Solar Light-to-Hydrogenated Organic Conversion

Hairus Abdullah
Editor

Solar Light-to-Hydrogenated Organic Conversion

Heterogeneous Photocatalysts

Springer

Editor
Hairus Abdullah
Department of Materials Science
and Engineering
National Taiwan University of Science
and Technology
Taipei, Taiwan

ISBN 978-981-99-8113-7 ISBN 978-981-99-8114-4 (eBook)
https://doi.org/10.1007/978-981-99-8114-4

This Springer imprint is published by the registered company Springer Nature Singapore Pte Ltd.
The registered company address is: 152 Beach Road, #21-01/04 Gateway East, Singapore 189721, Singapore

Preface

Sustainable work on photocatalytic chemical transformations shows promise for various industrial applications. Photocatalysis is a process that utilizes light energy to drive chemical reactions, and it has gained significant attention due to its potential to provide environmentally friendly and energy-efficient solutions. Some essential results on photocatalytic hydrogenation of nitrophenol, nitrobenzene, azobenzene, hydroxylation of benzene to phenol, oxygen reduction reaction to form H_2O_2, and glycerol valorization are discussed in this book.

Photocatalytic hydrogenation involves the reduction of nitro compounds to their corresponding amines using light energy. This process is vital because nitro compounds are often toxic and environmentally harmful. Using suitable photocatalysts, such as semiconducting materials like Zn(O,S), TiO_2, or other modified materials, can transform nitro compounds into less harmful products under mild conditions. The reaction conditions and catalyst selection greatly influence the efficiency of the transformation. The hydroxylation of benzene to phenol is a significant transformation in the chemical industry, as phenol is a versatile compound used in producing plastics, pharmaceuticals, and other chemicals. Traditional hydroxylation methods require high temperatures and pressures and often involve environmentally harmful reagents. Photocatalytic hydroxylation offers a milder alternative using light energy and suitable catalysts, potentially reducing the environmental impact of the process. In addition, photocatalytic oxygen reduction reaction involves the electrochemical reduction of oxygen to produce hydrogen peroxide (H_2O_2). H_2O_2 has various industrial applications, including as a bleaching agent and an oxidant. Traditional methods of H_2O_2 production involve energy-intensive processes. Photocatalytic ORR presents an alternative route where oxygen is reduced under light irradiation, potentially improving the energy efficiency and sustainability of H_2O_2 production. Furthermore, glycerol is a byproduct of biodiesel production and has limited direct applications. Photocatalytic transformations can convert glycerol into value-added chemicals through oxidation, hydrogenation, or dehydration. These transformations can enhance the economic viability of biodiesel production and reduce waste.

In all these cases, selecting a photocatalyst, reaction conditions, and light source is crucial in achieving efficient and selective transformations. Researchers continually optimize these factors to make photocatalytic processes more viable for industrial-scale applications. Although there has been progress in this field, it is essential to note that challenges remain, such as catalyst stability, scalability, and cost-effectiveness. However, the promising results suggest that photocatalytic chemical transformations could contribute to more sustainable and environmentally friendly industrial practices.

Taipei, Taiwan

Dr. Hairus Abdullah

Contents

Contributors

Hairus Abdullah Department of Industrial Engineering, Universitas Prima Indonesia, Medan, Indonesia

Mohamed Tarek Ahmed Department of Materials Science and Engineering, National Taiwan University of Science and Technology, Taipei, Taiwan

Bramantyo Bayu Aji Department of Materials Science and Engineering, National Taiwan University of Science and Technology, Taipei, Taiwan

Siti Khodijah Chaerun Department of Metallurgical Engineering, Institut Teknologi Bandung, Bandung, Indonesia;
Geomicrobiology-Biomining and Biocorrosion Laboratory, Microbial Culture Collection Laboratory, Biosciences and Biotechnology Research Center (BBRC), Institut Teknologi Bandung, Bandung, Indonesia

Lee Daeun Department of Materials Science and Engineering, National Taiwan University of Science and Technology, Taipei, Taiwan

Nurul Ika Damayanti Department of Chemical Engineering, National Taiwan University of Science and Technology, Taipei, Taiwan, ROC

Lusi Ernawati Department of Chemical Engineering, Institut Teknologi Kalimantan, Balikpapan, East Kalimantan, Indonesia

Shoeb Azam Farooqui Department of Materials Science and Engineering, National Taiwan University of Science and Technology, Taipei, Taiwan

Vivi Fauzia Department of Physics, Universitas Indonesia, Kampus UI Depok, Indonesia

Riski Titian Ginting Department of Electrical Engineering, Universitas Prima Indonesia, Medan, Indonesia

Sheng-Hsiang Hsu Department of Materials Science and Engineering, National Taiwan University of Science and Technology, Taipei, Taiwan

Andromeda Dwi Laksono Department of Materials and Metallurgical Engineering, Institut Teknologi Kalimantan, Balikpapan, East Kalimantan, Indonesia; Department of Materials Science and Engineering, National Taiwan University of Science and Technology, Taipei, Taiwan, ROC

Bobby Refokry Oeza Department of Materials Science and Engineering, National Taiwan University of Science and Technology, Taipei, Taiwan, ROC; Department of Material Engineering, Institut Teknologi Sepuluh Nopember Surabaya, Keputih Sukolilo, Surabaya, East Java, Indonesia

Ade Wahyu Yusariarta Putra Parmita Department of Materials and Metallurgical Engineering, Institut Teknologi Kalimantan, Balikpapan, East Kalimantan, Indonesia

Tran Nguyen Hoang Phan College of Natural Sciences, Can Tho University, Can Tho City, Vietnam

Abdul Qadir Department of Mechanical Engineering, National Taiwan University of Science and Technology, Taipei, Taiwan, ROC

Ulya Qonita Department of Applied Science and Technology, National Taiwan University of Science and Technology, Taipei, Taiwan

Tran Thi Bich Quyen Faculty of Chemical Engineering, College of Engineering, Can Tho University, Can Tho City, Vietnam; Nano-Electrochemistry Laboratory, Can Tho University, Can Tho City, Vietnam

Fadila Arum Rhamadani Department of Materials Science and Engineering, National Taiwan University of Science and Technology, Taipei, Taiwan

Salva Salshabilla Department of Materials Science and Engineering, National Taiwan University of Science and Technology, Taipei, Taiwan, ROC

Leonardo Togar Samosir Department of Physics, Universitas Indonesia, Kampus UI Depok, Indonesia

Albertus Jonathan Suciatmaja Department of Chemical Engineering, National Taiwan University of Science and Technology, Taipei, Taiwan

Diah Susanti Department of Materials and Metallurgical Engineering, Sepuluh Nopember Institute of Technology, Surabaya, Indonesia

Phan Thi Bao Tran Faculty of Economics, Dong Thap University, Cao Lanh City, Dong Thap Province, Vietnam

Phan Van Hoang Khang Department of Materials Science and Engineering, National Taiwan University of Science and Technology, Taipei City, Taiwan

Januar Widakdo Department of Physics, Universitas Indonesia, Jakarta, Indonesia

Ly Tho Xuan Department of Materials Science and Engineering, National Taiwan University of Science and Technology, Taipei City, Taiwan

Afandi Yusuf Department of Materials Science and Engineering, National Taiwan University of Science and Technology, Taipei, Taiwan, ROC

Chapter 1
Photocatalytic Reduction of Nitrophenol and Nitrobenzene with Zn Oxysulfide Semiconductor Without Using Reducing Agents

Lusi Ernawati, Andromeda Dwi Laksono, Ade Wahyu Yusariarta Putra Parmita, Diah Susanti, and Abdul Qadir

Abstract The content in this review paper discusses a photocatalytic process involving the reduction of nitrophenol and nitrobenzene using a Zn oxysulfide semiconductor. Importantly, this process does not rely on using external reducing agents. Instead, the Zn oxysulfide semiconductor serves as a catalyst, utilizing light energy to drive the reduction reactions of nitrophenol and nitrobenzene. This innovative approach demonstrates the potential of Zn oxysulfide as an efficient photocatalyst for environmentally friendly reduction reactions, contributing to sustainable and cleaner chemical processes. The use of a Zn oxysulfide semiconductor for the photocatalytic reduction of nitrophenol and nitrobenzene represents a greener and more sustainable approach to chemical transformations. Traditional reduction methods often require the use of external reducing agents that can be harmful to the environment. The other side, Zn oxysulfide exhibits significant photocatalytic properties, enabling it to harness light energy to drive chemical reactions. This property makes it a promising

L. Ernawati
Department of Chemical Engineering, Institut Teknologi Kalimantan, Soekarno Hatta Street Km. 15, Karang Joang, Balikpapan, East Kalimantan 76127, Indonesia

A. D. Laksono (✉) · A. W. Y. P. Parmita
Department of Materials and Metallurgical Engineering, Institut Teknologi Kalimantan, Soekarno-Hatta Street Km. 15, Karang Joang, Balikpapan, East Kalimantan 76127, Indonesia
e-mail: andromeda@itk.ac.id

A. D. Laksono
Department of Materials Science and Engineering, National Taiwan University of Science and Technology, Taipei 10672, Taiwan, ROC

D. Susanti
Department of Materials and Metallurgical Engineering, Sepuluh Nopember Institute of Technology, Surabaya 60111, Indonesia

A. Qadir
Department of Mechanical Engineering, National Taiwan University of Science and Technology, Taipei 10607, Taiwan, ROC

H. Abdullah (ed.), *Solar Light-to-Hydrogenated Organic Conversion*,
https://doi.org/10.1007/978-981-99-8114-4_1

candidate for various photocatalytic applications beyond nitrophenol and nitrobenzene reduction. The novelty of this approach lies in its ability to achieve reduction without the need for additional reducing agents. This idea simplifies the reaction process and eliminates the potential risks associated with using certain reducing agents. The specific focus on nitrophenol and nitrobenzene reduction underscores the significance of this research in addressing the reduction of nitroaromatic compounds. These compounds are toxic and have adverse environmental and health effects. The successful reduction of nitrophenol and nitrobenzene using Zn oxysulfide semiconductor opens up possibilities for other catalytic applications. Researchers might explore similar catalysts for various organic transformations, thereby expanding the scope of green catalysis. By utilizing a photocatalytic approach without reducing agents, the study contributes to reducing chemical waste and minimizing the environmental impact associated with traditional reduction processes. The findings in this work could potentially stimulate further research into optimizing Zn oxysulfide catalysts, exploring different reaction conditions, and investigating their performance in diverse catalytic reactions.

Keywords Nitrophenol · Nitrobenzene · Photocatalysis · Zn oxysulfide

1.1 Introduction

1.1.1 Nitrophenol

Photocatalysis is a subfield of chemistry that concerns itself with chemical reactions that occur when light is present along with a photocatalyst [1, 2]. A photocatalyst, being a semiconductor, accelerates the reaction rate simply by being there [3–5]. Photochemists have been intrigued by the electron and hydrogen transfers that occur in aromatic nitro compounds, and they have conducted a thorough investigation of the individual stages in the photoinduced reactions. Nevertheless, certain nitro compounds, such as nitrophenols, are more acidic than phenol itself. Nitrophenol refers to a class of chemical compounds that consist of a phenolic ring with one or more nitro groups (–NO_2) attached to it. The commonly utilized reaction route for converting the extremely hazardous 4-nitrophenol to 4-aminophenol involves chemical adsorption on the surface of the nanostructures and the transfer of surface-hydrogen species (Fig. 1.1).

The three main isomers of nitrophenol are:

1. 2-Nitrophenol: An isomer that has a nitro group attached to the ortho position (position 2) of the phenolic ring.
2. 3-Nitrophenol: The nitro group is attached to the meta position (position 3) of the phenolic ring.
3. 4-Nitrophenol: This isomer has the nitro group attached to the para position (position 4) of the phenolic ring.

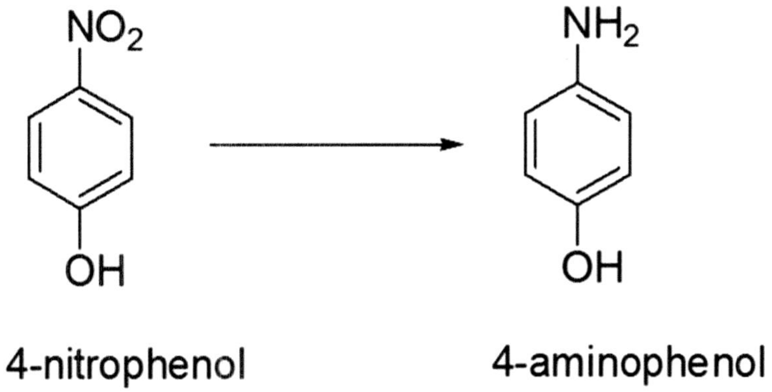

Fig. 1.1 Chemical reaction: p-nitrophenol to p-aminophenol

These compounds are usually yellow to brown in color and have various applications. They can be used as intermediates in producing dyes, pharmaceuticals, and pesticides. Some nitrophenols are also found as secondary metabolites produced by certain plants and fungi. Handling and disposing of nitrophenols properly is important since they are toxic to humans and the environment. Nitrophenols have several applications and uses in various industries.

Here are some common uses of nitrophenol compounds:

1. Chemical Intermediates: Nitrophenols are used as intermediates in synthesizing various chemicals, including dyes, pharmaceuticals, agrochemicals, and other organic compounds. They can undergo further chemical reactions to produce a wide range of products.
2. Dye Production: Nitrophenols are employed in producing dyes, particularly azo dyes. Azo dyes are extensively used in textile, printing, and coloring industries (Fig. 1.2).
3. Pesticides and Herbicides: Some nitrophenols, such as 2,4-dinitrophenol (DNP) and 2,4-dinitro-6-tert-butylphenol (DNBP), are used as active ingredients in pesticides and herbicides. These compounds can act as effective biocides against insects, weeds, and fungi.
4. Wood Preservation: Certain nitrophenols, such as pentachlorophenol (PCP), are used as wood preservatives to protect timber from decay, insect infestations, and fungal growth.
5. Laboratory Reagents: Nitrophenols can be used as reagents in chemical laboratories for various purposes, such as analytical testing, organic synthesis, and research experiments (Fig. 1.3).

It is worth noting that while nitrophenols have several industrial applications, some compounds in this class, like pentachlorophenol, have been phased out or restricted in many countries due to environmental and health concerns. Proper handling, use, and disposal of nitrophenols are crucial to minimize their impact on human health and the environment.

Fig. 1.2 The example of dye production: digital textile printing machine

Fig. 1.3 In an actual laboratory, there is an arrangement of jars and flasks containing reagent solutions of various colors

Here are some general guidelines for the safe handling and disposal of nitrophenols:

1. Storage: Store nitrophenols in a cool, dry, and well-ventilated area. Ensure they are kept in properly labeled and tightly sealed containers to prevent leaks or spills. Follow any specific storage requirements or recommendations provided by the manufacturer.
2. Personal Protective Equipment (PPE): When handling nitrophenols, use appropriate personal protective equipment such as gloves, safety goggles, and a lab coat or protective clothing to prevent direct contact with the skin, eyes, and clothing. Follow good laboratory or industrial hygiene practices.

Fig. 1.4 Hazard waste disposal

3. Spill Response: In case of a spill or leak, immediately contain the area to prevent the spread of the nitrophenol. Use suitable absorbent materials, such as spill control pads or inert absorbents, to soak up the spilled material. Dispose of the absorbent material as hazardous waste.
4. Disposal: Nitrophenols are typically considered hazardous waste due to their toxicity. It is important to follow local regulations and guidelines for removing hazardous chemicals. Contact our local waste management or environmental agency for specific disposal requirements and recommendations (Fig. 1.4).
5. Chemical Treatment: Nitrophenols can be chemically treated or neutralized to render them less harmful before disposal. Consult with a qualified chemist or waste management professional to determine appropriate treatment methods, if applicable.
6. Recycling and Reuse: Whenever possible, consider recycling or reusing nitrophenols rather than disposing of them. Check if there are any recycling programs or options available for hazardous chemicals in our area (Fig. 1.5).

Nitrophenols are resistant to both photoreduction and photo substitution reactions. This resistance can be attributed to the existence of a state with a significant contribution from charge transfer in their excited state [6]. Reducing nitro compounds is a considerable challenge due to their polluting nature. Of particular importance is the reduction of 4-nitrophenol and other nitroaromatics, which are artificial, toxic, and inhibitory. Industrial effluents frequently contain 4-nitrophenol as a persistent pollutant, whereas 4-aminophenol is highly commercially valuable as an intermediate for various industrial products such as agrochemicals, pharmaceuticals, and dyestuffs [7]. As a result, the fast reduction of hazardous 4-nitrophenol to 4-aminophenol (as shown in Fig. 1.1) using suitable catalysts has become a widely researched topic.

Fig. 1.5 Reuse, recycle, reduce

The reduction product of 4-nitrophenol, 4-aminophenol, has diverse applications such as a photographic developer for black and white films, a corrosion inhibitor, a drying agent, and an intermediate in the synthesis of analgesic and antipyretic drugs, particularly paracetamol [8].

There are two main pathways for removing 4-nitrophenol from a solution: oxidative and reductive. The objective of oxidative degradation is to completely oxidize 4-nitrophenol to CO_2, NO_3^-, and H_2O. However, this process is typically complicated and may involve several reactions, including ring opening, which makes it challenging to achieve complete decomposition. On the other hand, the reductive pathway aims to convert 4-nitrophenol into 4-aminophenol, which is a crucial intermediate in the production of dyes [9], photographic developers [10], and antipyretic drugs [11]. Although the reductive degradation may not be the final solution, the resulting 4-aminophenol is less toxic than 4-nitrophenol [12] and can be further degraded by microorganisms [13]. Unfortunately, the reduction of 4-nitrophenol typically requires high temperatures and hydrogen pressure [14], so a milder reduction route is highly desirable. Previous studies have commonly used an excess of $NaBH_4$ in the 4-nitrophenol reduction [15–20], but this is undesirable due to the corrosive and irritative nature of $NaBH_4$, as indicated in the material safety data sheet.

1.1.2 Nitrobenzene

Nitrobenzene (as shown in Fig. 1.6) is a type of nitroaromatic compound that is considered a major pollutant from human-made sources [21]. These chemicals are prevalent in the environment because they are commonly used as raw materials or intermediates in the production of various products, such as pharmaceuticals, wood preservatives, rubber chemicals, pigments, dyes, plastics, pesticides, fungicides, explosives, and industrial solvents [22]. However, nitrobenzene is toxic and can be harmful to humans. It is absorbed through the skin and can cause skin irritation, damage to the central nervous system, and adverse effects on the liver, kidneys, and blood (Fig. 1.7).

Handling nitrobenzene with proper safety precautions and using protective equipment when working with it is crucial. Nitrobenzene is persistent in the environment and contaminate water and soil if not properly managed. It is classified as hazardous and should be handled and disposed of according to local regulations and guidelines. Cleaning up wastewater contaminated by nitrobenzene is a challenging process because the presence of the nitro group in the aromatic ring increases the stability of these molecules, making them resistant to chemical and biological degradation.

Nitrobenzene can undergo reduction reactions to yield aniline under suitable conditions. Reduction of nitrobenzene is typically achieved using catalytic hydrogenation or other reducing agents. Catalytic hydrogenation to reduce nitrobenzene to aniline is a widely used method in both industrial and laboratory settings. Transition metals like Cu and Ni, as well as noble metals like Pt, Pd, and Au, are typically used as catalysts for this reaction [23–25]. However, this method requires high temperatures, high H_2 pressure, and longer reaction times to achieve satisfactory selectivity of

Fig. 1.6 Nitrobenzene with molecular formula $C_6H_5NO_2$

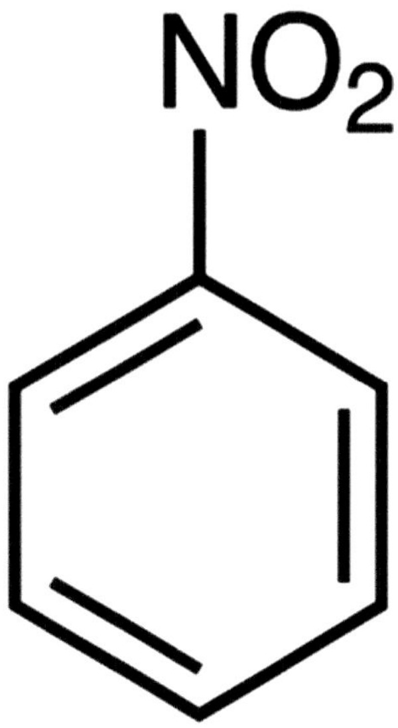

Fig. 1.7 Toxic safety sign

aniline [26]. In contrast, photochemical-induced reduction of nitrobenzene to aniline using a photocatalyst can be performed at room temperature, making it an attractive environmentally friendly option. Although this method has been explored in some literature [27–29], the conversion and selectivity rates are typically low, and TiO_2 is commonly used as the photocatalyst [30, 31], which only responds to UV light. Thus, to better use solar energy, it is necessary to develop visible-light-activated photocatalysts.

Several studies have reported that various oxidation processes, including O_3/H_2O_2, H_2O_2/UV, and H_2O_2/Fe^{2+}/UV [32–34], are capable of rapidly removing nitroaromatic from contaminated water sources. These processes are effective because they produce hydroxyl radical (·OH), which is a potent and non-selective oxidizing agent that can react with organic compounds to produce dehydrogenated or hydroxylated derivatives, and ultimately mineralize them into CO_2, water, and inorganic ions. Figure 1.8 illustrates a clear and concise explanation of the reaction sequence involved in the degradation of nitrobenzene. This scheme provides a feasible general pathway for the complete breakdown of nitrobenzene under acidic conditions using anodic oxidation. The sequence encompasses all the identified products from nitrobenzene in a simplified manner. The mechanism helps providing a detailed understanding of the degradation process of nitrobenzene in oxidation reactions.

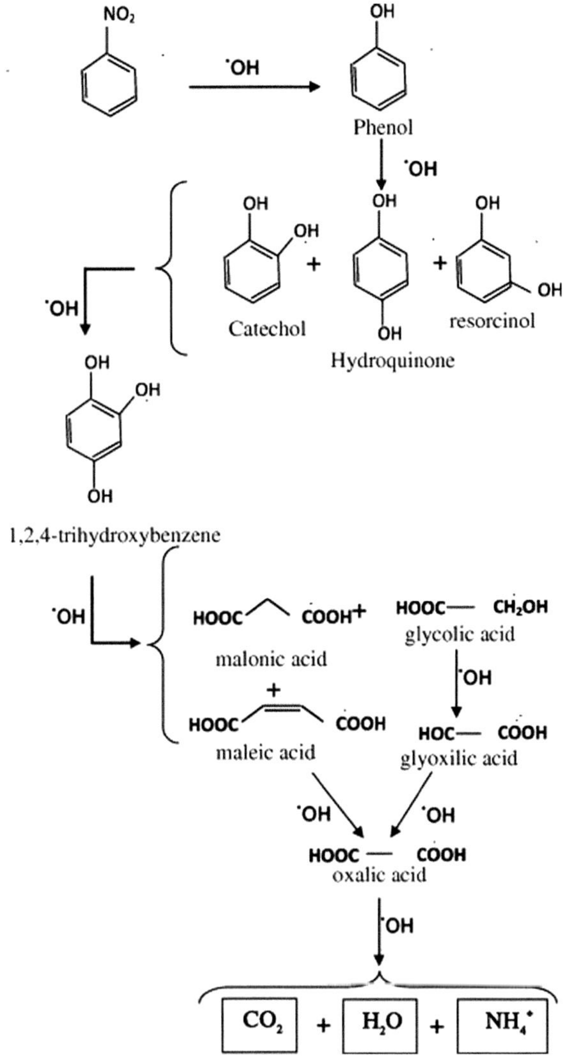

Fig. 1.8 This is a concise statement describing a suggested reaction pathway for the complete breakdown of nitrobenzene in an acidic aqueous medium. [21] Copyright 2023, Elsevier (License Number: 5540810535401)

1.1.3 Photocatalytic Reduction Mechanism of 4-NP to 4-AP

Photocatalytic reduction refers to the process where a photocatalyst utilizes light energy to drive a reduction reaction. In the case of 4-nitrophenol (4-NP) to 4-aminophenol (4-AP), the process typically involves the use of a photocatalyst such as semiconductor nanoparticles, commonly titanium dioxide (TiO_2), and an appropriate light source. The process of reducing 4-NP to 4-AP is significant and shows promise. There is a great deal of interest in developing new catalyst materials for this reduction process. The utilization of semiconductor catalysts such as ZnO, TiO_2, ZnS, CdS,

WO_3, and Fe_2O_3 in photocatalytic reduction of toxic organic effluents has emerged as a highly effective method for their treatment. Studies have been conducted on the photocatalytic oxidation of 4-NP using TiO_2. Langmuir–Hinshelwood-type pseudo-first-order kinetics were observed by Augugliaro et al. during the TiO_2-mediated photocatalytic degradation of 4-NP [35]. They discovered that the degradation rate of 4-NP decreased as the initial solution pH increased. In another study, researchers achieved the selective reduction of 4-NP to 4-AP by utilizing arginine-modified TiO_2. The process resulted in complete mineralization. The summarized redox pathways of 4-NP, including the rarely observed reductive pathway, are presented in Fig. 1.9. By modifying the surface of TiO_2 with arginine, the surface charge was altered, leading to an enhanced reduction of 4-NP at pH 9. This enhancement occurred because the positively charged amine group on the arginine side chain effectively formed complexes with the negatively charged 4-NP. The photocatalytic reduction specifically targeted the aromatic nitro-groups, and the reaction rate was primarily influenced by the half-wave reduction potential [36].

Graphene hybrids, silver-supported nanoporous iron oxide hybrids, nickel-palladium nanoparticles, and carbon-supported gold nanoparticles have been successfully employed as catalysts for reducing 4-NP to 4-AP in the presence of $NaBH_4$. These catalyst materials have demonstrated superior stability and reusability. However, their preparation methods are intricate and time-consuming. Researchers

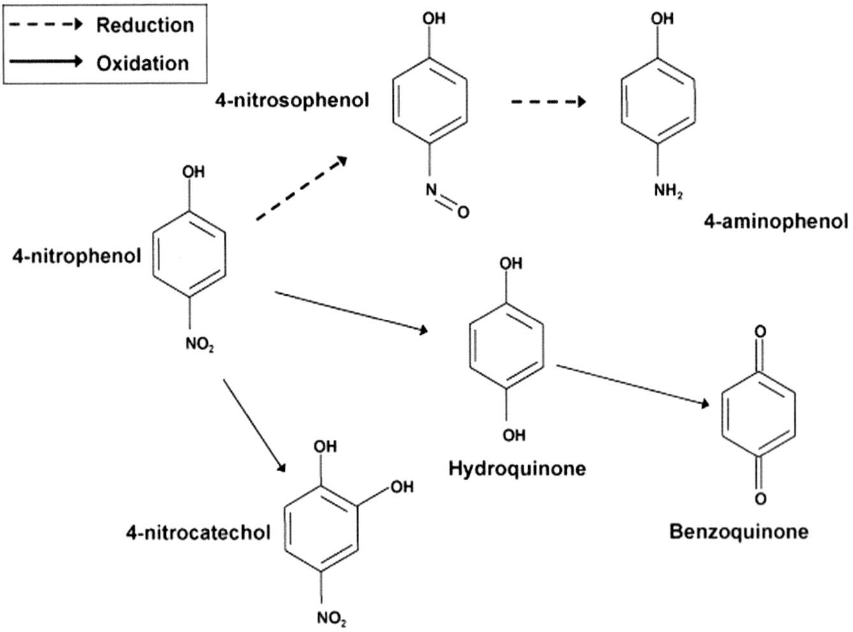

Fig. 1.9 The anticipated redox pathways of 4-NP. [36] Copyright 2023, Elsevier (License Number: 5553580107626)

have made significant efforts to develop an efficient catalyst that can reduce 4-NP to 4-AP under mild conditions. Metal–organic frameworks (MOFs), also known as metal–organic coordination polymers, have emerged as promising candidates. They offer adjustable properties, organic functionality, high thermal and mechanical stability, large pore size, high surface area, and straightforward synthesis [37–39]. In recent years, MOFs have been extensively explored in various fields, such as gas adsorption, separations, catalysis, sensors, photocatalysis, and drug delivery. The use of MOFs as photocatalysts is a recent area of exploration [40–42]. In a recent study, MOF named [Zn(BDC)(DMF)] was synthesized using the solvothermal method with ultrasonic irradiation. The synthesized MOF exhibited excellent photocatalytic reduction of 4-NP to 4-AP in the presence of $NaBH_4$ under direct sunlight irradiation, completing the reduction process within 10 min. Moreover, the catalyst maintained high catalytic activity even after 10 cycles (Fig. 1.10), with an efficiency exceeding 95%. Overall, this study suggests that the synthesized MOF [Zn(BDC)(DMF)] can be effectively utilized for the removal of toxic organic dyes from wastewater [43].

In a separate study conducted by Bekena et al. [44], they investigated novel magnesium-doped Zn(O,S) nanoparticles. Based on their findings, the following reaction mechanism is proposed for the photocatalytic reduction of 4-NP using magnesium-doped Zn(O,S). The photogenerated holes in the valence band (VB) move to the surface of the catalyst and react with water molecules, resulting in the formation of hydroxyl radicals. These hydroxyl radicals (·OH) quickly react

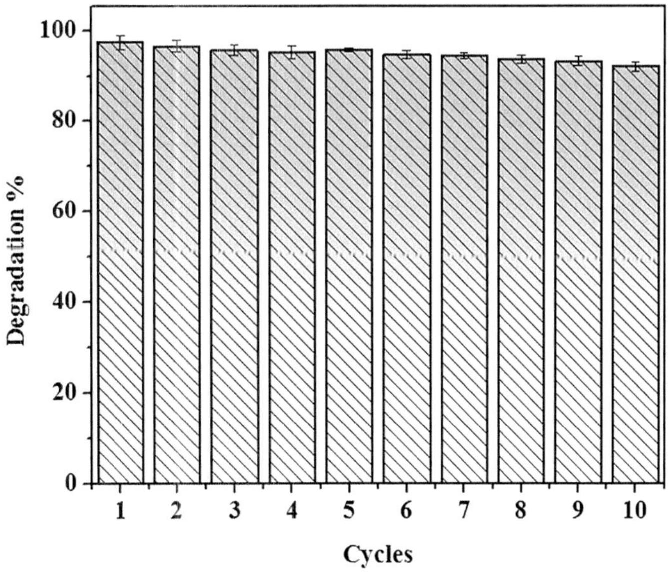

Fig. 1.10 The ability to reuse the synthesized MOF [Zn(BDC)(DMF)] for reducing 4-nitrophenol (4-NP) in the presence of $NaBH_4$. [43] Copyright 2023, Elsevier (License Number: 5553581030409)

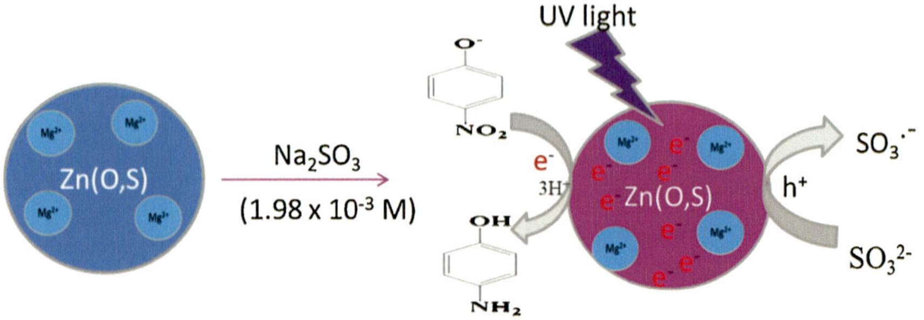

Fig. 1.11 A diagram illustrating the proposed reaction mechanism for the photocatalytic reduction of 4-NP to 4-AP in the presence of Na_2SO_3 using magnesium-doped Zn(O,S) nanoparticles. [44] Copyright 2023, Elsevier (License Number: 5553660700966)

with SO_3^{2-} ions present on the surface of the magnesium-doped Zn(O,S) nanoparticles, leading to an oxidation reaction that forms sulfate ions. This process reduces the number of charge carriers available for recombination with electrons. Simultaneously, the photogenerated electrons in the conduction band (CB) transfer to the adsorbed 4-nitrophenolate (4-NP_{ate}) ions, driving the reduction process. While a small amount of water is oxidized by the holes in the VB, the 4-NP_{ate} ions are reduced by the electrons in the CB. The overall reaction is thermodynamically favorable and can be visualized as shown in Fig. 1.11.

The photocatalytic reduction mechanism of 4-NP to 4-AP can be summarized as follows:

1. Photocatalyst Activation: The photocatalyst, usually TiO_2, is exposed to an appropriate light source, typically ultraviolet (UV) light. The light energy excites the electrons in the valence band of the photocatalyst to the conduction band, creating electron–hole pairs.
2. Adsorption: The 4-NP molecules adsorb onto the surface of the photocatalyst through weak interactions such as Van der Waals forces or hydrogen bonding. This adsorption allows the 4-NP molecules to come into contact with the photocatalyst and facilitates subsequent reaction steps.
3. Photogenerated Electron Transfer: The photogenerated electrons in the conduction band of the photocatalyst are transferred to the adsorbed 4-NP molecules, reducing them to 4-AP. This electron transfer process leads to converting the nitro group (–NO_2) in 4-NP to an amino group (–NH_2) in 4-AP.
4. Proton Transfer: Concurrently with the electron transfer, protons (H^+) from the surrounding solution are typically involved in the reduction process. Protonation occurs, where the transferred electrons react with the protons, further facilitating the reduction of the nitro group.
5. Desorption and Product Formation: Once the reduction reaction occurs, the resulting 4-AP molecules desorb from the surface of the photocatalyst. The 4-AP molecules are then collected and separated from the reaction mixture.

It is important to note that the photocatalytic reduction mechanism can vary depending on the specific photocatalyst, reaction conditions, and experimental setup. Various factors such as pH, temperature, light intensity, and catalyst morphology can influence the efficiency and kinetics of the photocatalytic reduction process. The photocatalytic reduction of 4-NP to 4-AP is of particular interest due to its potential application in environmental remediation and organic synthesis, as well as its relevance to the broader field of photocatalysis.

1.2 Zinc Oxysulfide-Based Photocatalysts

Zinc oxysulfide (ZnO_xS_y) is a semiconductor material that exhibits photocatalytic properties, meaning it can utilize light energy to drive chemical reactions. It is composed of zinc, oxygen, and sulfur atoms.

Here are some key points about zinc oxysulfide-based photocatalysts:

1. Photocatalytic Activity: Zinc oxysulfide has been studied for its photocatalytic activity in various applications. When exposed to light, it can generate electron–hole pairs in its semiconductor structure. These photogenerated charge carriers can participate in redox reactions, promoting the degradation of organic pollutants or facilitating other desirable chemical transformations.
2. Visible Light Response: One advantage of zinc oxysulfide is its ability to respond to visible light, which makes it attractive for applications where sunlight or visible light sources are available. This broader light absorption range compared to other photocatalysts like titanium dioxide (TiO_2) allows for more efficient solar energy utilization.
3. Environmental Applications: Zinc oxysulfide photocatalysts have been explored for ecological remediation purposes. They have been studied for degrading organic pollutants in water, such as dyes, pesticides, and pharmaceutical compounds. The photocatalytic activity of zinc oxysulfide can promote the breakdown of these pollutants into less harmful substances.
4. Photocatalytic Water Splitting: Another application of zinc oxysulfide-based photocatalysts is in the field of renewable energy. They have been investigated for their potential to split water molecules under light irradiation, generating hydrogen gas (H_2). Photocatalytic water splitting is a promising approach for clean and sustainable hydrogen production as a fuel source.
5. Synthesis and Modifications: Zinc oxysulfide can be prepared through various synthesis methods, including hydrothermal methods, sol–gel techniques, or solid-state reactions. Additionally, its properties can be tuned or enhanced through doping with other elements or surface functionalization.

Zinc oxysulfide-based photocatalysts are an active area of research due to their potential in environmental remediation, energy conversion, and other photocatalysis-related applications. However, it is essential to note that further research is still being conducted to optimize their performance, understand their mechanisms, and address any challenges associated with stability and efficiency.

ZnO and TiO_2 are extensively used as photocatalysts among semiconductor catalysts due to their exceptional activity, non-toxicity, chemical stability, cost-effectiveness, favorable optical and electrical properties, UV-driven behavior, and environmentally friendly nature [45]. Efforts have been made to enhance the performance of ZnO through methods such as metal and non-metal doping and the development of ZnO-based composites. One promising approach involves doping ZnO photocatalyst with transition metal ions, which reduces the band gap energy and enhances the separation of electrons and holes by creating electron traps [46]. This step enables the migration of holes towards the photocatalyst surface to degrade organic compounds [47]. Cu-doped ZnO nanoparticles were found to exhibit superior photodegradation efficiency compared to undoped ZnO, as confirmed by Fu et al. [48]. Additionally, Hairus et al. reported the successful UV light-driven photocatalytic reduction of 4-NP using a Zn(O,S)/Ga_2O_3 composite [49]. The process of photocatalytic hydrogen evolution reaction (HER) was observed to commence through water oxidation, resulting in the creation of oxygen vacancy sites on the catalyst surfaces. The heterojunction band diagram of the Zn(O,S)/Ga_2O_3 nanocomposite illustrates a typical straddling gap. This arrangement facilitates the transfer of electrons and holes from Ga_2O_3 to the conduction and valence bands of Zn(O,S), respectively. It is believed that defect states in the Ga_2O_3 conduction band play a role in generating photoinduced electrons by forming oxygen vacancies during the low-temperature process. Throughout the photoreaction, the hydrogen ions produced on the nanocomposite surfaces are utilized for the reduction of 4-NP, resulting in the formation of 4-AP. This reduction of 4-NP on the Zn(O,S)/Ga_2O_3 nanocomposite was confirmed through various means, including UV–VIS spectroscopy, high-performance liquid chromatography (HPLC) measurements, and the observation of a decrease in the amount of evolved hydrogen consumed during the 4-NP reduction. In conclusion, the Zn(O,S)/Ga_2O_3 nanocomposites can simultaneously generate hydrogen and remediate the toxic pollutant 4-NP in ethanol solution when exposed to low UV light illumination.

Research has been conducted on detoxifying 4-NP to 4-AP without a reduction agent [50]. In this study, it was discovered that Zn(O,S) doped with rare-earth lanthanum metal exhibited excellent potential for detoxifying 4-NP and forming less toxic 4-AP in a 10% ethanol solution without the need for $NaBH_4$ as a reduction agent. The introduction of La into the Zn(O,S) lattice enhanced the presence of adsorbed hydrogen ions on the catalyst surfaces, thereby promoting the reduction of 4-NP. Consequently, a solution containing 30 ppm of 4-NP could be reduced entirely within 2 h under the illumination of a low-intensity blacklight UV lamp when La-doped Zn(O,S) was present. The proposed mechanism, depicted in Fig. 1.12, illustrates the process. The formation of H^+ and the positive charge on the catalyst surface facilitated the attraction of 4-NP to the catalyst surfaces through electrostatic interactions. Simultaneously, the excess photogenerated electrons on the catalyst surfaces reduced the uptake of electrons by the adsorbed 4-NP molecules. Additionally, the

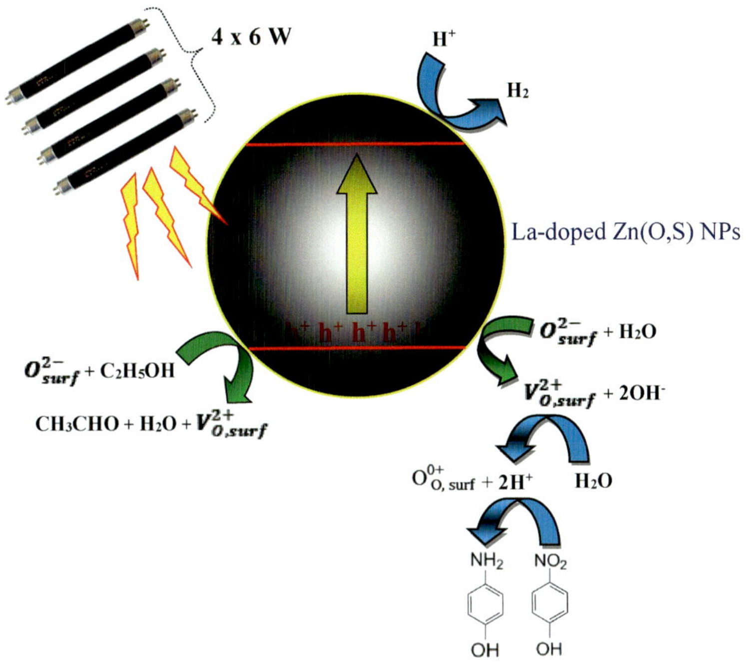

Fig. 1.12 The mechanism of photocatalytic detoxification involving the reduction of 4-NP to 4-AP using La-doped Zn(O,S) nanoparticles was observed under the illumination of a blacklight UV tube lamp with a power of 4 × 6 W. [50] Copyright 2023, Elsevier (License Number: 5570020310638)

presence of H^+ on the catalyst surfaces contributed to reducing 4-NP into 4-AP products. This research demonstrates a promising chemical synthesis approach not only for converting 4-NP to 4-AP but also for potentially catalyzing other hydrogenation reactions using the hydrogen evolution photocatalyst.

1.3 Modification of Zn Oxysulfide-Based Photocatalysts with Metal Dopants

Ho-doped Zn(O,S) catalysts for photocatalytic hydrogenation reaction (PHR) were successfully produced and studied. XRD and TEM examination revealed that Ho had been doped into the Zn(O,S) lattice in a relatively minor amount of about 4% throughout the preparation procedure. According to the characterization results from photoluminescence (PL), electrochemical impedance spectroscopy (EIS), and transient photocurrent (TPC), ZOS-2.5 Ho not only displayed the best photocatalytic HER of 18.581 mmol/g in 5 h but also a great PHR for 4-NP-to-4-AP conversion. The in-situ produced proton on catalyst surfaces was confirmed to be utilized to hydrogenate 4-NP to 4-AP without reducing chemicals during the PHR. It was proposed

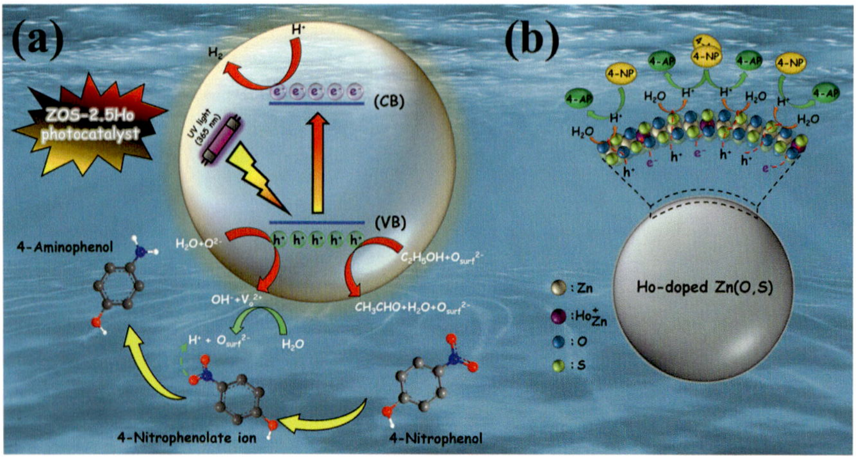

Fig. 1.13 (**a**) Photocatalytic reaction mechanism of 4-NP-to-4-AP conversion on ZOS-2.5Ho particle in ethanol solution under blacklight UV tube lamp illumination with (**b**) the schematic of molecular reaction during the PHR. Reprinted with permission from [51]. Copyright (2023) American Chemical Society

that occupying Zn sites by Ho in the Zn(O,S) lattice played essential roles not only in separating photocarriers and enhancing charge transfer but also in shortening the diffusing time of nitrophenolate ions to catalyst surfaces for subsequent PHR process. As a result, the work adds to a green and gentle hydrogenation method of chemical conversion and a viable strategy for industrial use (Fig. 1.13).

The improved photocatalytic activity of the Zn(O,S)/Ga_2O_3 nanocomposite was attributed to effective electron transfer between the Zn(O,S) and Ga_2O_3 phases, which produced a straddling gap to increase the amount of photo carriers in photo processes. The development of a Zn(O,S)/Ga_2O_3 heterojunction boosted the rate of hydrogen evolution by 30% as compared to a single Zn(O,S) phase. The hydrogen ions formed on nanocomposite surfaces during photoreactions can be used for 4-NP reduction to form 4-AP. UV–vis spectroscopy, HPLC studies, and the decreased amount of generated hydrogen used for 4-NP reduction indicated the reduction of 4-NP on Zn(O,S)/ Ga_2O_3 nanocomposite. Finally, it is concluded that the formation of Zn(O,S)/ Ga_2O_3 nanocomposites can simultaneously generate hydrogen and remediate toxic pollutant of 4-NP in ethanol solution under low UV light illumination (0.088 mW/cm^2 or approximately 1/40-fold UV light intensity of sunlight).

Sn(II)-doped Zn(O,S) catalysts have been produced, studied, and used to reduce 4-NP. This study effectively converted 4-NP to 4-AP by using a hydrogen-evolved catalyst in ethanol solution under low-intensity UV light illumination to replace the use of excess $NaBH_4$. It was discovered that 5% Sn(II) precursor was adequate to dope the Zn(O,S) catalyst, allowing for successful 4-NP conversion to 4-AP with decreased impedance resistance and better photoinduced charge separation. Water and ethanol oxidation produced oxygen vacancy sites required for H^+ production during the photocatalytic reaction. Because the oxidation reaction in ethanol solution was more substantial than in water, more oxygen vacancy sites were created in ethanol solution to have more H ions to decrease 4-NP faster. Furthermore, it is anticipated that the current work will provide an idea to the catalysis field for future research into hydrogen-evolution-assisted reactions not only for 4-NP reduction but also for chemical synthesis reactions.

The hydrogen evolved Sn(II)-doped Zn(O,S) photocatalyst was produced at 90 °C by precipitation and then used to totally decrease 30 ppm 4-NP to 4-AP in 3 h under 4 × 6 W UV lamp illumination (0.088 mW/cm^2) without employing $NaBH_4$ as a hydrogen source. Pure Zn(O,S) nanoparticles did not demonstrate 4-NP reduction behavior, but Zn(O,S) with Sn(II) cation doping did, as evidenced by UV–vis absorbance and HPLC data. Sn(II) doped into Zn(O,S) is critical in increasing the number of photoinduced hydrogen ions for reducing 4-NP to 4-AP on catalyst surfaces. The detoxification of 4-NP to 4-AP without $NaBH_4$ via a photocatalytic technique is considered a green chemical conversion in this work. The experimental findings demonstrated that 30 ppm hazardous 4-NP was completely transformed to helpful 4-AP with lower toxicity in 2 h, as evidenced by a particular peak shift as indicated by UV–vis absorbance spectra and high-performance liquid chromatography (HPLC) measurements. The decreased amount of hydrogen evolved from the photocatalytic process on La-doped Zn(O,S) NPs in the presence of 4-NP demonstrated that the created hydrogen was used as a reducing agent during the 4-NP-to-4-AP conversion. This work exhibited photocatalytic detoxification of 4-NP to 4-AP and postulated and clarified an appropriate mechanism based on experimental results.

The Mn-doped Zn(O,S) was effectively synthesized at low temperatures using a simple technique with varying Mn concentrations. Mn inclusion in the Zn(O,S) host lattice not only boosted absorbance but also lowered photocarrier electron–hole recombination, as shown by DRS and PL spectra, respectively. Furthermore, Mn may greatly reduce charge transfer resistance. The 10% Mn-doped to Zn(O,S) host lattice was found to have the best characteristics, and it could completely reduce the 30 ppm 4-NP to 4-AP after a 2-h photoreaction without requiring $NaBH_4$ as a reducing agent. The hydrogen ion was proposed to involve the 4-NP reduction to 4-AP, in which the hydrogen ion and electron replaced the oxygen in the amino (NO_2) group to form the nitro (NH_2) group. The capability of Mn-doped Zn(O,S) to reduce the 4-NP was contributed by the synergetic effect of the high absorbance, low charge recombination, and fast charge transfer. Substituting Mn in the side of Zn in the Zn(O,S) host lattice also weakened the oxygen bonding to easily form the oxygen vacancy.

Using a low temperature (95 °C) precipitation process, cerium and gallium codoped Zn(O,S) with various quantities of Ce have been successfully synthesized and studied. The positively charged substitutional defects of $Ce_{Zn}{}^{+}$, $Ce_{Zn}{}^{2+}$, and $Ga_{Zn}{}^{+}$ that promote the e^-/h^+ separation and charge transfer are thought to cause increased photocatalytic activity. In a 10% ethanol solution, 15Ce-5 Ga-Zn(O,S) produced the most hydrogen at 7130 mol/g/h. Additionally, the oxygen vacancies have greatly aided the evolution of hydrogen and the chemical conversion of nitrobenzene and azobenzene. The catalyst demonstrated good reusability and stability for practical

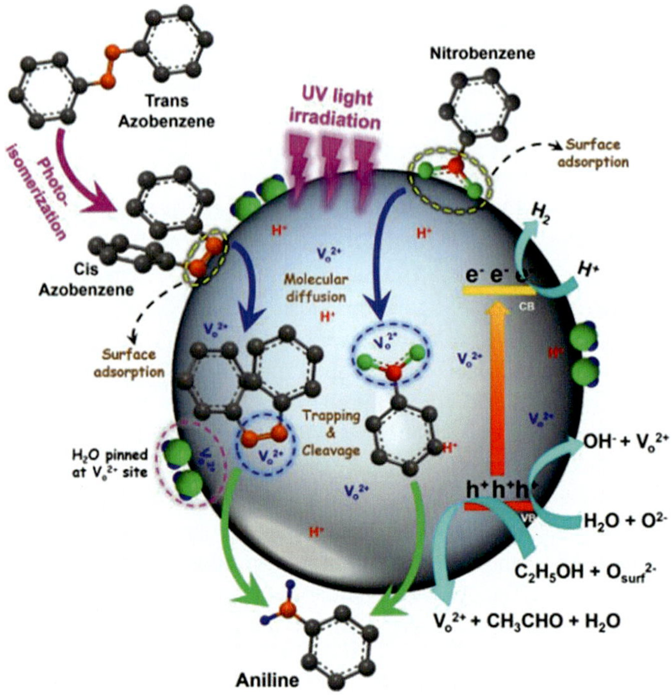

Fig. 1.14 Schematic drawing of HER and PHR mechanisms. [52] Copyright 2023, Elsevier (License Number: 5618570340359)

application. In general, this work reveals the possible use of cerium and gallium codoped Zn(O,S) as photocatalysts for hydrogen evolution and chemical conversion of harmful organic pollutants like azobenzene and nitrobenzene into a beneficial product of aniline (Fig. 1.14).

It has been proved that the loading of nickel hydroxide, zinc hydroxide, and nickel sulfide on the $Cd_{0.3}Zn_{0.7}S$ or $Cd_{0.3}Zn_{0.7}S$/FTO surface increased the photocatalytic activity and short-circuit current density. The growth of power conversion efficiency was retarded by the co-catalyst's presence since it caused the open-circuit voltage and fill factor to drop. The results of the electrochemical impedance spectroscopy showed that the electron lifetime was increasing and the photoanode resistivity was decreasing. For the photoelectrodes made using two different techniques—SILAR and drop-casting—the differences in target parameters were seen in the following consequences: $Cd_{0.3}Zn_{0.7}S$/FTO; ZnOH-20/FTO; NiOH-0.06/FTO; and NiS-0.3/FTO. A similar outcome was fixed by the improvement in photocatalytic activity. For the $Cd_{0.3}Zn_{0.7}S$/FTO modified by various nickel- and zinc-containing compounds, a link between the photocatalytic activity and photoelectrochemical parameters was initially established (Fig. 1.15).

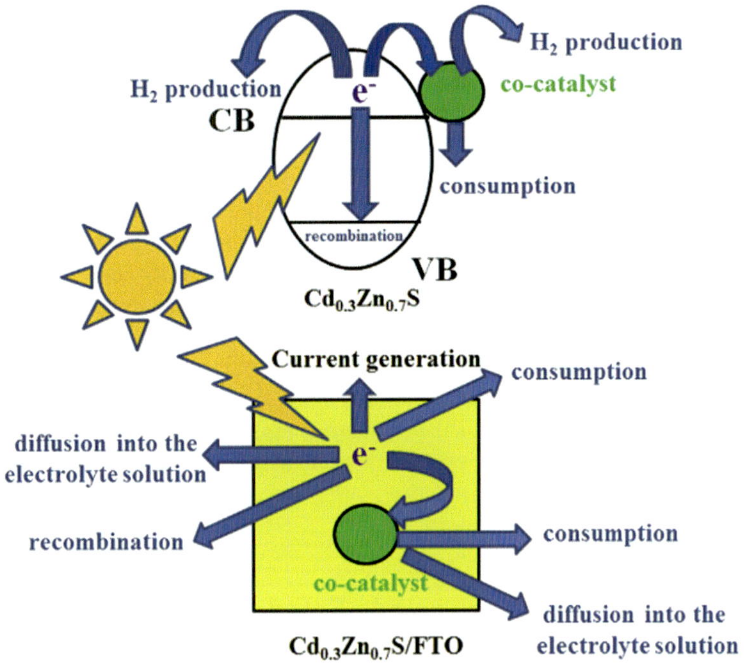

Fig. 1.15 Scheme of the photocatalytic H_2 production and light energy conversion processes. [53] Copyright 2023, Elsevier (License Number: 5618570690168)

1.4 Material Characterization of Zn Oxysulfide Photocatalysts

Material characterization of Zn oxysulfide photocatalysts involves the comprehensive analysis and understanding of the physical, chemical, and structural properties of Zn oxysulfide compounds used as photocatalysts. These materials are crucial in various photocatalytic applications, including environmental remediation, energy conversion, and sustainable chemical processes. The characterization process helps researchers gain insights into the catalyst performance, efficiency, and potential applications.

Techniques like X-ray diffraction (XRD) are used to determine the crystal structure of Zn oxysulfide. This information is vital for understanding the arrangement of atoms in the material and how it influences its photocatalytic properties. Scanning electron microscopy (SEM) and transmission electron microscopy (TEM) provide high-resolution images of the catalyst's surface, allowing researchers to study particle size, shape, and surface features. Energy-dispersive X-ray spectroscopy (EDX) or X-ray photoelectron spectroscopy (XPS) can identify the elemental composition of the photocatalyst, confirming the presence of zinc, oxygen, and sulfur in the oxysulfide. UV–visible spectroscopy helps determine the material's absorption properties in the ultraviolet and visible light regions. This information is crucial for understanding the wavelength range over which the catalyst can effectively absorb light energy. Techniques like UV–visible spectroscopy also allow researchers to calculate the material band gap, which provides insights into its ability to absorb photons and generate electron–hole pairs. Brunauer–Emmett–Teller (BET) analysis using gas adsorption helps determine the surface area and porosity of the material. This information is important for evaluating the catalyst's active surface sites available for catalytic reactions. Photoluminescence spectroscopy provides information about the recombination of photo-generated electron–hole pairs, which is crucial for understanding the efficiency of charge separation and utilization. Techniques like cyclic voltammetry and impedance spectroscopy can offer insights into the catalyst's electron transfer kinetics and charge transport properties. The long-term stability and durability of the catalyst can be evaluated through accelerated aging tests and prolonged exposure to photocatalytic reactions. To assess their effectiveness, the synthesized Zn oxysulfide photocatalysts are tested for their catalytic activity in specific reactions, such as water splitting or organic pollutant degradation.

1.5 Application of Zn Oxysulfide-Based Materials in Reducing Nitrophenol and Nitrobenzene

The highly stable 4-NP is one of the toxic pollutants from industrial wastes that may threaten environmental and human health. As a result, remediating the toxic 4-NP using an effective method without inheriting another toxic material to the environment was considerably required. Fundamentally, an oxidative or a reductive pathway can remove 4-NP from the solution.

Nitrobenzene is one of the most well-known nitroaromatic compounds commonly present as a precursor to manufacture pesticides, polymers, explosives, and dyestuffs. The U.S. Environmental Protection Agency classifies nitrobenzene as a hazardous pollutant with a threshold limit of about one ppm. Moreover, the nitro group attached to the benzene ring causes delocalization of π-electrons to overcome its charge deficiency and make it very stable and resistant to oxidative degradation. Thus, several methods were developed to detoxify nitrobenzene for environmental remediation. Likewise, aniline can also be obtained via reduction of nitrobenzene [54].

The synergistic effect is expected from the combination of Zn(O,S) solid solution and heterojunction concept by forming a Zn(O,S)/Ga_2O_3 nanocomposite. Ga_2O_3 was selected due to its suitable conduction band potential for water reduction to achieve a better photocatalytic activity. After the simultaneous photo excitation on Zn(O,S) and Ga_2O_3 nanoparticles, the excited electron in the conduction band from one phase will transfer to the other phase near the interfaces between them. As a result, this process will increase the amount of photoelectrons, which promote photocatalytic activities on catalyst surfaces. The Zn(O,S)/Ga_2O_3 nanocomposite was synthesized in one pot with a simple precipitation method at 90 °C and normal pressure. The formation of Ga_2O_3 was easily occurred by adding a little hydrazine monohydrate into a thioacetamide solution at 90 °C. Different amounts of Ga precursor were used to obtain various amounts of Ga_2O_3 in Zn(O,S)/Ga_2O_3 nanocomposites to find a suitable composition to optimize the hydrogen evolution rate. The photocatalyst with the best performance was further utilized to reduce 4-NP to 4-AP without using any reducing agents [49]. The list of applications for Zn oxysulfide-based materials in reducing nitrophenol and nitrobenzene can be seen in Table 1.1.

According to Fig. 1.16, the photocatalytic HER begins with water oxidation, creating oxygen vacancy sites on the catalyst surfaces due to surface oxygen anions. Moreover, the presence of ethanol in the solution further enhances the formation of oxygen vacancy sites. These active oxygen vacancy sites play a vital role in facilitating water reduction to produce H_2 and surface oxygen anions.

In Fig. 1.17, based on the reaction mechanisms, both photocatalytic HER and photocatalytic hydrogenation reaction (PHR) compete for the same generated proton on the catalyst during their respective processes. The successful occurrence of PHR relies on the duration that the adsorbed proton remains before being reduced to hydrogen gas. If the proton is reduced and released as hydrogen gas, the availability of protonated catalyst surfaces decreases, leading to a reduction in PHR. Notably, Ho doping has been observed to enhance the oxygen vacancy sites, electrical conductivity, and photoresponse compared to undoped Zn(O,S) as confirmed through XPS, EIS, and TPC analyses.

Figure 1.18 illustrates the kinetic mechanism underlying the hydrogen evolution-involved reaction and the reduction of 4-nitrophenol to 4-aminophenol. When the catalyst is exposed to light energy exceeding 3.6 eV, photocarrier holes and electrons are generated in the valence and conduction bands, respectively. These electrons

Table 1.1 Applications of Zn oxysulfide-based materials for nitrophenol and nitrobenzene reduction

Types of Zn oxysulfide-based materials	Method of preparation	Uses	References
Zn(O,S)/Ga_2O_3	One pot with a simple precipitation method	Reducing 4-nitrophenol	[49]
Ho-doped Zn(O,S)		Reducing 4-nitrophenol	[51]
Mn-doped Zn(O,S) nanoparticle		Reducing 4-nitrophenol	[55]
La-doped Zn(O,S) NPs		Reducing 4-nitrophenol	[50]
Zn, Sn (O,S) NPs		Reducing 4-nitrophenol	[56]
Mg-doped Zn(O,S) NPs	Chemical precipitation process	Reducing 4-nitrophenol	[44]
Y-doped Zn (O, S)		Reducing 4-nitrophenol	[57]
Co-doped Zn (O, S) nanocatalyst		Reducing nitrobenzene	[54]

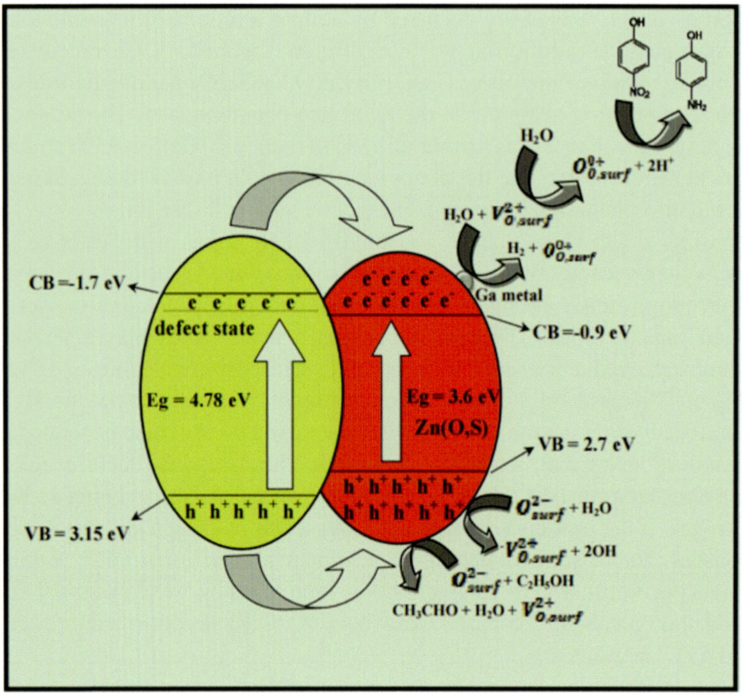

Fig. 1.16 A typical straddling gap of Zn(O,S)/Ga_2O_3 heterojunction with Fermi level alignment [49]

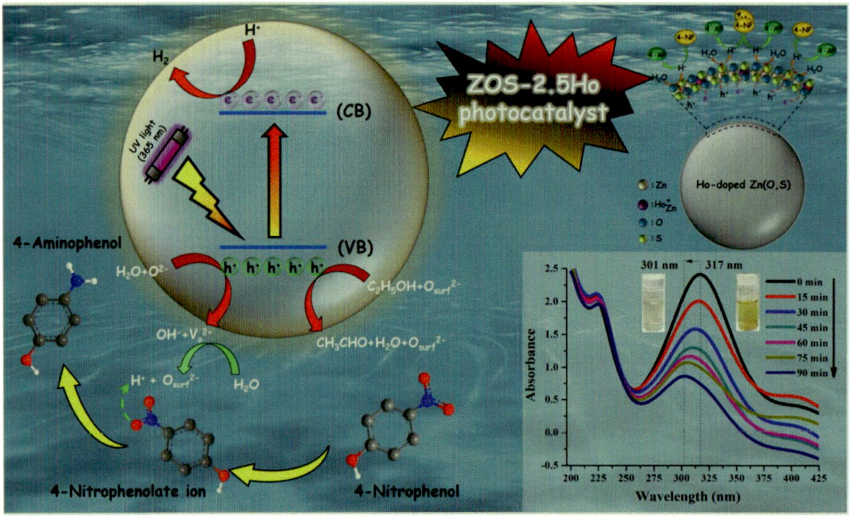

Fig. 1.17 The photocatalytic reaction mechanism for converting 4-NP to 4-AP using ZOS-2.5Ho particles in an ethanol solution under blacklight UV tube lamp illumination and illustration of the molecular reaction during the photocatalytic hydrogenation process [51]

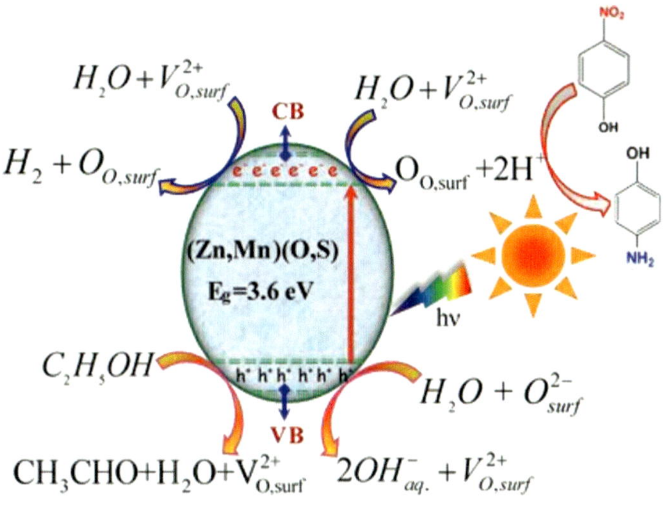

Fig. 1.18 Diagram illustrating the mechanistic process of 4-NP reduction using Mn-doped Zn(O,S) [55]

and holes must migrate to the catalyst surface to participate in the chemical reaction before they recombine. The reduction of 4-nitrophenol commences with the oxidation reaction of water, creating oxygen vacancies. Subsequently, water undergoes reduction via these oxygen vacancies, forming hydrogen ions (H^+). It proposes that these hydrogen ions take part in reducing 4-nitrophenol to 4-aminophenol by replacing the oxygen in the nitro (NO_2) group with an amino (NH_2) group facilitated by hydrogen and electrons. Moreover, incorporating Mn is suggested to weaken the bonding and increase the surface's reactivity. Mn-doped Zn(O,S) with the optimal composition has significantly enhanced its photocatalytic activity.

Figure 1.19 shows the detailed photocatalytic reduction mechanism of converting 4-NP to 4-AP. After the photoexcitation, electrons and holes will separate and diffuse on catalyst surfaces to interact with adsorbed chemical species for reduction and oxidation reactions, respectively. The role of ethanol was crucial for 4-NP reduction; it proposed oxygen vacancy sites as the product of water and ethanol oxidation by photogenerated holes.

As shown in Fig. 1.20, a mechanism involving molecular adsorption phenomena on Y_5-Zn(O,S) nanocatalyst surfaces is depicted in Fig. 1.20a, b for 4-NP and AB, respectively. In Fig. 1.20a, the diagram illustrates that 4-NP molecules containing nitro (NO_2) and hydroxyl (–OH) groups are adsorbed and immobilized at oxygen vacancies (active sites) on Y_5-Zn(O,S). Subsequently, the conversion of 4-NP to 4-AP occurs when the NO_2 groups of 4-NP attach to the oxygen vacancy sites. However, after nitro is reduced to amino groups, the remaining trapped hydroxyl groups of 4-NP obstruct the active sites of the nanocatalyst, leading to a decrease in the HER process.

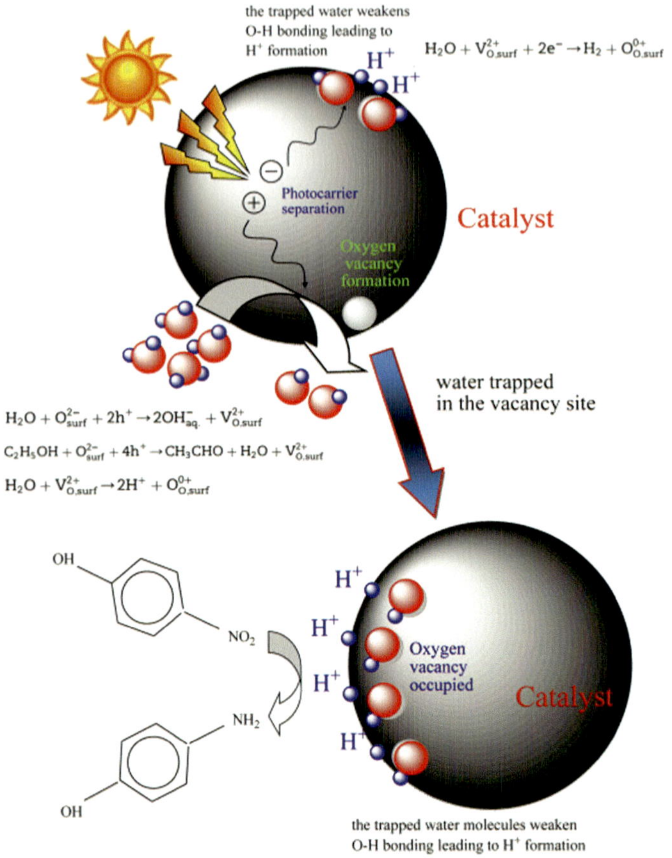

Fig. 1.19 The photocatalytic reduction mechanism of 4-nitrophenol to 4-aminophenol involves the weakening of the O–H bond on the catalyst surface, resulting in the formation of H+ ions

All mechanism discusses various aspects of the photocatalytic reduction of 4-NP and nitrobenzene using Zn oxysulfide-based materials. The highly stable and toxic nature of 4-NP raises environmental and health concerns, prompting the need for effective remediation methods. Different pathways, oxidative or reductive, can remove 4-NP from the solution. Nitrobenzene, another hazardous pollutant commonly used in various industries, necessitates detoxification methods for environmental remediation. Combining Zn(O,S) solid solution and heterojunction concept with Ga_2O_3 results in a synergistic effect, enhancing photocatalytic activity. Zn(O,S)/Ga_2O_3 nanocomposite is synthesized through a simple precipitation method. The optimal composition of Mn-doped Zn(O,S) nanocatalyst shows significantly improved photocatalytic activity. The mechanisms of 4-NP reduction with Y_5-Zn(O,S) nanocatalyst involve molecular adsorption phenomena at oxygen vacancies on the catalyst's surface, leading to 4-AP formation. The presence of ethanol in the solution enhances the formation of oxygen vacancies, crucial for the HER and

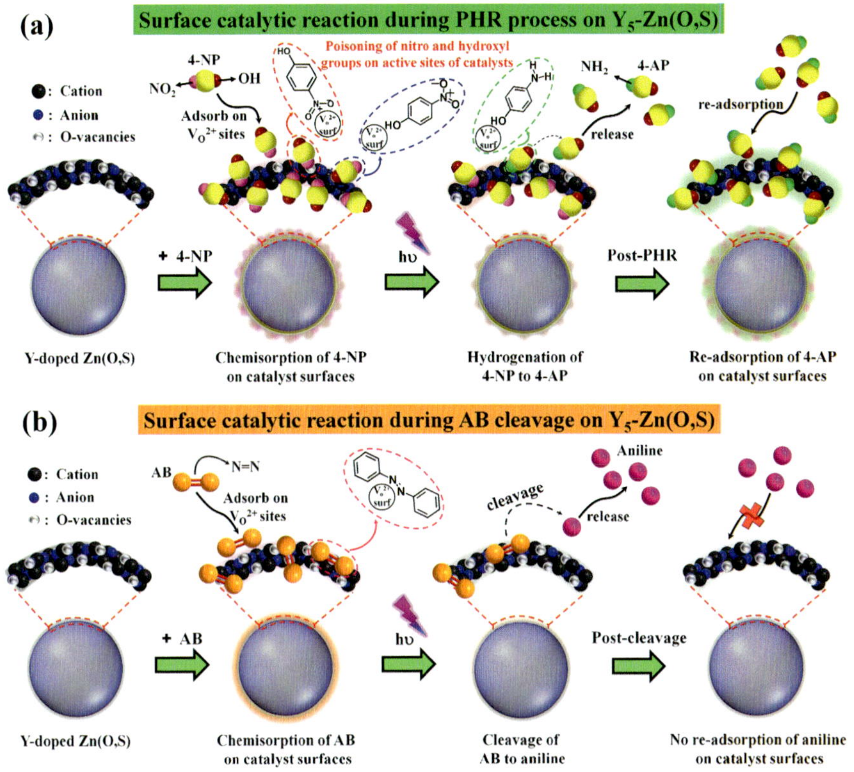

Fig. 1.20 The mechanisms involving molecular adsorption phenomena during the PHR of **a** 4-NP and **b** the cleavage reaction of AB utilizing Y_5-Zn(O,S) nanocatalyst [57]

PHR. Incorporating Mn weakens bonding and enhances surface reactivity, further improving the photocatalytic activity. Additionally, Fig. 1.20 provides a mechanism for molecular adsorption on Y_5-Zn(O,S) nanocatalyst surfaces during 4-NP and AB reduction, affecting the HER process.

1.6 Performances and Mechanisms of Catalytic Materials

Performances and mechanisms of catalytic materials on photocatalytic reduction of nitrophenol and nitrobenzene with Zn oxysulfide semiconductor without using reducing agents refer to the evaluation of the effectiveness and underlying processes involved in using Zn oxysulfide semiconductor materials as catalysts for the photocatalytic reduction of nitrophenol and nitrobenzene. Notably, this process does not rely on external reducing agents, showcasing the potential of green and sustainable catalysis. Here are the critical parameters of catalytic materials:

1. Catalytic Efficiency: The performance of the Zn oxysulfide catalysts is assessed based on their ability to facilitate the reduction of nitrophenol and nitrobenzene under light irradiation. This involves measuring reaction rates, conversion efficiency, and selectivity towards desired products.
2. Kinetics: The reaction kinetics provide insights into the rate of reaction, which helps researchers understand the catalyst's activity and the factors influencing the reaction speed.
3. Stability: The long-term stability of the catalysts is examined to determine whether they maintain their performance over extended periods of photocatalytic reactions.
4. Reusability: The possibility of reusing the Zn oxysulfide catalysts for multiple cycles of photocatalytic reactions is explored to understand their durability and effectiveness.

The study of the performances and mechanisms of Zn oxysulfide semiconductor catalytic materials in the photocatalytic reduction of nitrophenol and nitrobenzene without external reducing agents contributes to our understanding of sustainable catalysis, cleaner chemical processes, and potential applications in environmental remediation and green chemistry.

1.7 Conclusion and Prospective Application

The prospective application of photocatalytic reduction of nitrophenol and nitrobenzene using Zn oxysulfide semiconductor without reducing agents has several potential advantages.

1. Environmentally friendly: The use of photocatalysis eliminates the need for additional reducing agents, which are often toxic or harmful to the environment. By harnessing light energy, this approach offers a greener alternative for reducing nitrophenol and nitrobenzene, minimizing hazardous waste generation (Fig. 1.21).
2. Selectivity and mild reaction conditions: Photocatalytic reduction allows for precise control over reaction conditions, such as pH and temperature, which can influence the selectivity of the reduction reactions. This method can selectively convert nitrophenol and nitrobenzene into desired amino derivatives, minimizing the formation of undesired by-products.
3. Wide range of applications: The photocatalytic reduction process can be applied to various nitroaromatic compounds beyond nitrophenol and nitrobenzene. This approach holds promise for treating different pollutants in industrial wastewater or contaminated sites, as well as in synthesizing valuable intermediates in pharmaceutical or chemical industries (Fig. 1.22).

Fig. 1.21 Environmental friendly is better

Fig. 1.22 Wind turbine solar panel applications

4. Renewable energy utilization: Using sunlight as the energy source for photocatalysis offers the advantage of using renewable energy. By harnessing solar energy, the photocatalytic reduction process becomes more sustainable and energy-efficient.
5. Catalyst stability and recyclability: Zn oxysulfide semiconductor photocatalysts have shown good stability and recyclability, allowing for multiple reaction cycles without significant activity loss. This characteristic contributes to the economic feasibility of the process and reduces the need for frequent catalyst replacement.

6. Synergistic effects: The photocatalytic reduction process can have additional benefits beyond pollutant degradation. For example, the generation of reactive oxygen species during the reaction can contribute to removing other organic contaminants or pathogens present in the system, leading to a more comprehensive treatment process.

It is important to note that while the photocatalytic reduction of nitrophenol and nitrobenzene using Zn oxysulfide semiconductor without reducing agents shows promise, further research and development are still needed to optimize the process parameters, enhance the efficiency, and scale it up for practical applications. Nevertheless, the prospective advantages make this approach attractive for sustainable pollutant remediation and synthesis of valuable compounds.

In conclusion, the photocatalytic reduction of nitrophenol and nitrobenzene using Zn oxysulfide semiconductor without reducing agents holds great potential as an environmentally friendly and selective approach for treating nitroaromatic compounds. This method harnesses light energy to initiate reduction reactions, eliminating the need for additional reducing agents that may be toxic or harmful to the environment. The process offers advantages such as green chemistry, mild reaction conditions, wide applicability, renewable energy utilization, catalyst stability, and synergistic effects. However, further research and development are still required to optimize the process parameters and scale it up for practical applications. Nonetheless, the prospective benefits make this approach attractive for sustainable pollutant remediation and the synthesis of valuable compounds.

References

1. Laksono AD, Damastuti R, Amanah NL, Assa MH, Cheng Y, Ernawati L et al (2023) Photocatalytic and adsorptive removal of liquid textile industrial waste with carbon-based nanomaterials. In: Abdullah H (ed) Photocatalytic activities for environmental remediation and energy conversion. Springer Nature Singapore, Singapore, pp 1–73
2. Ernawati L, Yusariarta AW, Laksono AD, Wahyuono RA, Widiyandari H, Rebeka R et al (2021) Kinetic studies of methylene blue degradation using $CaTiO_3$ photocatalyst from chicken eggshells. J Phys: Conf Ser 1726(1):012017. https://doi.org/10.1088/1742-6596/1726/1/012017
3. Ameta R, Solanki MS, Benjamin S, Ameta SC (2018) Photocatalysis. In: Ameta SC, Ameta R (eds) Advanced oxidation processes for waste water treatment. Academic Press, pp 135–175
4. Cheng C, Liang Q, Yan M, Liu Z, He Q, Wu T et al (2022) Advances in preparation, mechanism and applications of graphene quantum dots/semiconductor composite photocatalysts: a review. J Hazard Mater 424:127721. https://doi.org/10.1016/j.jhazmat.2021.127721
5. Miseki Y, Sayama K (2019) Photocatalytic water splitting for solar hydrogen production using the carbonate effect and the Z-scheme reaction. Adv Energy Mater 9(23):1801294. https://doi.org/10.1002/aenm.201801294
6. Brezová V, Blažková A, Šurina I, Havlínová B (1997) Solvent effect on the photocatalytic reduction of 4-nitrophenol in titanium dioxide suspensions. J Photochem Photobiol, A 107(1):233–237. https://doi.org/10.1016/S1010-6030(96)04577-7

7. Liu L, Chen R, Liu W, Wu J, Gao D (2016) Catalytic reduction of 4-nitrophenol over Ni-Pd nanodimers supported on nitrogen-doped reduced graphene oxide. J Hazard Mater 320:96–104. https://doi.org/10.1016/j.jhazmat.2016.08.019
8. Zhao P, Feng X, Huang D, Yang G, Astruc D (2015) Basic concepts and recent advances in nitrophenol reduction by gold- and other transition metal nanoparticles. Coord Chem Rev 287:114–136. https://doi.org/10.1016/j.ccr.2015.01.002
9. Corbett JF (1999) An historical review of the use of dye precursors in the formulation of commercial oxidation hair dyes. Dyes Pigm 41(1):127–136. https://doi.org/10.1016/S0143-7208(98)00075-8
10. Lunar L, Sicilia D, Rubio S, Pérez-Bendito D, Nickel U (2000) Degradation of photographic developers by Fenton's reagent: condition optimization and kinetics for metol oxidation. Water Res 34(6):1791–1802. https://doi.org/10.1016/S0043-1354(99)00339-5
11. Chang Y-C, Chen D-H (2009) Catalytic reduction of 4-nitrophenol by magnetically recoverable Au nanocatalyst. J Hazard Mater 165(1):664–669. https://doi.org/10.1016/j.jhazmat.2008.10.034
12. Razo-Flores E, Donlon B, Lettinga G, Field JA (1997) Biotransformation and biodegradation of N-substituted aromatics in methanogenic granular sludge. FEMS Microbiol Rev 20(3–4):525–538. https://doi.org/10.1111/j.1574-6976.1997.tb00335.x
13. Afzal Khan S, Hamayun M, Ahmed S (2006) Degradation of 4-aminophenol by newly isolated *Pseudomonas sp.* strain ST-4. Enzyme Microbial Technol 38(1):10–13. https://doi.org/10.1016/j.enzmictec.2004.08.045
14. Du Y, Chen H, Chen R, Xu N (2004) Synthesis of p-aminophenol from p-nitrophenol over nano-sized nickel catalysts. Appl Catal A 277(1):259–264. https://doi.org/10.1016/j.apcata.2004.09.018
15. Narayanan RK, Devaki SJ (2015) Brawny silver-hydrogel based nanocatalyst for reduction of nitrophenols: studies on kinetics and mechanism. Ind Eng Chem Res 54(4):1197–1203. https://doi.org/10.1021/ie5038352
16. Huang C, Ye W, Liu Q, Qiu X (2014) Dispersed Cu_2O octahedrons on h-BN nanosheets for p-nitrophenol reduction. ACS Appl Mater Interfaces 6(16):14469–14476. https://doi.org/10.1021/am5037737
17. Noh J-H, Meijboom R (2015) Synthesis and catalytic evaluation of dendrimer-templated and reverse microemulsion Pd and Pt nanoparticles in the reduction of 4-nitrophenol: the effect of size and synthetic methodologies. Appl Catal A 497:107–120. https://doi.org/10.1016/j.apcata.2015.02.039
18. Fang X, Liu Z, Hsieh M-F, Chen M, Liu P, Chen C et al (2012) Hollow mesoporous aluminosilica spheres with perpendicular pore channels as catalytic nanoreactors. ACS Nano 6(5):4434–4444. https://doi.org/10.1021/nn3011703
19. Mohamed MM, Al-Sharif MS (2013) Visible light assisted reduction of 4-nitrophenol to 4-aminophenol on Ag/TiO_2 photocatalysts synthesized by hybrid templates. Appl Catal B 142–143:432–441. https://doi.org/10.1016/j.apcatb.2013.05.058
20. Hernández-Gordillo A, Arroyo M, Zanella R, Rodríguez-González V (2014) Photoconversion of 4-nitrophenol in the presence of hydrazine with AgNPs-TiO_2 nanoparticles prepared by the sol–gel method. J Hazard Mater 268:84–91. https://doi.org/10.1016/j.jhazmat.2013.12.069
21. Rabaaoui N, Moussaoui Y, Allagui MS, Ahmed B, Elaloui E (2013) Anodic oxidation of nitrobenzene on BDD electrode: variable effects and mechanisms of degradation. Sep Purif Technol 107:318–323. https://doi.org/10.1016/j.seppur.2013.01.047
22. Uberoi V, Bhattacharya SK (1997) Toxicity and degradability of nitrophenols in anaerobic systems. Water Environ Res 69(2):146–156. https://doi.org/10.2175/106143097X125290
23. Corma A, Concepción P, Serna P (2007) A different reaction pathway for the reduction of aromatic nitro compounds on gold catalysts. Angew Chem Int Ed 46(38):7266–7269. https://doi.org/10.1002/anie.200700823
24. Wang J, Yuan Z, Nie R, Hou Z, Zheng X (2010) Hydrogenation of nitrobenzene to aniline over silica gel supported nickel catalysts. Ind Eng Chem Res 49(10):4664–4669. https://doi.org/10.1021/ie1002069

25. Lee S-P, Chen Y-W (2000) Nitrobenzene hydrogenation on Ni–P, Ni–B and Ni–P–B ultrafine materials. J Mol Catal A: Chem 152(1):213–223. https://doi.org/10.1016/S1381-1169(99)00298-8
26. Li H, Zhao Q, Wan Y, Dai W, Qiao M (2006) Self-assembly of mesoporous Ni–B amorphous alloy catalysts. J Catal 244(2):251–254. https://doi.org/10.1016/j.jcat.2006.08.025
27. Flores SO, Rios-Bernij O, Valenzuela MA, Córdova I, Gómez R, Gutiérrez R (2007) Photocatalytic reduction of nitrobenzene over titanium dioxide: by-product identification and possible pathways. Top Catal 44(4):507–511. https://doi.org/10.1007/s11244-006-0098-2
28. Chen S, Zhang H, Yu X, Liu W (2010) Photocatalytic reduction of nitrobenzene by titanium dioxide powder. Chin J Chem 28(1):21–26. https://doi.org/10.1002/cjoc.201090030
29. Mahdavi F, Bruton TC, Li Y (1993) Photoinduced reduction of nitro compounds on semiconductor particles. J Org Chem 58(3):744–746. https://doi.org/10.1021/jo00055a033
30. Rychnovsky SD, Vaidyanathan R, Beauchamp T, Lin R, Farmer PJ (1999) AM1-SM2 calculations model the redox potential of nitroxyl radicals such as TEMPO. J Org Chem 64(18):6745–6749. https://doi.org/10.1021/jo990636c
31. Dai H-P, Shiu K-K (1998) Voltammetric behavior of alizarin red S adsorbed on electrochemically pretreated glassy carbon electrodes. Electrochim Acta 43(18):2709–2715. https://doi.org/10.1016/S0013-4686(97)10185-2
32. Jiang P, Zhou J, Zhang A, Zhong Y (2010) Electrochemical degradation of p-nitrophenol with different processes. J Environ Sci 22(4):500–506. https://doi.org/10.1016/S1001-0742(09)60140-6
33. Yuan S, Tian M, Cui Y, Lin L, Lu X (2006) Treatment of nitrophenols by cathode reduction and electro-Fenton methods. J Hazard Mater 137(1):573–580. https://doi.org/10.1016/j.jhazmat.2006.02.069
34. Zhang W, Xiao X, An T, Song Z, Fu J, Sheng G et al (2003) Kinetics, degradation pathway and reaction mechanism of advanced oxidation of 4-nitrophenol in water by a UV/H_2O_2 process. J Chem Technol Biotechnol 78(7):788–794. https://doi.org/10.1002/jctb.864
35. Augugliaro V, López-Muñoz MJ, Palmisano L, Soria J (1993) Influence of pH on the degradation kinetics of nitrophenol isomers in a heterogeneous photocatalytic system. Appl Catal A Gener 101(1):7–13. https://doi.org/10.1016/0926-860X(93)80133-B
36. Ahn W-Y, Sheeley SA, Rajh T, Cropek DM (2007) Photocatalytic reduction of 4-nitrophenol with arginine-modified titanium dioxide nanoparticles. Appl Catal B 74(1):103–110. https://doi.org/10.1016/j.apcatb.2007.01.016
37. Chang L, Li J, Duan X, Liu W (2015) Porous carbon derived from metal–organic framework (MOF) for capacitive deionization electrode. Electrochim Acta 176:956–964. https://doi.org/10.1016/j.electacta.2015.07.130
38. Aguado S, Canivet J, Farrusseng D (2010) Facile shaping of an imidazolate-based MOF on ceramic beads for adsorption and catalytic applications. Chem Commun 46(42):7999–8001. https://doi.org/10.1039/C0CC02045A
39. Eddaoudi M, Kim J, Rosi N, Vodak D, Wachter J, O'Keeffe M et al (2002) Systematic design of pore size and functionality in isoreticular MOFs and their application in methane storage. Science 295(5554):469–472. https://doi.org/10.1126/science.1067208
40. Grau-Crespo R, Aziz A, Collins AW, Crespo-Otero R, Hernández NC, Rodriguez-Albelo LM et al (2016) Modelling a linker mix-and-match approach for controlling the optical excitation gaps and band alignment of zeolitic imidazolate frameworks. Angew Chem Int Ed 55(52):16012–16016. https://doi.org/10.1002/anie.201609439
41. Bordiga S, Lamberti C, Ricchiardi G, Regli L, Bonino F, Damin A et al (2004) Electronic and vibrational properties of a MOF-5 metal-organic framework: ZnO quantum dot behaviour. Chem Commun 20:2300–2301. https://doi.org/10.1039/b407246d
42. Mahata P, Madras G, Natarajan S (2007) New photocatalysts based on mixed-metal pyridine dicarboxylates. Catal Lett 115(1–2):27–32. https://doi.org/10.1007/s10562-007-9067-z
43. Samuel MS, Bhattacharya J, Parthiban C, Viswanathan G, Pradeep Singh ND (2018) Ultrasound-assisted synthesis of metal organic framework for the photocatalytic reduction of 4-nitrophenol under direct sunlight. Ultrason Sonochem 49:215–221. https://doi.org/10.1016/j.ultsonch.2018.08.004

44. Bekena FT, Abdullah H, Kuo D-H, Zeleke MA (2019) Photocatalytic reduction of 4-nitrophenol using effective hole scavenger over novel Mg-doped Zn(O, S) nanoparticles. J Ind Eng Chem 78:116–124. https://doi.org/10.1016/j.jiec.2019.06.029
45. Zang Z (2018) Efficiency enhancement of ZnO/Cu_2O solar cells with well oriented and micrometer grain sized Cu_2O films. Appl Phys Lett 112(4). https://doi.org/10.1063/1.5017002
46. Saleh R, Djaja NF (2014) Transition-metal-doped ZnO nanoparticles: synthesis, characterization and photocatalytic activity under UV light. Spectrochim Acta - Part A: Mol Biomol Spectrosc 130:581–590. https://doi.org/10.1016/j.saa.2014.03.089
47. Kumar SG, Devi LG (2011) Review on modified TiO_2 photocatalysis under UV/visible light: Selected results and related mechanisms on interfacial charge carrier transfer dynamics. J Phys Chem A 115(46):13211–13241. https://doi.org/10.1021/jp204364a
48. Fu M, Li Y, Wu S, Lu P, Liu J, Dong F (2011) Sol-gel preparation and enhanced photocatalytic performance of Cu-doped ZnO nanoparticles. Appl Surf Sci 258(4):1587–1591. https://doi.org/10.1016/j.apsusc.2011.10.003
49. Abdullah H, Gultom NS, Kuo DH (2017) A simple one-pot synthesis of a Zn(O, S)/Ga_2O_3 nanocomposite photocatalyst for hydrogen production and 4-nitrophenol reduction. New J Chem 41(21):12397–12406. https://doi.org/10.1039/c7nj02505j
50. Abdullah H, Susanto Gultom N, Kuo D-H (2019) Synthesis and characterization of La-doped Zn(O, S) photocatalyst for green chemical detoxification of 4-nitrophenol. J Hazard Mater 363:109–118. https://doi.org/10.1016/j.jhazmat.2018.09.030
51. Abdullah H, Gultom NS, Shuwanto H, Kebede WL, Kuo D-H (2020) Self-protonated Ho-doped Zn(O, S) as a green chemical-conversion catalyst to hydrogenate nitro to amino compounds. ACS Appl Mater Interfaces 12(39):43761–43770. https://doi.org/10.1021/acsami.0c12695
52. Sisay G, Abdullah H, Kuo D-H, Lakew W, Shuwanto H, Fentie S (2021) Zn-Ce-Ga trimetal oxysulfide as a dual-functional catalyst: hydrogen evolution and hydrogenation reactions in a mild condition. Appl Surf Sci 563:150383. https://doi.org/10.1016/j.apsusc.2021.150383
53. Markovskaya DV, Gribov EN, Kozlova EA, Kozlov DV, Parmon VN (2020) Modification of sulfide-based photocatalyst with zinc- and nickel-containing compounds: correlation between photocatalytic activity and photoelectrochemical parameters. Renew Energy 151:286–294. https://doi.org/10.1016/j.renene.2019.11.030
54. Shuwanto H, Gultom NS, Abdullah H, Kuo D-H (2020) Environmentally benign photoreactions for hydrogen production and cleavage of N=N bond in azobenzene over co-doped Zn(O, S) nanocatalyst: the role of in situ generated H+. ACS Appl Energy Mater 3(12):12692–12702. https://doi.org/10.1021/acsaem.0c02668
55. Susanto Gultom N, Abdullah H, Kuo D-H (2018) Convenient synthesis of Mn-doped Zn (O, S) nanoparticle photocatalyst for 4-nitrophenol reduction. J Phys: Conf Ser 1007(1):012061. https://doi.org/10.1088/1742-6596/1007/1/012061
56. Abdullah H, Kuo D-H (2019) Utilization of photocatalytic hydrogen evolved (Zn, Sn)(O, S) nanoparticles to reduce 4-nitrophenol to 4-aminophenol. Int J Hydrogen Energy 44(1):191–201. https://doi.org/10.1016/j.ijhydene.2018.04.036
57. Shuwanto H, Abdullah H, Kuo D-H, Gultom NS (2021) Surface active sites of Y-doped Zn(O, S) for chemisorption and hydrogenation of azobenzene and nitroaromatic compounds under light via self-generated proton. Appl Surf Sci 552:149508. https://doi.org/10.1016/j.apsusc.2021.149508

Chapter 2
Photoreactions on Hydrogen Production and Cleavage of Azo Bond in Azobenzene Over Metal Oxide and Sulfide Nanocatalysts in a Mild Condition

Phan Van Hoang Khang, Ly Tho Xuan, Tran Nguyen Hoang Phan, Tran Thi Bich Quyen, Phan Thi Bao Tran, Hairus Abdullah, and Riski Titian Ginting

Abstract The rapid global population during this decade is instigating a substantial surge in demand for global energy resources. Nevertheless, the depletion of fossil fuels, including petroleum, coal, and natural gas, is progressively increasing. Furthermore, the detrimental environmental pollution from conventional energy resources poses a significant challenge. For those reasons, research on sustainable energy sources such as H_2 production by photocatalysts is becoming a hot topic due to its eco-friendliness and sustainability. The discharge of an enormous quantity of untreated azobenzene into the environment gives rise to many significant environmental issues and complications. In order to address these concerns and maximize

P. Van Hoang Khang · L. T. Xuan
Department of Materials Science and Engineering, National Taiwan University of Science and Technology, Number 43, Section 4, Keelung Road, Da'an District, Taipei City 10607, Taiwan

T. N. H. Phan
College of Natural Sciences, Can Tho University, 3/2 Street, Ninh Kieu District, Can Tho City 90000, Vietnam

T. T. B. Quyen
Faculty of Chemical Engineering, College of Engineering, Can Tho University, 3/2 Street, Ninh Kieu District, Can Tho City 90000, Vietnam

Nano-Electrochemistry Laboratory, Can Tho University, Room 218, CTU Hi-Tech Building, 3/2 Street, Ninh Kieu District, Can Tho City 900000, Vietnam

P. T. B. Tran
Faculty of Economics, Dong Thap University, Number 783, Pham Huu Lau Street, Cao Lanh City, Dong Thap Province, Vietnam

H. Abdullah
Department of Industrial Engineering, Universitas Prima Indonesia, Medan, Indonesia

R. T. Ginting (✉)
Department of Electrical Engineering, Universitas Prima Indonesia, Medan, Indonesia
e-mail: titiangt@unprimdn.ac.id

H. Abdullah (ed.), *Solar Light-to-Hydrogenated Organic Conversion*,
https://doi.org/10.1007/978-981-99-8114-4_2

economic advantages, there has been considerable scientific interest in employing metal oxide and sulfide nanocatalysts under mild conditions for the hydrogenation of azobenzene, leading to the production of aniline and hydrazobenzene. This approach is highly efficient and reproducible, making it a significant focus research field for researchers. In this chapter, we would like to summarize the requirements of H_2 production and organic chemicals conversion. Besides, we also discuss promising photocatalysts for H_2 production and hydrogenation of azobenzene, such as TiO_2, Zn(O,S), CdS, and $NaNbO_3$. To enhance our comprehension of H_2 production and the hydrogenation process of azobenzene, the possible mechanisms and detailed performance analyses was discussed. Additionally, we provide an overview of potential application prospects for future research.

Keywords Photocatalyst System · Photoreaction · Mechanism · Azobenzene Reduction · Hydrogenation of Azobenzene

2.1 Introduction

Considering the escalating global population, there has been a substantial surge in energy consumption requirements for daily living and production activities in recent years. Nonetheless, conventional energy sources like petroleum, natural gas, and coal are undergoing depletion. Furthermore, these fossil fuels are prominent contributors to climate change and the greenhouse effect. Renewable resources such as wind, sunlight, geothermal heat, and wave energy hold great promise as viable alternatives, with sunlight being particularly noteworthy due to its affordability, eco-friendliness, and limitless availability. Consequently, scientists and industry have a significant surge of interest in researching hydrogen (H_2) production and the hydrogenation of organic compounds into valuable compounds using photocatalysts.

Under photoexcitation, the incident light triggers the generation of electron–hole pairs. Subsequently, the charge carriers become actively involved in photo-reduction processes, leading to the formation of H_2 gas or the disruption of organic linkages, ultimately resulting in the production of valuable compounds. Hairus et al. (2021) successfully synthesized a composite of In-doped Zn(O,S)/Zn-doped $In(OH)_3$-xS_x through a one-step hydrothermal method and investigated its capacity for H_2 evolution under visible light. The yield of the H_2 evolution reaction (HER) was found to be approximately 730 μmol/g·h. In a separate study, Sethupathi et al. (2021) introduced a novel material, Ni-doped $ZnIn_2(O,S)_4/In(OH)_3$ heterojunction composites for HER. They explored the influence of sulfur dosages on the photocatalytic activity. The HER process demonstrated an impressive yield of up to 2100 μmol/g·h. Furthermore, Hairus et al. (2022) fabricated an environmentally friendly composite using Zn(O,S) and banana peel for HER under solar light irradiation. The composite material exhibited great promise as a candidate for HER, achieving a H_2 gas yield of up to 9232 μmol/g for 5 h, which was 1.7 times higher than that of pristine Zn(O,S).

This material presents a novel approach to utilizing bio-waste for the generation of green fuel.

On the other hand, wastewater from industrial textile processes can have severe adverse effects on the environment and human health [4]. The textile industry is one of the largest water-consuming industries globally, with significant amounts of wastewater generated from dyeing, printing, and finishing processes [5]. This wastewater is often discharged into aquatic environments such as rivers and lakes, leading to pollution of these resources. The wastewater can contain many kinds of toxic compounds such as dyes, heavy metals, and solvents that can have serious ecological consequences [6]. These pollutants can affect the water quality, reducing oxygen levels and killing aquatic life, disrupting the ecosystem [7]. In addition, the discharge of untreated or poorly treated wastewater can contaminate the soil, leading to reduced agricultural productivity and posing a risk to human health [4]. This contaminated water can also seep into underground water sources, leading to long-term contamination of drinking water sources, which can lead to health problems such as cancer, liver, and kidney damage [8]. The pollution of water bodies from textile industry wastewater can also have economic implications, as it can affect tourism and other industries that rely on clean water sources [9]. The improper disposal of textile industry waste into the water system further exacerbates environmental pollution, contributing to the accumulation of non-biodegradable materials and causing long-term ecological harm [4, 8]. The textile industry uses many different kinds of chemicals and dyes as well as pigments. Among them, azobenzene (AB) is a common pollutant found in textile wastewater [10–13]. AB can be released into wastewater in various forms, including unreacted AB, the traces in reactive azo dyes, non-reactive azo dyes, and AB-based pigments [14–17]. AB generally consists of two aromatic rings connected by a diazo group (–N = N–), which gives its characteristic shape [14, 18]. AB is also a nonpolar molecule due to the symmetrical arrangement of its atoms, which makes it miscible in aqueous medium but immiscible in organic solvents [19]. From a chemical standpoint, AB demonstrates high thermal and chemical stability, displaying resilience against degradation processes [20].

AB is highly toxic to human health and the environment. Exposure to AB has been linked to various health hazards, including skin and eye irritation, respiratory problems, and cancer [14, 17, 18, 21]. Furthermore, azobenzene (AB) is classified as a hazardous air and water pollutant, with the potential to persist in the environment and cause adverse effects on aquatic ecosystems [21]. The treatment of AB-contaminated wastewater poses challenges due to its toxic nature and persistence in the environment [21]. Several biological treatment approaches, such as biodegradation and bio-sorption, have been employed for the removal of AB from wastewater [22]. Additionally, techniques like activated carbon adsorption and membrane filtration can be utilized to eliminate AB from water sources [23]. However, despite the effectiveness of these methods, they can generate secondary waste, necessitating further treatment steps.

Aniline (AN) and hydrazobenzene (HB) are two important chemical compounds that play significant roles in various industrial processes and everyday life. AN is commonly used as a precursor to various chemical compounds, including dyes, pharmaceuticals, and rubber additives [14]. It is also used in the manufacturing of polyurethane foams and insulation materials [24]. HB, on the other hand, is used primarily as a reducing agent and a precursor to other chemicals such as azo dyes, pharmaceuticals, and herbicides [25]. In addition to their industrial applications, both AN and HB have been found to have potential biological activities, with AN showing antimicrobial and antitumor properties, while HB is being investigated for its anti-inflammatory and antioxidant properties [26, 27].

However, the production of AN and HB by traditional methods can exist several drawbacks [28]. Firstly, these methods often require the use of toxic and expensive metal catalysts, which can lead to environmental pollution and increased production costs [29–33]. Secondly, traditionally synthetic methods can result in unwanted by-products, which can increase waste generation and reduce the yield of the desired product. Thirdly, the traditional methods may require high temperatures and pressure, increasing energy consumption and production costs. Therefore, the use of photocatalytic hydrogenation of AB (PHA) using nano-catalysts, such as Au [34], Mo [35], Fe complexes [36], Bi_2MoO_6 with $(NH_4)_2C_2O_4$ [31, 37], TiO_2 [18, 38, 39], $NaNbO_3$ [25], Fe_2O_3 [40], $BiVO_4$ [41, 42], WO_3 [41, 43, 44], Ag_3PO_3 [45], g-C_3N_4 [46], Zn(O.S) [1, 3, 14–16, 47], CdS [38, 48–50] and so on to produce AN and HB offers a potential solution to these negative sides, as it provides a more sustainable, cost-effective, and selective novel.

The PHA offers several advantages over traditional methods. Firstly, photocatalysis is a green and sustainable technology that has no requirements toxic metals that generate toxic secondary wastes. Secondly, photocatalysis can achieve high selectivity in producing AN and HB, leading to minimal waste generation and reduced costs. Thirdly, using photocatalytic materials such as TiO_2 is relatively cost-effective and abundant, making the process economically viable. Furthermore, the photocatalytic hydrogenation can be proceeded under mild reaction conditions, minimizing energy consumption and increasing the efficiency of the process. Therefore, the PHA provides a promising novel to the production of AN and HB with significant economic and environmental benefits.

Because the photocatalytic hydrogenation reaction (PHR) using nano-catalysts is a novel approach and highly promising to convert AB to AN and HB, or decompose organic wastes [51]. Thus, this field is received tremendous attention among researchers. Undoubtedly, as science and technology continue to advance, numerous novel approaches have been proposed for the synthesis of photocatalytic materials. These methods include hydrothermal, solvothermal, ultrasonic irradiation, mechanothermal, sol–gel, microwave irradiation, solid-state reactions, and co-precipitation methods. Ho et al. (2006) reported low-temperature hydrothermal synthesized of S-doped TiO_2 photocatalysts. The well-defined nano-crystalline of S-doped TiO_2 could be achieved under hydrothermal conditions, where TiS_2 and HCl served as the precursors [52]. The correlation between TiS_2 concentration and band energy was confirmed by UV–vis spectrometer. When TiS_2 concentration, the band energy becomes narrower. Yang et al. (2000) synthesized CdS nanorods (NRs) via solvothermal route using thiourea and cadmium nitrate as precursors in ethylenediamine [53].

The formation of CdS NRs was studied by TEM, SAED, XRD, SAED, and IR techniques. The results confirmed that CdS NRs was formed along c-axis. Reza et al. (2017) used the sol–gel method to synthesize ZnO nanoparticles (NPs) and study the effects of time and ultrasonic irradiation power [54]. Yin et al. (2003) synthesized N-doped rutile TiO_2 by utilizing the ball milling method, followed by calcination at 400 °C [55]. This material showed two absorption peak at 400 and 540 nm in UV–vis spectra, simultaneously demonstrated the excellent photo-activity for NO oxidization under visible light. Olga et al. (2020) reported a novel to fabricate ZrO_2-ZnO NPs with various dosages of ZrO_2 by utilizing a combine method (precipitation and microwave irradiation) was examined by XRD, TEM, EDS, and FTIR [56]. It was found that concentration of ZrO_2 10% was the optimal value for the photocatalytic reaction.

Many efforts have been devoted to fabricate the new materials as well as attempting better understanding about their properties and performance. Hairus et al. (2019) synthesized successfully the Dy-doped Zn(O,S) NPs with different Dy ratios, and further used for PHR [57]. To better understand their properties, the structural, morphological, electrochemical, and optical analyses were carried out systematically. It is found that Dy-doped Zn(O,S) NPs containing 10% Dy can convert completely 60 ppm AB to AN in 6 h under UV light. Another example, Hardy et al. (2020) fabricated Co-doped Zn(O,S) solid solution with various Co doping [15]. These materials promise as a high potential candidate for H_2 production as well as hydrogenation reaction of contaminants. The structure of Co-based catalysts was investigated by SAED, XRD, and DRS. The result showed that a Co dosage of 2.5% is the optimal value for the doping as well as suitable for HER. This material performed desirable yield for photo-evolved H_2 with 27,000 and 1565 μmol/g under UV light and sunlight, respectively. Moreover, incorporating Co of 2.5% also exhibited a decent efficiency for PHR either to break the azo bonds in AB or to reduce NO_2 groups in nitrobenzene (NB). AB and NB were converted completely during 30 and 120 min, respectively. Hairus et al. (2021) fabricated Ni-CdS NPs with different ratios of Ni, and applied then to cleave the azo bond of AB and methyl orange (MO) under visible light [58]. The optimal doping ratio of Ni CdS was 2%, in which the conversion AB and MO can reach the highest yield. Furthermore, this material also exhibited the high potential possibility for industrial application due to reusability up to 100 cycles.

On the other hand, the studies about kinetic mechanism are also carried out systematically. Julia et al. (2009) carried out the photo-reduction of NB to AN in an alcoholic media as well as investigating the effects of reaction conditions on kinetics [59]. An appropriate model about the effects of catalyst concentration, light intensity, initial concentration, as well as temperature was considered. Xu et al. (2022) investigated TiO_2/Ce_2S_3 S-scheme heterojunction photocatalysts for PHR with H_2O [60]. Furthermore, they also proposed the mechanism using the density functional theory calculation to elucidate the electron transfer from Ce_2S_3 to TiO_2 due to the differences in Fermi level.

Recently, machine learning, artificial intelligence, and computational science also were applied in the photocatalytic field. Daniels et al. (2021) [61] studied the selectivity azo(xy) versus AN in the nitroarene reduction over intermetallic such as Pd_2Sn, Pd_3Sn_2, Pd_2Ge, PtSn, Pd_3Pb, and Pd_xSn_{1-x} in experiment and using machine learning algorithm. In this study, "machine-learned potentials using a deep neural network" were applied to identify the crucial reaction species on the various types of nanocatalysts and can be used to study the selectivity further. Accordingly, Pd_2Ge, Pd_2Sn, Pd_3Sn_2, and PtSn NPs favor for azo(xy) formation, compared to that of AN. Regarding Pd_3Pb NPs and Pd_xSn_{1-x} nanoalloys, AN formation to azo(xy) formation is preferable. Wang et al. (2019) used "support vector machines and quadratic regression" to study the influences of silanol groups and experimental factors on hydrogenation of NB [62]. Roy et al. (2021) reported a high promising approach using the combination of "machine learning-aided high-throughput screening" and microstructure model to determine the new active photocatalysts for CO_2 conversion to methanol [63]. They found seven new active photocatalysts, including CuCoNiZn-based tetrametallic (two types), CuNiZn-based trimetallic (three types), and CuCoZn-based trimetallic alloys (two types) by computational simulations.

The PHA is a promising technique that has gained significant attention because of its excellent selective conversion. AB is a widely used organic compound that can be converted into HB and AN by photocatalytic hydrogenation using suitable photocatalysts and light sources. This process has several advantages, such as mild reaction conditions, high efficiency, and selectivity, making it an attractive alternative to traditional methods. Moreover, HB and AN are versatile intermediates that can be further converted into useful compounds such as pharmaceuticals, agrochemicals, and dyes. Therefore, PHA has great potential as a green and sustainable approach to produce value-added chemicals. Despite the growing number of annual publications on PHA and the significant future potential of this process, there is currently a lack of comprehensive literature reviews of addressing the photo-evolution of H_2 and the cleavage of azo bonds in azobenzene (AB) using metal oxide and sulfide nanocatalysts under mild conditions. Based on these considerations, this chapter aims to delve deeply into the aforementioned issues and provide an in-depth discussion. The requirements of photocatalytic H_2 production and organic chemical conversion are displayed in Sect. 2.2. The promising photocatalysts such as TiO_2, Zn(O,S), CdS, and $NaNbO_3$ are focused analytically in Sect. 2.3. The applications also are listed in Sect. 2.4. To better understand photo-reduction, we also list the performances and mechanisms of catalytic materials in Sect. 2.5 and Sect. 2.6, respectively. Finally, Sect. 2.7 provides a comprehensive overview of the conclusions drawn from the study and outlines the future prospects and potential applications.

2.2 Requirements of Hydrogen Production and Organic Chemical Conversion

Hydrogen, being an environmentally friendly and promising renewable energy source, holds great potential for the future; not only does it possess a higher energy density compared to traditional fuels [65] but hydrogen-based fuels are also preferable due to their combustion process producing only water as a by-product, which is a topic of increasing interest in both basic and practical researches. This characteristic has attracted a lot of attention in both fundamental and applied research. However, conventional techniques for hydrogen production often rely on the consumption of fossil minerals and artificial chemicals, resulting in carbon dioxide emissions [66]. Consequently, there is an urgent need for the development of straightforward, eco-friendly, and efficient processes that can generate affordable emission-free hydrogen energy. A notable breakthrough in this regard is photocatalytic H_2O splitting, which employs semiconductor particles in a solution to produce H_2 and O_2 when exposed to sunlight. The conceptual representation of this innovation is depicted in Fig. 2.1. Ultimately, this process yields a hydrogen supply that is environmentally benign, renewable, and free from harmful by-products.

$$\mathrm{H_2O} \rightarrow \mathrm{H_2} + 0.5\mathrm{O_2}\delta G^0 = 238\mathrm{kJ/mol} \tag{2.1}$$

$$\text{photocatalysts} \overset{hv>Eg}{\rightarrow} \text{photocatalysts}^* + \mathrm{H}^+ + e^- \tag{2.2}$$

$$\mathrm{H_2O} + 2h^+ \rightarrow 0.5\mathrm{O_2} + 2\mathrm{H}^+(+0.82\mathrm{VvsNHE, pH} = 7) \tag{2.3}$$

$$2\mathrm{H}^+ + 2\mathrm{e}^- \rightarrow \mathrm{H_2}(-0.41\mathrm{VvsNHE, pH} = 7) \tag{2.4}$$

When illuminated by sunlight or light from another energy source, the photocatalysts can initiate photochemistry, creating an electron–hole pair. In order to begin the oxidation reaction, the valence band (VB) and conduction band (CB) must be over the threshold of water oxidation potential and hydrogen evolution potential, respectively (Eqd. 2.3–2.4). The process of photocatalytic water splitting, as evidenced in Eqs. 2.1–2.4, encompasses three key stages. The outcome of this approach yields an eco-friendly source of hydrogen without any toxic by-products, but achieving simultaneous production of H_2 and O_2 through the single photocatalysts. In addition, the photocatalysts must meet specific energy and band structure criteria. Therefore, a viable approach is to combine co-catalysts, and numerous catalyst designs have been developed to achieve this objective. Among these, heterogeneous photocatalysts stand out as they offer remarkable chemical stability in liquid environments, recyclability, robustness, low toxicity, and cost-effectiveness.

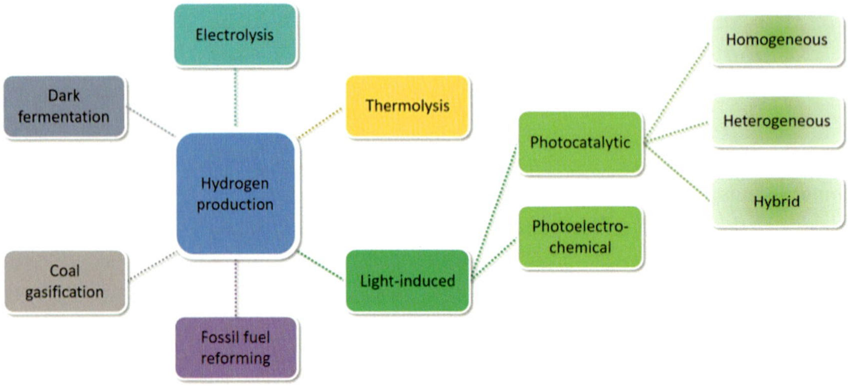

Fig. 2.1 The novels for H_2 generation and solar-induced H_2 evolution (reproduced with permission Copyright 2023, MDPI) [64]

Nevertheless, achieving the simultaneous generation of H_2 and O_2 from a single photocatalyst (p-cats) presents a significant challenge. Thus, the utilization of co-catalysts has emerged as a common approach. Notably, heterogeneous photocatalysts exhibit desirable characteristics such as high chemical stability in liquid environments, good recyclability, robustness, low toxicity, and cost-effectiveness. Commonly, the interest in metal sulfides (MS) as semiconducting materials has garnered considerable attention due to their exceptional properties. Various binary MS compounds, including SnS_2, PbS, ZnS, CdS, MoS_2, In_2S_3, NiS/NiS_2, Cu_2S, Bi_2S_3, and CoS_2, as well as their derivatives and heterostructures, have found extensive use in diverse applications [67, 70–74]. Figure 2.2 provides an overview of commonly employed MS compounds for photocatalytic H_2 production. These compounds have been extensively investigated due to their narrow bandgap, favorable electron transfer properties, and enhanced chemo-selectivity, rendering them highly effective as photocatalysts.

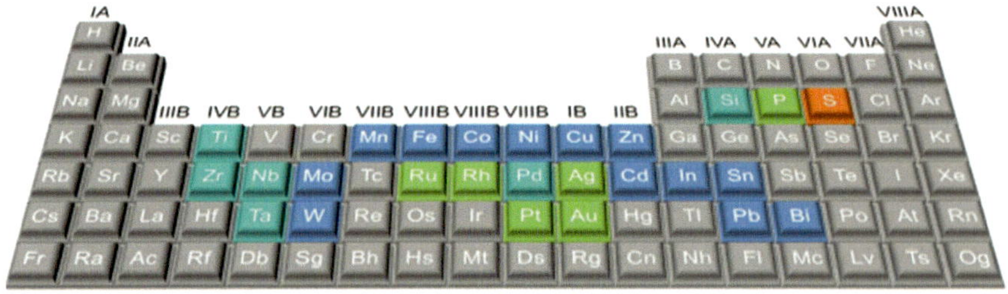

Fig. 2.2 Most common chemical elements for fabrication of (blue color) metal sulfide compounds/ (cyan color) oxides/(green color) dopant agents for photocatalytic hydrogenation (Reproduced with permission. Copyright 2023, MDPI) [67–69]

Moreover, the quantum size effects can give rise to desirable properties, including enhanced charge transfer rates and prolonged excited state lifetimes [75, 76]. In the past decade, there has been significant progress in enhancing solar hydrogen generation through the utilization of metal sulfide (MS) semiconductor photocatalysts [77–79]. Studies have demonstrated that ternary sulfides such as $CuInS_2$ and $CuGaS_2$ exhibit high selectivity and photocatalytic activity, particularly when combined with titanium dioxide (TiO_2) [80, 81]. Current research efforts are focused on exploring MSs with reduced dimensionalities, such as 2D nanosheets, anisotropic structures, and flakes, which have shown promising results in hydrogen production. Furthermore, the functionalization of these low-dimensional structures with metal dopants or co-catalysts (as depicted in Fig. 2.2) is proving to be an effective strategy for enhancing their photoactivity, thereby emerging as a significant area of research [82].

The requirements for an exceptional photocatalyst are outlined in Fig. 2.3, as detailed in references [64, 83]. In addition to these considerations, safety and cost factors are of utmost importance, particularly in industrial applications. To address these challenges, researchers have been actively investigating alternative approaches that prioritize sustainability, safety, and cost-effectiveness while minimizing energy consumption, as evidenced by referenced [84–88]. Furthermore, achieving high chemo-selectivity under environmentally benign conditions is crucial for ecologically friendly chemical conversions.

The stability of hydrogen production and hydrogenation processes is a critical aspect of their development. Additionally, there is a preference for employing

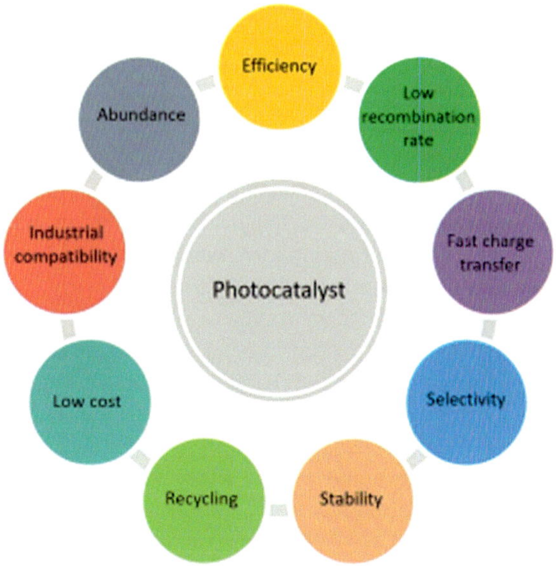

Fig. 2.3 General requirements for designing highly efficient photocatalytic H_2 production (reproduced with permission. Copyright 2023, MDPI) [64]

mild conditions due to their advantages of lower energy consumption and cost-effectiveness. Hardy et al. (2020) [15] used Co – Zn(O,S) nanoparticles as a photocatalyst to convert AB and NB to AN under mild conditions. This photocatalyst shows good reusability with five cycles without lowering the efficiency [15].

Figure 2.4 presents an overview of the general strategies commonly employed in addition to specific strategies for metal sulfide-based photocatalysts. The current focus lies in designing new morphologies and modifying particles, including hollow structures, one-dimensional nanowires, porous morphologies, nanorods (NRs), zero-dimensional quantum dots, and other innovative approaches, to enhance the performance of the hydrogen evolution reaction (HER). Notably, mesoporous structures have shown favorable characteristics for facilitating the efficient transport of reactants and reaction products [89]. Furthermore, the presence of a porous structure enhances the absorption of sunlight by promoting multiple reflections of light waves within the porous network [90]. Moreover, the porous structure offers numerous active sites and a larger specific surface area, thereby facilitating the photocatalytic reaction. This can be achieved by designing suitable architectures with low-dimensional nanostructures. It is worth noting that the recombination rate of electrons and holes in semiconductor photocatalysts is approximately 2×10^7 fs, which is three orders of magnitude faster compared to oxidation–reduction reactions (10^{10}–10^{11} fs) [91]. Hence, single-component photocatalysts are typically insufficient for the hydrogen evolution reaction (HER). To address this limitation, the utilization of metal sulfide (MS) metal oxide co-catalysts has emerged as a promising and effective approach. This type of co-catalyst reduces the bandgap energy and recombination rate of electrons and holes, resulting in enhanced photoactivity for HER.

In addition, the emergence of quantum size effects, discrete and trap states formed on quantum particles, and size reduction has shown beneficial effects on photoreactions. These effects facilitate charge transport, accelerate charge transfer, and reduce recombination rates [92–95]. Moreover, integrating metal sulfide (MS) photocatalysts with plasmonic metals has garnered significant attention in recent years. In such cases, photo-generated electrons can be transferred to adjacent metals due to the lower fermi level, enhancing photocatalytic performance.

The photocatalytic reaction involves oxidation and reduction processes, which are facilitated by generating photo-induced holes and electrons, respectively. The presence of OH* radicals produced through photosynthesis can further enhance these processes and effectively oxidize various hazardous chemical compounds [96–98]. Furthermore, with appropriate valence band (VB) and conduction band (CB) positions, the photo-generated charges can promote water oxidation and proton reduction, leading to the production of hydrogen gas. [57, 87, 88, 99–102].

In the present day, photocatalysts offer two significant advantages for hydrogenation in the liquid phase: safety and purity. Traditional catalytic hydrogenation reactions often rely on reducing substances such as $NaBH_4$ and NH_4HCO_2 as proton sources [103–106]. However, these compounds are neither environmentally friendly nor safe. To address these limitations, photocatalysts are a viable alternative, providing a more sustainable and better approach [107–109].

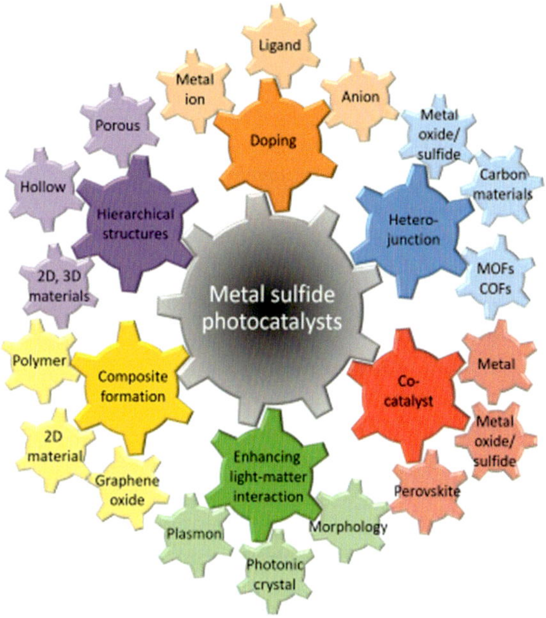

Fig. 2.4 Strategies for improving HER of MS-based photocatalysts (Reproduced with permission. Copyright 2023, MDPI) [64]

Several studies have suggested that the cleavage of the (–N = N–) bond in AB occurs at a slow conversion rate [17, 18] Previous research indicates that AB exhibits higher selectivity for HB conversion compared to AN. Specifically, Pt/TiO_2 achieved only 9% AN conversion after 1 h, which increased to 19.2% after 3 h [86]. Shiraishi also reported that the PHR of AB favors HB over AN [51]. Notably, commonly used photocatalysts such as $BiVO_4$, WO_3, Fe_2O_3, g-C_3N_4, and Ag_3PO_4 showed negligible conversion of AB to AN, with yields below 0.1% [42, 110]. Consequently, further investigations are necessary to enhance the reaction efficiency by modifying potential photocatalysts capable of hydrogen evolution, while considering the potential environmental impact of these compounds.

2.3 Promising Photocatalysts

2.3.1 TiO_2

Pd nanoparticle supported on TiO_2 (Pd/TiO_2 NPs) is a popular photocatalytic system due to the excellent properties of TiO_2 and the catalytic activity of Pd for various reactions. TiO_2 is a wide-bandgap semiconductor that exhibits strong absorption of UV light and generates reactive oxygen species upon irradiation, which can promote various oxidation and reduction reactions. The incorporation with Pd can increase TiO_2 photocatalytic efficiency, making them effective for various reactions, including the synthesis of secondary amines. Pd/TiO_2 NPs have demonstrated remarkable efficacy in the eco-friendly secondary amine synthesis from AB and alcohols under UV irradiation at room temperature. This approach provides a viable and sustainable alternative to conventional techniques requiring higher temperatures and/or hazardous chemicals [111].

$Pd(NO_3)_2$ was impregnated on P25 TiO_2 to create the Pd_x/TiO_2 NPs. Figure 2.5 reveals that the Pd/TiO_2 NPs are spherical in shape [111]. The spectra obtained through DRS analysis of the catalysts revealed a distinctive even absorption pattern at around 430 nm (Fig. 2.6). This is explained by the light diffraction caused by the Pd NPs [111]. As presented in Table 2.1, the Pd_2/TiO_2 catalyst demonstrates its effectiveness in the synthesis of diverse secondary amines. When non-aromatic alcohols are reacted with AB, high yields (99%) of the corresponding secondary amines are achieved, as observed with benzyl alcohol (entry 4). The catalyst also performs well in large-scale reactions, although longer reaction times are necessary (entry 5). It is worth noting that reducing the catalyst quantity and lowering the light intensity result in prolonged reaction times for secondary amine synthesis, emphasize the significance of these parameters for efficient production. Moreover, the Pd_2/TiO_2 catalyst proves successful in converting AB with various substituents (–CH_3, –OCH_3, –OH, –CN, and –COOH groups) into secondary amines with yields of at least 80%, while maintaining the substituents intact. However, the reaction with halogen-substituted AB is unsuccessful due to dehalogenation of the H–Pd species (entry 14) [112]. Therefore, these findings demonstrate that the Pd_2/TiO_2 catalyst has a broad applicability in the synthesis of secondary amines, with only a few cases of failure attributed to subsequent reactions of the substituents [111].

To summarize, the utilization of Pd/TiO_2 NPs system enables the production of secondary amines through a consecutive photocatalytic and catalytic reaction pathway initiated by photoexcitation. Despite the challenge of efficiently consuming a substantial amount of H-atoms from alcohols, this approach offers several advantages, including the use of recyclable and recoverable heterogeneous catalysts, mild reaction conditions, and the utilization of waste AB. As a result, this method holds significant promise as an environmentally friendly and efficient approach for the one-step synthesis of secondary amines [111].

2.3.2 *Zn(O, S)*

Zn(O, S) is a promising nanocatalyst for various photoreactions, including hydrogen production and azo bond cleavage in AB under mild conditions. This nanocatalyst is the combination of ZnO and ZnS, which work together to enhance its photo-activity. Recent studies have shown that Zn(O, S) nanocatalysts exhibit remarkable performance in promoting these photoreactions, which are of great interest in renewable energy and chemical synthesis. Additionally, the mild reaction conditions of Zn(O, S) suitable for attractive practical applications. The use of metal-based oxide and sulfide nanocatalysts for photoreactions is a rapidly growing field of research, and Zn(O, S)

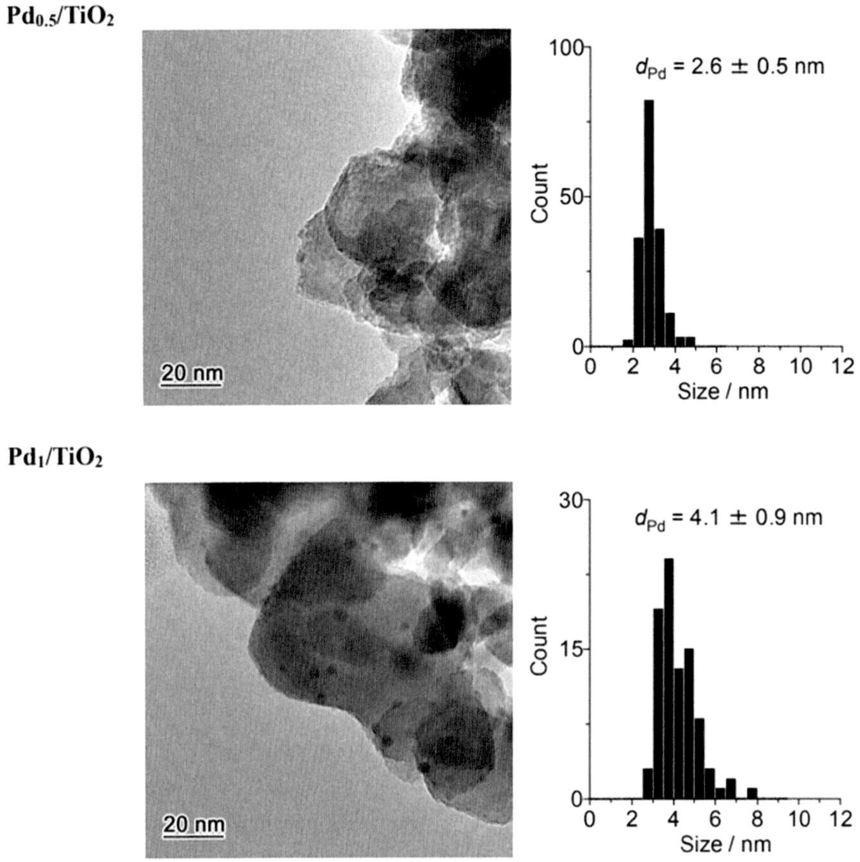

Fig. 2.5 The size distribution and TEM images of Pd NPs on the catalysts. (reproduced with permission. Copyright 2023, Royal Society of Chemistry)

Fig. 2.6 Absorption of Pd_x/TiO_2 measured by diffuse reflectance spectroscopy (DRS). (reproduced with permission. Copyright 2023, Royal Society of Chemistry)

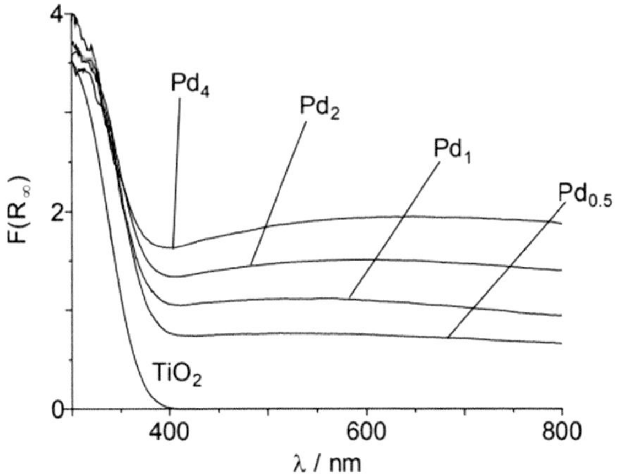

Table 2.1 The hydrogenation of AB over Pd_2/TiO_2 NPs for the synthesis of secondary amines

Run	Alcohol	R–N=N–R	*t*/h	Product	Yield[c]/%
1	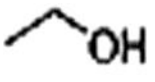	R = 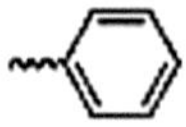	1	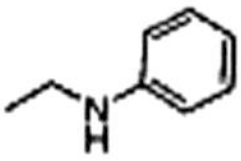	99
2		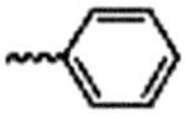	1.5	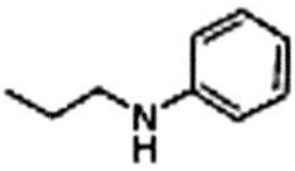	99
3		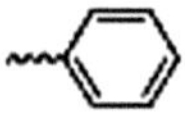	1.5	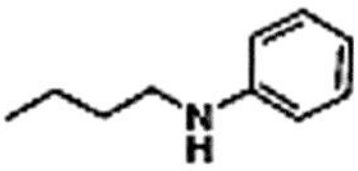	99
4	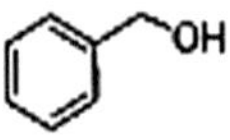	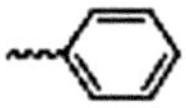	1.5	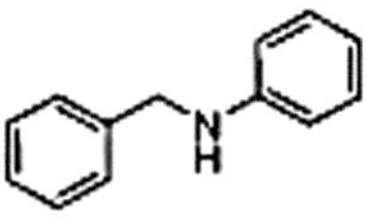	98
5			24		91
6			6		98
7			5		94
8		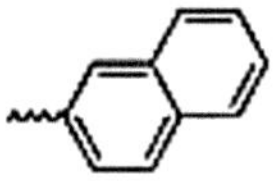	16	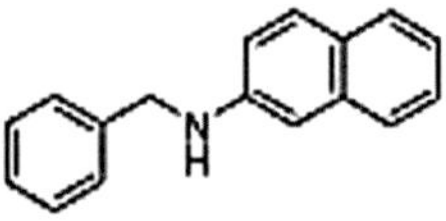	80
9			2	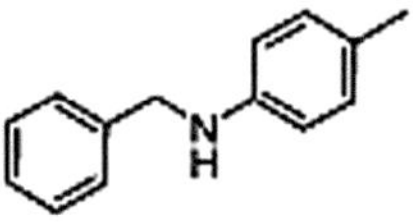	90
10		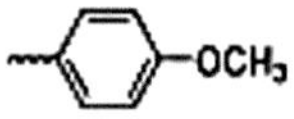	4	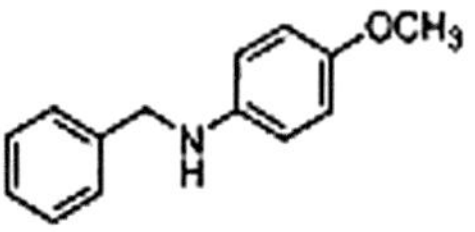	95
11		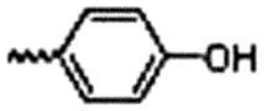	8	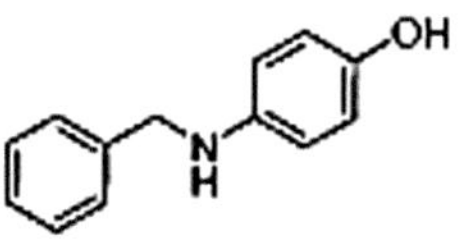	90
12		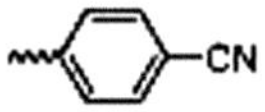	10	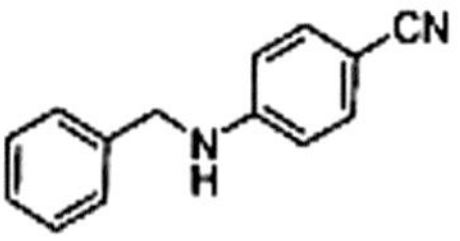	87

(continued)

Table 2.1 (continued)

Run	Alcohol	R–N=N–R	*t*/h	Product	Yield^c/%
13		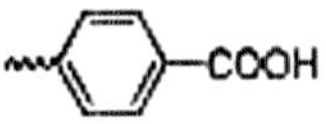	7	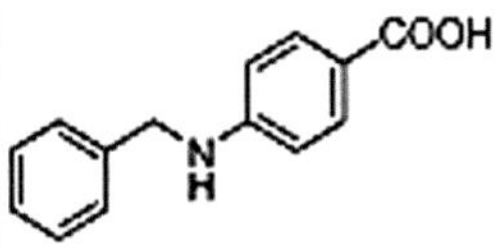	91
14		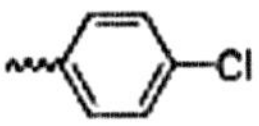	5	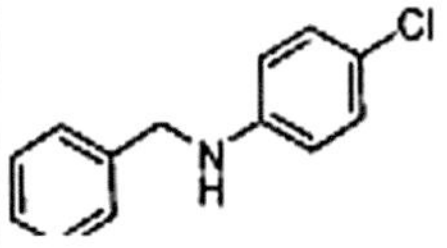	38^h

is a particularly promising candidate for further exploration [14, 100, 102, 113, 114]. For instance, the Zn(O, S) solid solution co-doped with varying concentrations of Co (0–10%) has been shown to be highly effective in facilitating the restoration of the natural environment and the production of clean energy. XRD, DRS, and selected area electron diffraction (SAED) analyses were utilized to confirm the synthesis of a compound containing oxygen and sulfur in Zn(O, S) [15].

Figure 7a illustrates the XRD patterns of CZ-*x* nanocatalysts with different Co concentrations. The Bragg peaks of CZ-0 can be observed in this figure at $2\theta = 48.8°$ (220), 57.6° (331), and 29.5° (111) that are positioned amidst the cubic ZnO and ZnS. CZ-x peak shifts result from oxygen atom occupation in the sulfur (anion) locations of the ZnS crystal lattice modification. Figure 7b shows a gradual shift of the peak towards slightly higher 2θ values, indicating the incorporation of Co-doping in the cation site of the Zn(O, S) nanocatalysts [3, 99, 115]. Furthermore, it is found that the optimal concentration of Co dopant is 2.5%, as indicated by the maximum shift observed due to Co doping at this level. However, for CZ-5 and CZ-10 nanocatalysts, less efficient Co doping was observed at the cationic site of the Zn(O, S) framework. This can be attributed to excessive deposition of Co on the catalyst surfaces [15]. Additionally, the diffraction pattern confirmed that the prepared CZ *x* nanocatalysts did not contain any secondary phases.

Figure 2.8 shows the corresponding images obtained from using FE-SEM to observe the shape of CZ-*x* nanocatalysts with varying Co concentrations. Interestingly, the images revealed that all CZ-*x* nanocatalysts had similar morphologies, irrespective of their Co concentration. Based on Scherrer calculation, the size of nanoparticle was approximately 2.5 nm. Nonetheless, owing to agglomeration, as evident from the FE-SEM images, aggregation of the nanoparticles resulted in the formation of larger particles with sizes ranging from 100 to 250 nm [15].

Figure 2.9 presents the mapping of the elements results of the CZ-2.5 nanocatalysts and the percentages of O, S, Zn, and Co components in CZ-*x* nanocatalysts. Uniform distribution of all components can be observed in the CZ-*x* nanocatalysts. The atomic percentages of Co incorporated in Zn(O, S) were increased with the amount of Co

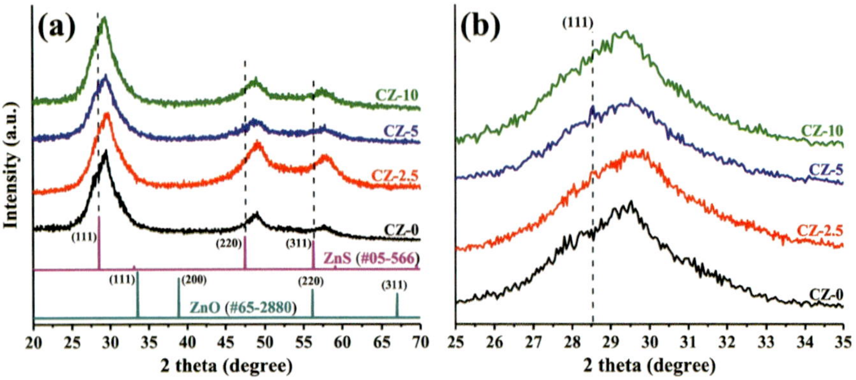

Fig. 2.7 **a** XRD patterns of Co-doped Zn(O, S) denoted as CZ-x nanocatalysts, where x represents Co concentration (Reproduced with permission. Copyright 2023, American Chemical Society)

Fig. 2.8 FE-SEM pictures of CZ-*x* nanocatalysts with **a** 0% Co, **b** 2.5% Co, **c** 5% Co, and **d** 10% Co at the magnification of 3×10^5 times (scale bar = 100 nm) (Reproduced with permission Copyright 2023, American Chemical Society)

(c)

Samples	Elements (atomic %)			
	Zn	O	S	Co
CZ-0	34.84	21.03	44.13	0
CZ-2.5	39.68	29.12	30.84	0.36
CZ-5	43.15	30.75	25.53	0.57
CZ-10	44.32	28.33	26.54	0.81

Fig. 2.9 **a** FE-SEM pictures of CZ 2.5 nanocatalysts; **b** its chemical mapping of S, Co, Z, and O and; **c** percentages of Co-doped Zn(O, S) nanocatalysts. (Reproduced with permission. Copyright 2023, American Chemical Society)

precursor while preparing the nanocatalysts, ranging from 0 to 0.36%, 0.57%, and 0.81% [15].

FEG-TEM was employed to further examine the structure of the CZ-2.5 nanocatalysts at the nano level. Figure 10a, b displays TEM images with high resolution, the results from XRD and SEM were in agreement with the size of nanoparticles found to be approximately 2.5 nm. Based on Fig. 10c, multiple inter-planar d-spacing was observed between ZnO and ZnS (111) planes of CZ-2.5 nanocatalysts. The corresponding values of the interatomic d-spacing were found to be 2.72, 2.84, and 2.96 Å. Figure 10d demonstrates the polycrystalline nature of the CZ-2.5 nanocatalysts as evidenced by multiple broad diffraction rings observed in the SAED patterns. The d-spacing of 3.11 and 2.63 Å was observed in the inner and outer diffraction patterns, respectively, which correspond to the ZnS (111) and ZnO (111) d-spacing. The presence of oxysulfide is also confirmed by the wide ring patterns [15].

Based on XPS analysis, CZ-2.5 surface composition and valence state were verified by examining XPS spectra of S, Zn, Co, and O components and their respective fittings, as depicted in Fig. 2.11. The binding energies at 1045.8 and 1022.9 eV are assigned to Zn $2p_{3/2}$ and Zn $2p_{1/2}$, respectively, as indicated in Fig. 11a [14, 113]. Moreover, Fig. 11b illustrates that the Co 2p XPS spectrum had a relatively low intensity, which can be attributed to the incorporation of a small quantity (2.5%) of

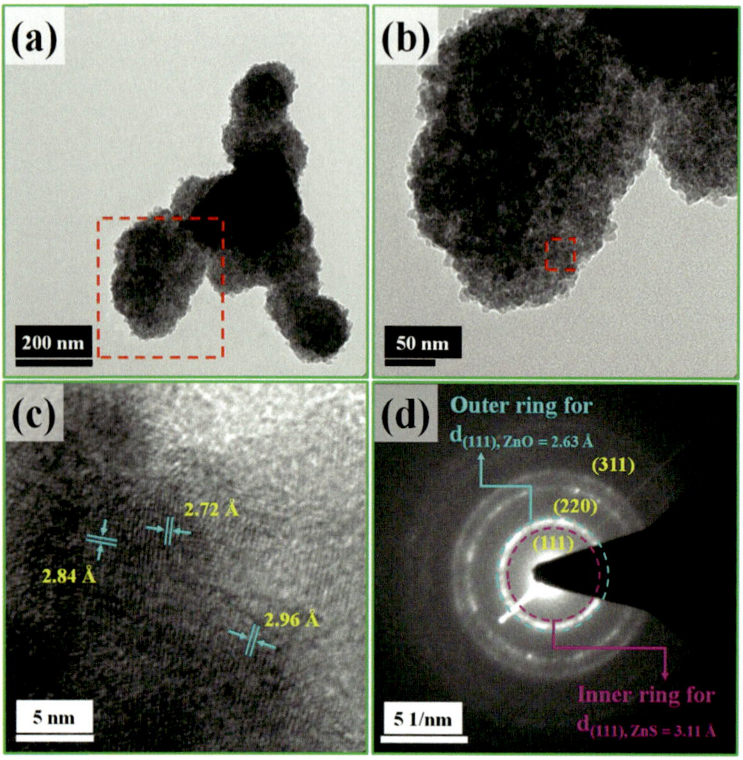

Fig. 2.10 **a**, **b** FEG-TEM images at various magnifications with **c** an image with high resolution to show its lattice structure and **d** its respective SAED (Reproduced with permission. Copyright 2023, American Chemical Society)

Co dopant into the Zn(O,S) matrix. Figure 11b shows the peaks of Co in the XPS spectrum, which is consistent with the previous studies [14, 116]. Figure 11c shows the feature peaks for O 1s at 531.2, 532.3, and 533.3 eV corresponding to O_{ads}, O_v, and O_L, respectively [17, 117, 118]. In Fig. 11d, the S $2p_{3/2}$ (162 eV) and $2p_{1/2}$ (162.8 eV) energy levels were identified as the peaks corresponding to sulfide species and metal-sulfur bonds, respectively. These findings are consistent with the research [119]. The doping of Co into the Zn(O, S) lattice, forming Co-sulfide and Co-oxide chemical bonds, was also confirmed by the XPS results, As previously stated in the XRD as well as XPS analyses [15].

The XPS spectra provide quantitative information of the elemental composition of the CZ-x nanocatalysts can be measured by integrating the area of the curves that have been fitted. The ratio of cations to anions in the CZ-2.5 nanocatalysts, as confirmed by XPS surface composition analysis, is in a stoichiometric proportion, as presented in Table 2.2. The significant presence of oxygen vacancies in CZ-2.5 can be attributed to the incorporation of cobalt atoms, as it generates vacancies in Zn sites. The formation of voids in the oxide lattice of the catalyst is compensated by the negative charge of

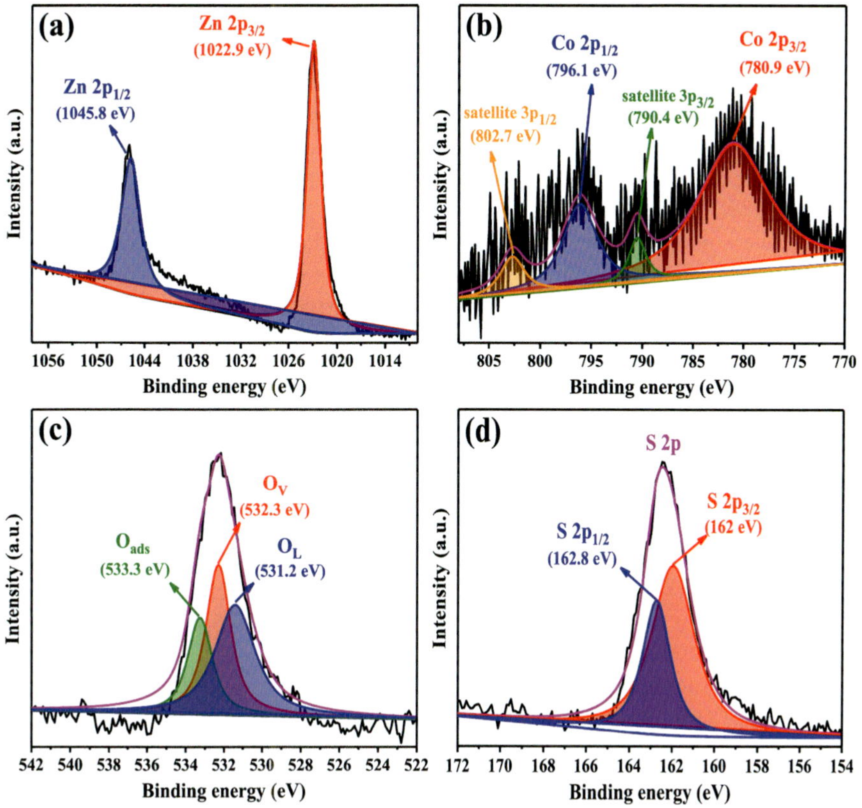

Fig. 2.11 The XPS spectra of O 1s, Co 2p, Zn 2p, and S 2p orbitals were obtained for the CZ-2.5 nanocatalysts (Reproduced with permission. Copyright 2023, American Chemical Society)

adsorbed oxygen species (O_{ads}). Considering its equilibrium charge state, the CZ-2.5 nano-sized catalyst will exhibit consistent performance in promoting photocatalytic reactions and demonstrate durability [14]. This suggests that the CZ-2.5 nanocatalysts have the potential for practical applications in photocatalysis.

In order to investigate the electrochemical properties of CZ-*x* nanocatalysts, this study employed both transient photocurrent (TPC) analyses and electrochemical impedance spectroscopy (EIS). The semicircle CZ-*x* nanocatalyst Nyquist plots in an equivalent Randles circuit were used to fit the data in Fig. 12a. Upon measuring

Table 2.2 Summarized of elemental composition of CZ-2.5 nanocatalysts obtained from XPS data

Elemental composition (%)						
O_L	O_v	O_{ads}	Zn	Co	S	Total
12.00	10.82	10.31	35.92	3.13	27.82	100

Reproduced with permission. Copyright 2023, American Chemical Society

CZ-0, it exhibited a resistance of approximately 6,521 Ω, but after being doped with 2.5% Co (CZ-2.5), the impedance significantly reduced to 2532 Ω, whereas further ascending the Co content to nanocatalysts with 5 and 10% Co doping resulted in an increase in the values of electrical resistance for charge transport decreased to 4414 and 5845 Ω, respectively. The CZ-2.5 nanocatalysts demonstrated the smallest electrical resistance, suggesting the most efficient charge transfer during photocatalysis among the tested CZ-x nanocatalysts. Additionally, transient photocurrent (TPC) analyses were performed to the photo-response of the CZ-*x* nanocatalysts, which were recorded under light alteration settings at a rate of 25 MHz for five cycles as illustrated in Fig. 12b [15]. TPC analysis is an effective technique to investigate the separation of electrons and holes generated by light absorption by measuring the photo-response of the catalyst under illumination. The CZ-2.5 nanocatalysts showed the highest photo-induced current with a measured value of about 80 μA, while the photo-response readings of CZ-10, CZ-5, and CZ-0 n-cats varied between 30 and 40 μA, and this enhancement in photocurrent can be attributed to the low charge transfer impedance of CZ-2.5, as determined from EIS testing [15].

DRS was used to measure the light-reflecting ability of CZ-x nanocatalysts and assess their optical properties. Figure 12c, d depicts the DRS spectra of the CZ-x nanocatalysts and the corresponding Tauc plots. The CZ-0 nanocatalysts initially had an absorbance peak at 316 nm, but the introduction of 2.5 and 5% Co doping caused the absorbance peaks shifted towards lower wavelengths. Increasing the Co concentration to 10% resulted in a bathochromic shift, indicating that higher amounts of Co dopant. The DRS analysis showed that CZ-*x* nanocatalysts were effective as UV-light-active photocatalytic materials [15].

$$(ahv)^2 = k\left(E_g - hv\right) \tag{2.5}$$

Through Tauc plot analysis (Eq. 2.5) of diffuse reflectance spectra (DRS), the band gap energy (Eg) values of CZ-*x* nanocatalysts were determined range from 3.52 to 3.64 eV, without significant variation observed with varying Co dopant amounts. These value are within the range of the bandgap energies of ZnO (3.2 eV) and ZnS (4.24 eV), indicating the generation of Zn(O,S) solid solution, as previously mentioned [120]. The experiments involving EIS, TPC, and DRS led to the conclusion that doping Zn cation sites with 2.5% Co content was the most effective approach for achieving optimum performance in the Zn(O, S) structure.

The photocatalytic capability of CZ-*x* nanocatalysts for hydrogen production was explored using a 10% aqueous ethanol solution under UV and sunlight. Figure 2.13 shows the quantity of hydrogen generated by CZ-*x* nanocatalysts under low-power UV light. After 5 h of illumination, CZ-2.5 exhibited the best performance compared to those without and with higher concentrations of Co dopant. Accordingly, both EIS and TPC analyses supported this conclusion [15]. Based on the above analysis, the CZ-x nanocatalysts were used for hydrogen evolution reaction (HER) application. The CZ-2.5 sample demonstrates excellent HER of 313 μmol/g h and a total amount of 1565 μmol/g under sunlight, whereas the hydrogen production reduced to 1410 μmol/g and 282 μmol/g h at 5% Co concentration. CZ-0 and CZ-10 displayed

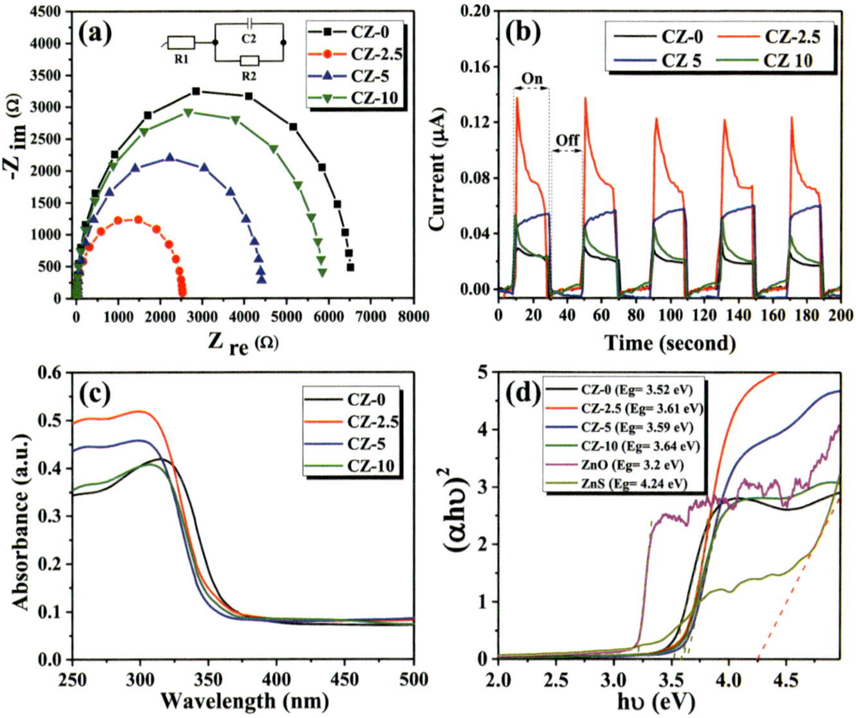

Fig. 2.12 **a** The inserted equivalent Randles circuit was used to fit the Nyquist plot in electrochemical impedance spectra (EIS) data, **b** the photocurrent response of CZ-*x* nanocatalysts with varying Co contents was analyzed using transient measurements, **c** variation of Co amount in CZ-*x* nanocatalysts measured by DRS and **d** Tauc plot for band gap energy of CZ-*x* (Reproduced with permission. Copyright 2023, American Chemical Society)

lower hydrogen evolution rates of 837 and 922 μmol/g h, respectively. It is worth mentioning that the CZ-*x* nanocatalysts demonstrated a higher hydrogen generation rate under UV light based on the DRS results. BET analysis indicated that CZ-2.5 had an improved surface area of 150 m^2/g, a mean pore diameter of 2.01 nm, and a total pore volume of 0.07 cm^3/g, representing significant advancements compared to prior research [120].

In conclusion, the optimal Co doping concentration for the Zn cation sites in the Zn(O, S) structure was 2.5%, as confirmed by EIS, TPC analyses, and HER experiments. The ability of CZ-*x* nanocatalysts to promote hydrogenation reactions was assessed in PHR, involving the cleavage of AB and reduction of NB to AN under UV and under solar irradiation, without the need for extra reducing agents. CZ-2.5 was found to exhibit excellent performance due to low resistance, high photo-induced current, and rapid hydrogen evolution rates, making it the ideal choice for use in the PHR. The UV–vis spectra shown in Fig. 2.14 illustrate the results of the PHR of AB and NB to AN under both UV and sunlight. The peak detected at 315 nm in Fig. 14a corresponds to the AB molecule, which undergoes cleavage under UV

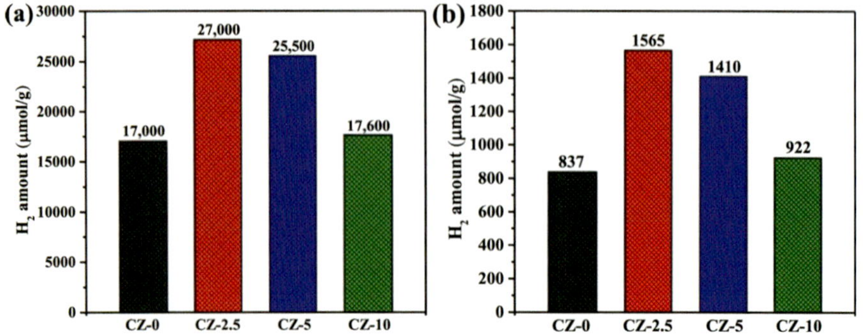

Fig. 2.13 Under various conditions, CZ-x nanocatalysts generated different amounts of hydrogen. **a** Ultraviolet light and **b** illumination with sunlight (Reproduced with permission. Copyright 2023, American Chemical Society)

light. Upon 15 min of UV light illumination, the AB peak was diminished and a new peak characteristic of AN emerged at 240 nm. The strength of the AN peak rises with the extension of the duration of illumination up to 30 min. However, it was observed that extending the PHR time to 45 and 60 min did not result in a further increase in peak intensity, suggesting that the majority of AB in the solution had already been transformed into AN. The UV–vis spectra of the NB solution during the PHR process over a duration of 0–120 min are shown in Fig. 14b. The peak corresponding to NB at 268 nm consistently decreased as the UV light exposure progressed from 0 to 120 min, while the peak absorbance of AN increased proportionally. The production of AN was further confirmed by the chromatogram shown in Fig. 2.15 [15].

A gradual degradation of the AB peak at 315 nm, accompanied by an increase in the intensity of the AN peak at 240 nm, was observed in Fig. 14c. However, due to lower visible light energy than UV light, the cleavage reaction of the azo bond in AB to AN took a total of three hours as opposed to the UV light, which required only 6 min. Similarly, Fig. 14d illustrates the solar light-induced PHR took 5 h to completely decrease NB to AN, which is significantly longer than the time required under UV irradiation. This phenomenon can be explained by the fact that the sunlight (AM 1.5G) has a relatively low intensity of UV radiation, which may result in longer reaction times for both AB and NB [15].

This section provides a detailed explanation of the mechanisms underlying the PHR and HER of AB using the CZ-x catalyst, which involves cleaving the azo bond in AB and reducing the nitro group in NB to amino group to produce AN (Fig. 2.16). Typically, exposing the CZ-2.5 semiconductor nanocatalysts to light with a suitable wavelength generates holes and electrons in VB and CB, respectively (e_{CB}^- and h_{VB}^+), as illustrated in Eq. 2.6 [15]. Exposure of the CZ-2.5 semiconductor nanocatalysts produces electrons and holes in the valence and conduction bands (e_{CB}^- and h_{VB}^+), leading to the generation of oxygen vacancies (V_O^{2+}) on its surface. Oxidation of water and ethanol with h_{VB}^+ generates V_O^{2+}, which holds onto water molecules, weakening their O–H bond, thus producing protons that accumulate on the

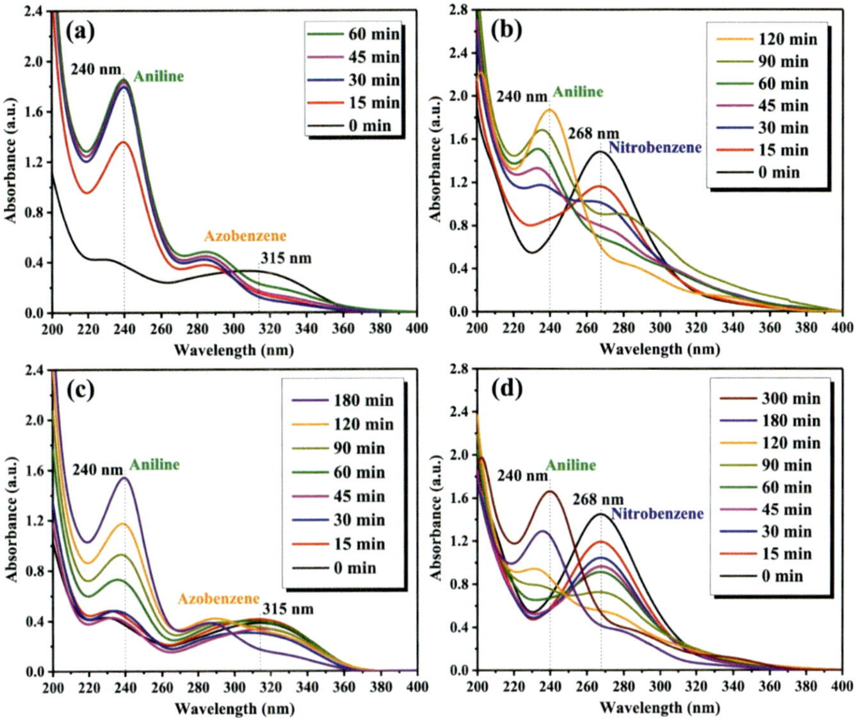

Fig. 2.14 The conversion of photocatalytic hydrogenation reaction (PHR) was analyzed using UV–vis spectra of **a** AB and **b** under UV light, nitrobenzene was converted to AN and **c** AB and **d** employing CZ-2.5 in a 10% ethanol solution, nitrobenzene was converted to AN under solar irradiation) (Reproduced with permission. Copyright 2023, American Chemical Society)

catalyst surface. The accumulation of protons is utilized for both hydrogen evolution (Eq. 2.8) and AB and NB are hydrogenated to produce AN (Eq. 2.7) [15].

$$CZ \quad 2.5 \overset{hv}{\rightarrow} e_{CB}^{-} + h_{VB}^{+} \tag{2.6}$$

$$H_2O + O_{surface}^{2-} + 2h^{+} \rightarrow 2OH_{aq.}^{-} + V_{O}^{2+} \tag{2.7}$$

$$C_2H_5OH + O_{surface}^{2-} + 4h^{+} \rightarrow CH_3CHO + H_2O + V_{O,}^{2+} \tag{2.8}$$

$$H_2O + V_{O}^{2+} \rightarrow 2H^{+} + O_{O}^{0} \tag{2.9}$$

$$2H^{+} + 2e^{-} \rightarrow H_2 \tag{2.10}$$

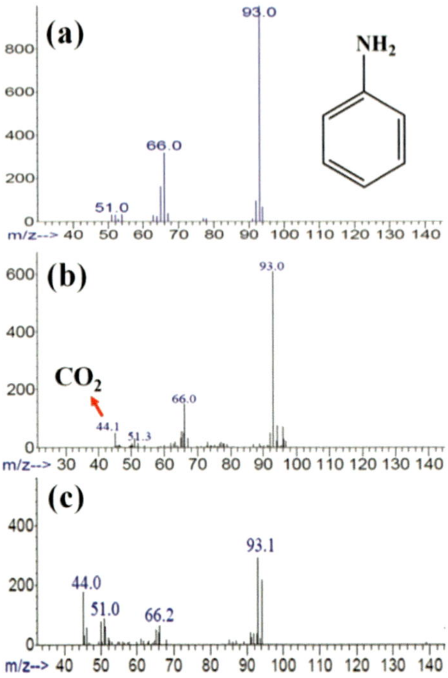

Fig. 2.15 Chromatogram of **a** AN common and the resulting AN after PHR from **b** AB and **c** NB over CZ-2.5 nanocatalysts (Reproduced with permission. Copyright 2023, American Chemical Society)

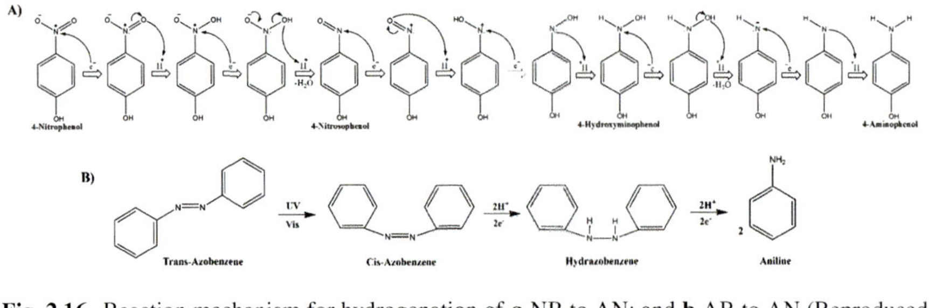

Fig. 2.16 Reaction mechanism for hydrogenation of **a** NB to AN; and **b** AB to AN (Reproduced with permission. Copyright 2023, American Chemical Society)

The proposed mechanisms for the evolution of hydrogen in an ethanol solution and the conversion of AB and NB to AN by the CZ-2.5 nanocatalysts are depicted in Fig. 2.17, providing a visual representation of the hypothesized processes [14]. The initial step under UV light illumination involved the photo-isomerization of trans-AB to cis-AB, followed by the adsorption of AB or NB molecules onto the catalyst surfaces through electrostatic forces. In this system, the presence of 10% ethanol was

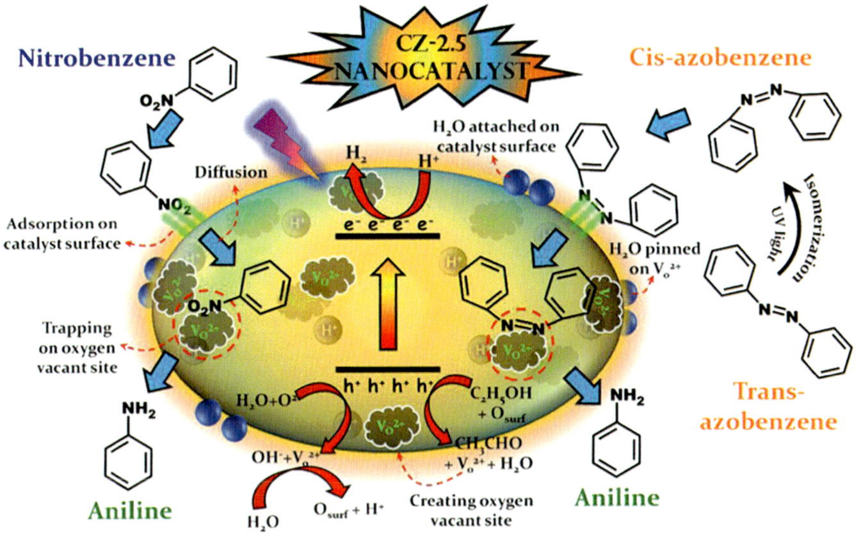

Fig. 2.17 Diagram showing the HER and PHR processes for cleaving the N–N bond in AB and reducing the nitro group in NB to AN in the presence of the CZ-2.5 nanocatalysts (Reproduced with permission. Copyright 2023, American Chemical Society)

crucial for molecular solvation due to the low solubility of stable trans-AB compound [19]. Ethanol also acted as a hole scavenger. It is worth noting that the PHR reaction may occur in a cyclic process, where the active sites of the nanocatalysts become available again for the conversion of another nitrobenzene or AB molecule after the production of AN. As a result, the hydrogenation reaction of nitrobenzene or AB to AN can continue as long as there are NB or AB molecules present in the solution [15].

The reusability of Zn(O, S) catalysts is an important aspect of their practical application and cost-effectiveness. The CZ-2.5 nanocatalysts were evaluated for its ability to maintain stability and be reused multiple times. The catalyst showed excellent performance, with hydrogen evolution rates of over 90% maintained after each cycle, as depicted in Fig. 18a. It is noteworthy that the CZ-2.5 catalyst underwent a color change from white to greyish-black during each cycle, which can be attributed to the formation of oxygen vacancies. However, during the subsequent washing and drying process, the vacancies were filled with oxygen again, indicating that the CZ-2.5 powder can be effectively reused for further experiments [14]. CZ-2.5 nanocatalysts stability was assessed through cyclic voltammetry (CV) analysis to assess its electrochemical performance towards the ferro/ferricyanide couple in a KCl electrolyte solution for up to 200 cycles. In Fig. 18b, distinct cathodic and anodic peaks were observed, with the cathodic peak appearing at 0.39 V and the anodic peak appearing at 0.11 V. The corresponding currents for the cathodic and anodic peaks were −0.84 mA and 0.98 mA, respectively. These results demonstrate the remarkable stability of the CZ-2.5 nanocatalysts in the electro-oxidation and electro-reduction of the redox

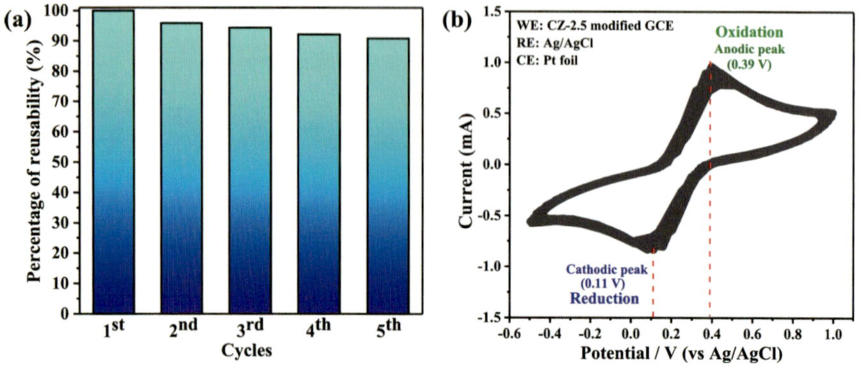

Fig. 2.18 **a** Five HER test runs to determine CZ-2.5 stability. **b** A 200-cycle cyclic voltammogram (CV) of a working electrode with CZ-2.5 showing its cathodic and anodic peaks (Reproduced with permission. Copyright 2023, American Chemical Society)

couple [121, 122]. The stability of the CZ-2.5 nanocatalysts was demonstrated by its consistent performance over 200 cycles of redox reactions [15].

2.3.3 CdS

The potential of CdS as a nanocatalyst for photoreactions, such as hydrogen production and the cleavage of azo bonds in AB, was investigated under mild conditions, along with other nanocatalysts composed of metal oxides and sulfides. The incorporation of Ni into CdS can enhance its photocatalytic efficiency in hydrogenation and azo-bond cleavage reactions. Ni acts as a co-catalyst, promoting the separation of electron–hole pairs and reducing recombination, thereby improving the efficiency of charge transfer. Furthermore, Ni can also act as an active site for the adsorption and activation of reactant molecules, facilitating catalytic reactions.

In recent years, Ni-doped CdS catalysts have attracted much interest for their potential in photocatalysis. In this context, it is crucial to understand the structural properties of these catalysts, including the integration of Ni dopants into the CdS lattices. XRD analysis was performed on Ni-doped CdS samples with different Ni concentrations to determine the phases present in the synthesized catalyst powders. CdS was investigated as the potential nanocatalysts for photoreactions, including hydrogen production and cleavage of azo bonds in AB, under mild conditions, along with other metal oxide and sulfide nanocatalysts. Moreover, the incorporation of Ni with CdS is thought to increase the photocatalytic efficiency of CdS in hydrogenation and azo-bond cleavage reactions. Ni acts as a co-catalyst, which promotes the separation of electron–hole pairs and reduces recombination, leading to increased charge transfer efficiency. Additionally, Ni can also serve as an active site for the adsorption of reactant molecules and their activation, facilitating the catalytic reactions [15, 97, 114, 124].

From Fig. 2.19, it can be clearly seen that the XRD pattern of Ni-CdS catalysts with varying Ni dopant contents (0, 2, 5, and 10%) compared to the standard pattern of hexagonal CdS [123]. The comparison of the diffraction patterns provides insight into the crystal composition of the Ni-CdS catalysts and the effect of Ni dopants on the lattice parameters. The XRD analysis revealed that the primary peaks of Ni-CdS corresponded to hexagonal CdS. As the Ni contents increased, a peak displacement towards higher angles was observed at (110), indicating that the presence of incorporated Ni had been confirmed into CdS lattices with smaller lattice parameters at (110) phase. The XRD analysis of Ni-doped CdS samples demonstrated successful integration of Ni dopant into CdS lattice due to (110) peak shifts towards higher angles with increasing Ni content. The Scherrer equation revealed that the sizes of Ni-CdS crystalline particles were around 11.8, 16.6, 10.1, and 11.3 nm, respectively, for the 0, 2, 5, and 10% Ni-doped CdS. Therefore, it can be concluded that a catalyst consisting of only one phase and doped with Ni was obtained [124].

The particle sizes of Ni-CdS catalysts in the range of 10–20 nm are shown in Fig. 20a. This size range was determined to be favorable for catalytic performance. The EDS analysis confirmed the close-to-stoichiometric ratios of Cd and S in the Ni-CdS catalysts, suggesting that the CdS catalyst was in a homogenous phase, as supported by XRD analysis. The Ni content in the catalysts increased with increasing Ni dopant, the low amount of Ni incorporated into the CdS lattice is possibly attributable to the synthesis being carried out at a low temperature. The

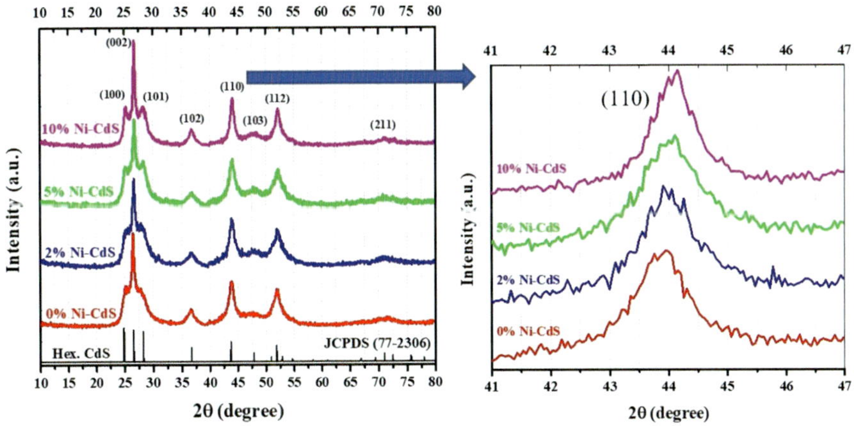

Fig. 2.19 The patterns of various materials X-ray diffraction (XRD) Ni-doped CdS catalysts showed a shift in the peak at (110) when compared to the hexagonal CdS JCPDS file with reference number 77–2306, which served as the baseline pattern (Reproduced with permission. Copyright 2023, Wiley-VCH GmbH)

EDS data in Fig. 20b showed a good correlation with ICP analysis to examine Ni concentration [97, 114].

One important method for investigating the light absorption characteristics of photocatalysts is using DRS. Figure 21a depicts the absorbance intensity of 0, 5, and 10% Ni-doped CdS, where the absorbance is observed from 200 to 580 nm; however, the light absorption intensity of 2% Ni-CdS sample is decreased. The direct bandgap of CdS materials can be determined by fitting the spectrum of diffuse reflectance as shown in Fig. 21b. Based on the Tauc plot, the calculated bandgap values were 2.22, 2.16, 2.22, and 2.24 eV for 0, 2, 5, and 10% Ni-CdS, respectively. The absorbance edges indicated that the bandgap values did not vary significantly among different Ni dopant concentrations [125, 126].

TPC analysis was performed to evaluate the response of the catalysts to incident light. In Fig. 2.22, the use of Ni-CdS catalysts with varying Ni concentrations

(A)

(B)

Samples	Elements (atomic %)		
	Cd	S	Ni
0% Ni-CdS	52.56 ± 0.59	47.44 ± 1.01	0 ± 0.05
2% Ni-CdS	52.24 ± 1.36	47.39 ± 1.84	0.37 ± 0.77
5% Ni-CdS	51.27 ± 0.74	47.54 ± 0.35	1.19 ± 0.52
10% Ni-CdS	50.05 ± 0.41	47.91 ± 0.87	2.03 ± 0.64

Fig. 2.20 **A** SEM images of CdS catalysts doped with (a) 0% Ni, (b) 2% Ni, (c) 5% Ni, and (d) 10% Ni; **B** EDS analyses of Ni-CdS (Reproduced with permission. Copyright 2023, Wiley-VCH GmbH)

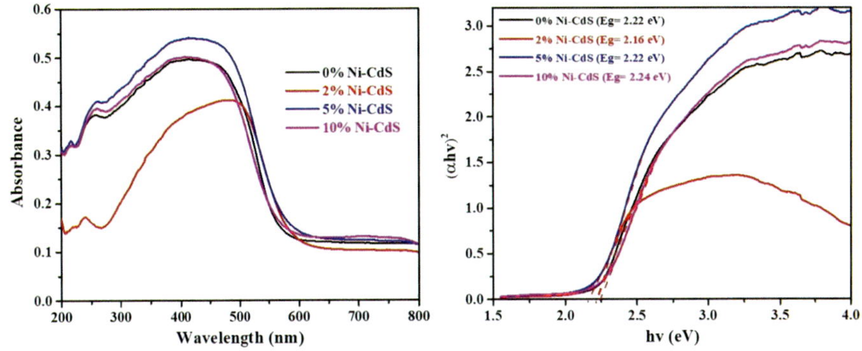

Fig. 2.21 DRS (on the right) and Tauc plots (on the left) of the CdS catalysts with varying amounts of Ni dopant (Reproduced with permission. Copyright 2023, Wiley-VCH GmbH)

demonstrated that the glassy carbon electrode (GCE) modified with 2% Ni-CdS yielded higher photo-induced current. This indicates efficient charge transfer and good photocatalytic activity for 2% Ni-CdS sample [127].

Additionally, important analysis of electrochemical impedance spectroscopy (EIS) was performed in a 1 M KCl electrolyte to corroborate the findings with TPC analysis. The electrical characteristics of the catalysts, especially their charge transfer behavior during electrochemical measurements, were examined using EIS analysis as depicted in Fig. 2.23. The charge transfer resistivities of Ni-CdS samples

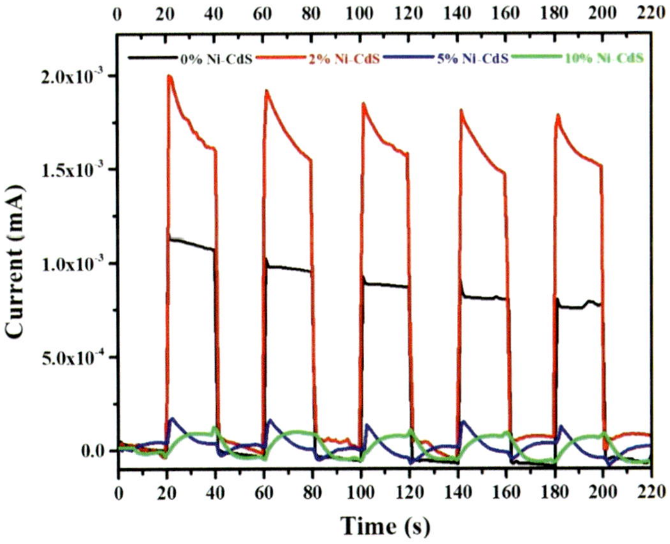

Fig. 2.22 Depicts the photocurrent transients of Ni-doped CdS catalysts with different percentages of Ni under illumination. (Reproduced with permission. Copyright 2023, Wiley-VCH GmbH)

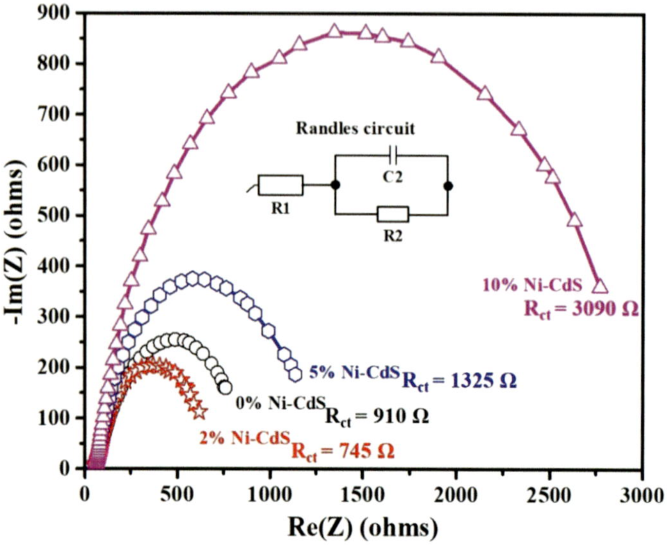

Fig. 2.23 Equivalent Randles circuit fitted spectra of Nyquist plots for Ni-CdS with varying Ni dopant concentrations. (Reproduced with permission. Copyright 2023, Wiley-VCH GmbH)

with varying Ni contents were measured, and the results indicated that 2% Ni-CdS with lower charge transfer resistance. This observation was found to be consistent with the TPC analysis, which suggested that 2% Ni-CdS had excellent photocatalytic activity. Based on the EIS and TPC analyses, it can be concluded that 2% Ni-CdS is a promising catalyst for photocatalytic reactions [124].

2.3.4 *$NaNbO_3$*

The synthesis of HB and its derivatives is crucial for the development of medications targeting rheumatoid arthritis and Alzheimer's diseases [128, 129]. Traditional methods for producing HB involve chemical or electrochemical reduction of nitrobenzene or AB, or hydrogenation catalysis using Pd or Raney Ni catalysts [130–134]. However, an ongoing challenge in this process is the excessive reduction of HB, leading to the formation of aniline [135]. To achieve selective hydrogenation, it is necessary to explore novel catalytic systems capable of activating the N=N bond without breaking the N–N bond. Previous studies have shown the effectiveness of $NaNbO_3$ as a photoactive agent for dye degradation, water splitting, and CO_2 reduction. Research findings indicate that $NaNbO_3$ fibers exhibit remarkable specificity in the hydrogenation process of AB to HB. The preparation of $NaNbO_3$ fibers involves additional steps such as centrifugation, washing, drying, and calcination to obtain a well-crystallized product [136].

The synthesis of $NaNbO_3$ fibers started with subjecting Nb_2O_5 to hydrothermal synthesis under basic conditions, resulting in the formation of $Na_2Nb_2O_6$ fibers (Fig. 24a). At a temperature of 700 °C, a thoroughly crystalline $NaNbO_3$ fiber (with space group Pbcm) was synthesized, as indicated by X-ray diffraction (XRD) analysis (Fig. 24b). The fibrous structure of $NaNbO_3$ is approximately 200 nm and lengths varying from 2 to 12 μm (Fig. 24c) [25]. $NaNbO_3$ fibers exhibit remarkable efficiency for PHA process when assisted by isopropanol as a reducing agent. The catalyst demonstrates high activity and selectivity while also demonstrating exceptional separation capabilities. Importantly, under the same reaction conditions, $NaNbO_3$ fibers do not exhibit any activity in breaking the –N–N– bond in HB (Table 2.3) [25].

Despite the potential of perovskite niobates like $NaNbO_3$ as photocatalysts for hydrogen production, their photocatalytic activity is limited due to a significant band gap and rapid recombination of electron–hole pairs generated by photon [137]. Various approaches have been employed to enhance their activity, including surface modification with Pt nanoparticles and doping with other elements. One promising strategy involves constructing fully solid-state Z-scheme heterojunction systems, such as the recently developed CdS/Pt/$NaNbO_3$ composites, which have shown significant potential in improving the photocatalytic efficiency of $NaNbO_3$-based materials for hydrogen production under visible light. Furthermore, the catalytic efficiency of $NaNbO_3$-based materials has been shown to be enhanced by CdS quantum

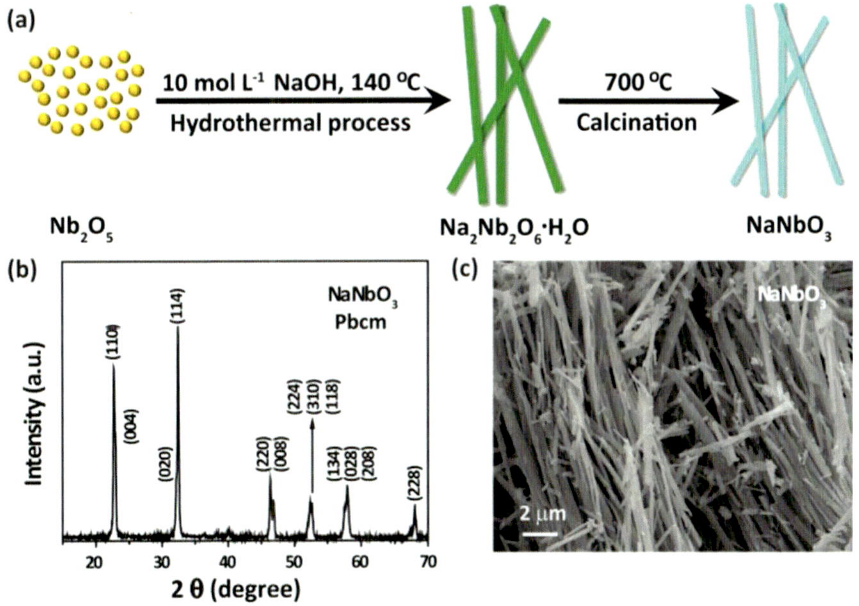

Fig. 2.24 **a** Schematic illustration of $NaNbO_3$ fiber synthesis route. **b** XRD pattern and **c** SEM image of $NaNbO_3$ micro-fibers. (Reproduced with permission. Copyright 2023, Elsevier)

Table 2.3 The catalytic performance of the prepared $NaNbO_3$ nanofibers

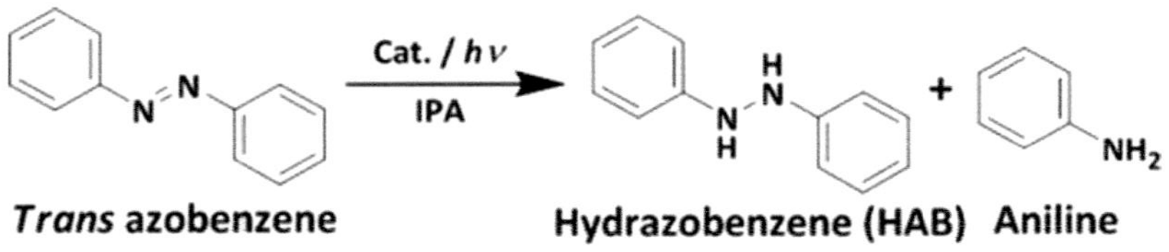

Entry	Catalyst	t (h)	$Yield_{HAB}$ (%)	$Sel._{HAB}$ (%)	$Sel._{Aniline}$ (%)
1	No catalyst	12	n.d.	n.d.	n.d.
2	$NaNbO_3$	12	92.7	100	n.d.
3[a]	$NaNbO_3$	12	n.d.	n.d.	n.d.
4[b]	$NaNbO_3$	12	–	–	<1
5	Carbon nitride	12	11.8	>99	<1
6	TiO_2 (P 25)	0.8	3.5	3.5	96.5

Reproduced with permission. Copyright 2023, Elsevier

dots (CdS QDs) [138, 139]. In this section, $NaNbO_3$ nanorods (NRs) were synthesized using a precipitation method, followed by the fabrication of Pt-doped $NaNbO_3$ NRs through photochemical deposition [140]. The proportions of Pt to N-$NaNbO_3$ by weight are held constant at varying amounts. Then, CdS QDs are covalently bound to the exterior of Pt/N-$NaNbO_3$ NRs through a hydrothermal synthesis, resulting in various compositions denoted as 12% CdS/2% Pt/$NaNbO_3$, 18% CdS/2% Pt/$NaNbO_3$, and 24% CdS/2% Pt/$NaNbO_3$ (Fig. 25a) [141].

Figure 25b displays the XRD patterns of the $NaNbO_3$ components. The diffraction patterns of the $NaNbO_3$ NRs coincide well with the orthorhombic crystal phase of $NaNbO_3$. On the other hand, the enlarged view (Fig. 25b) also shows that the intensity of (111) crystal plane of CdS located at 26.5° is very weak, implying the successful construction of the CdS/Pt/N-$NaNbO_3$ NRs with good crystallinity. The sentence states that the Raman spectrum shown in Fig. 25c contains peaks that are attributed to different lengths of the Nb–O bonds in the stretching mode of NbO_6 octahedral [142–144]. These peaks provide evidence for the successful synthesis of $NaNbO_3$.

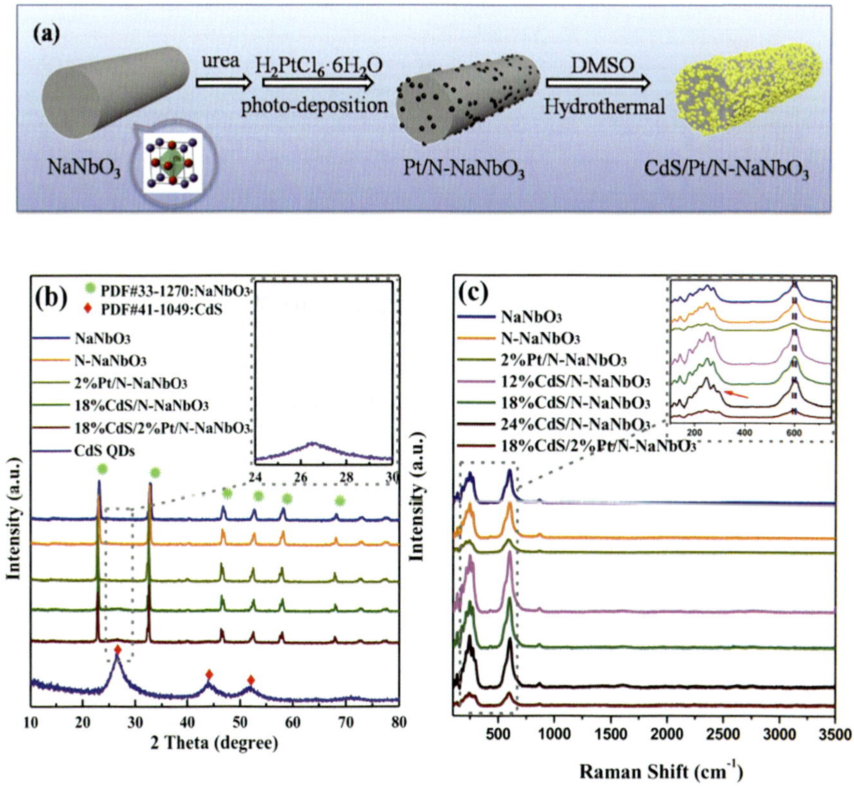

Fig. 2.25 **a** Synthesis and configuration of the CdS/Pt/N-$NaNbO_3$ NRs via the photochemical deposition and hydrothermal process, **b** XRD patterns, and **c** Raman spectra of the synthesized samples. (Reproduced with permission. Copyright 2023, Elsevier)

The stretching mode is a type of vibrational mode in which the bond lengths in the material change due to vibrations caused by an external force. Figure 26a shows TEM image with low loading of CdS QDs on the solid surface of N-$NaNbO_3$ NRs without the assistance of Pt NPs as a "bridge".

Figure 25d and Fig. 26c depict the attachment of CdS QDs to N-doped $NaNbO_3$ surfaces using Pt as a bridge and partially attached to the surface of $NaNbO_3$. There is a difference between CdS/N-$NaNbO_3$ NRs and Pt/N-$NaNbO_3$ NRs in terms of CdS QD loading and distribution. The Pt/N-$NaNbO_3$ NRs have a higher amount of CdS QDs loaded on the rods, with the CdS QDs completely covering the Pt NPs. Additionally, agglomerated CdS QDs are present on the surface of the Pt/N-$NaNbO_3$ NRs. The lattice distance of 0.35 nm allocated to the characteristic (111) plane of CdS in the hetero-interface of the 18%CdS/2%Pt/N-$NaNbO_3$ NRs in Fig. 26e and f confirms the successful production of the composite, which is shown by the red

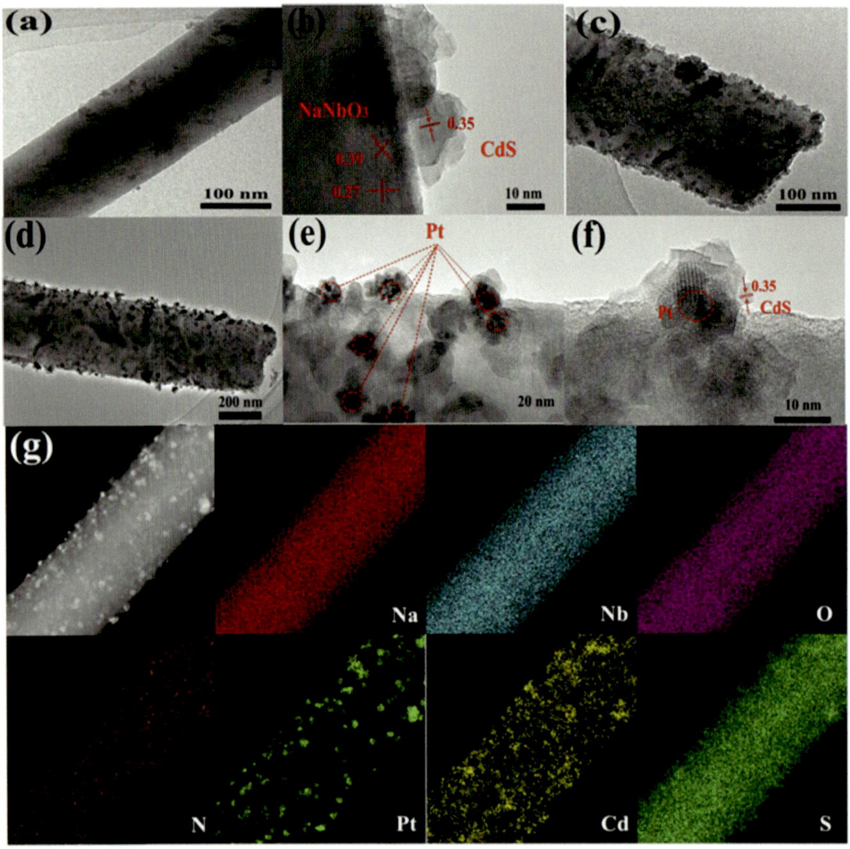

Fig. 2.26 **a** TEM and **b** HRTEM images of 18%CdS/N-$NaNbO_3$, **c**, **d** TEM and **e**, **f** HRTEM images of 18% CdS and 2% Pt co-doped $NaNbO_3$ NRs, **g** EDS analysis of 18%CdS/2%Pt/N-$NaNbO_3$ nanorods. (Reproduced with permission. Copyright 2023, Elsevier)

circles around the Pt NPs. According to Fig. 2.27, the usual diameter of pure CdS QDs is around 5 nm, so this is comparable to the size of CdS QDs that are generated on $NaNbO_3$ NRs in the field, indicating that the CdS particle size is mostly constant both before and after composite creation [141].

The quantity of Pt NPs present has an impact on the hydrogen generation efficiency, and an ideal amount is necessary to preserve the robust bridged heterojunction configuration and the function of Pt NPs as a bridge connecting CdS QDs and N-$NaNbO_3$. Efficient hydrogen generation through the Z-scheme junction is influenced by the presence of an optimal amount of Pt NPs, lower or higher Pt concentration can weaken the junction and decrease the efficiency of hydrogen generation. These findings highlight the importance of the Pt NPs in facilitating the strong bridged heterojunction structure between CdS quantum dots and nitrogen-doped $NaNbO_3$ [141].

Figure 28a–e shows the XPS survey and narrow scan was used to investigate the chemical state of the $NaNbO_3$ composites with Cd, S, Na, Nb, O, and N, indicating the successful doping of nitrogen and the bridging role of Pt NPs [145, 146]. The Pt

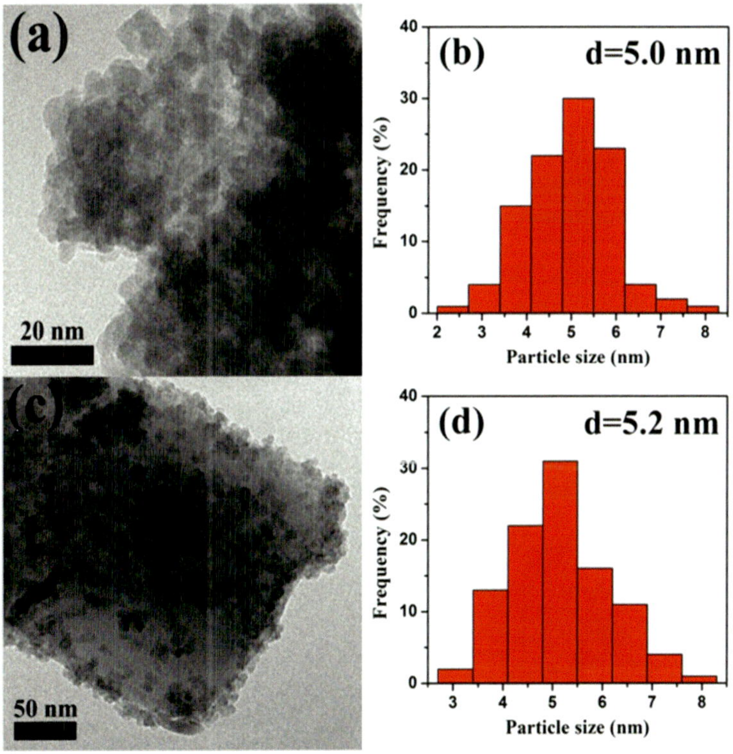

Fig. 2.27 **a** TEM image and **b** nanoparticle size distribution of CdS QDs, **c** TEM image and **d** size distribution of 18%CdS/2%Pt/N-$NaNbO_3$ NPs. (Reproduced with permission. Copyright 2023, Elsevier)

concentration was relatively low (inset Fig. 28a), hence it is consistent with the findings of SEM analysis since Pt NPs were completely covered by CdS QDs. The N 1s spectra exhibited a weak signal for the N peak, possibly due to the low concentration of nitrogen dopants and close to the binding energy (B.E.) of Cd 3d characteristic peak. The peak of S $2p_{3/2}$ and S $2p_{1/2}$ centered at 161.3 and 162.6 eV, respectively, as well as the higher B.E. peaks at 404.7 and 411.5 eV correlated with $Cd_{5/2}$ and $Cd_{3/2}$, respectively were in accordance with those reported for CdS, confirming that CdS was present in the composite. The presence of chemical state Nb^{5+} in $NaNbO_3$ was revealed by the Nb 3d spectra [147]. The XPS analysis validated the efficient deposition of CdS QDs and Pt NPs onto the N-$NaNbO_3$ surface. It was also observed that the Pt NPs played a crucial role as a bridge between the CdS QDs and $NaNbO_3$ NRs, creating a closely interconnected bridged heterojunction configuration.

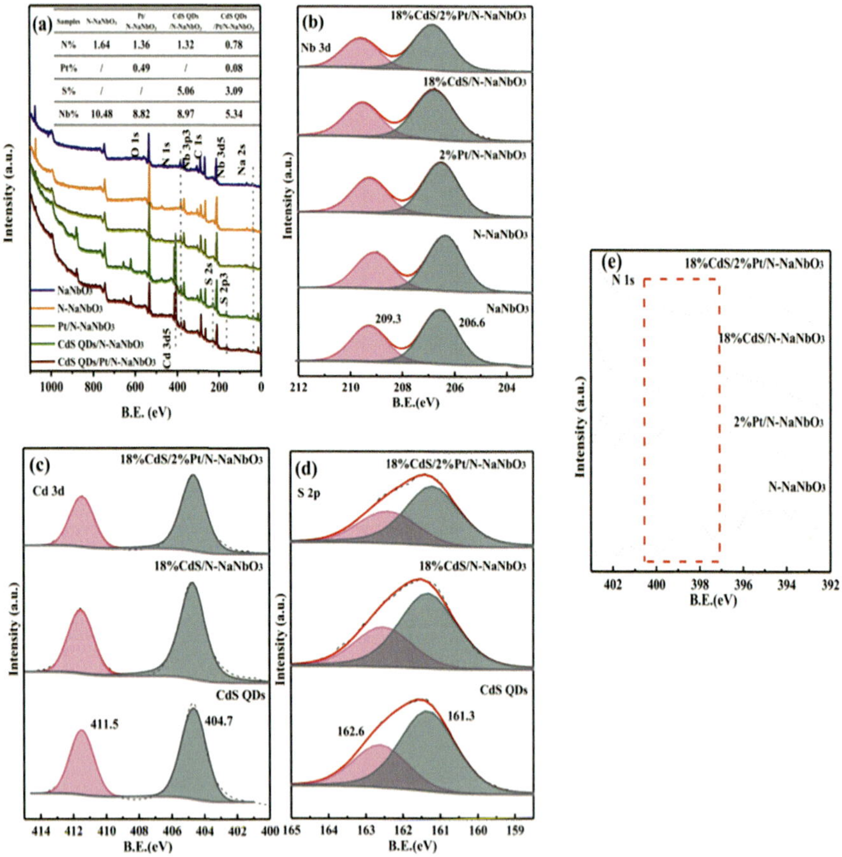

Fig. 2.28 XPS spectra of the synthesized samples for **a** wide scan and narrow scan of **b** Nb 3d, **c** Cd 3d, **d** S 2p, and **e** N 1s. (Reproduced with permission. Copyright 2023, Elsevier)

The CdS/Pt/N-NaNbO_3 photocatalysts demonstrate enhanced photochemical conversion under visible light (wavelengths > 420 nm) and simulated sunlight exposure. The optimum concentration ratio is determined through screening of photocatalytic activity, with 18% CdS QDs and 2% Pt yielding a maximum level of activity. The highest photocatalytic activity was observed for the 18%CdS/2%Pt/N-NaNbO_3 NRs, as depicted in Fig. 2.29, where the rate of hydrogen generation for various combinations is presented. This composite achieves a remarkable increase to 14,218 μmol/g h under visible light and 29,297 μmol/g h under simulated sunlight irradiation, which is the highest reported value for NaNbO_3-based photocatalysts. The presence of Pt as a bridge is essential for forming the bridged hetero-structure and achieving a controllable composition. It profoundly affects the Z-scheme junction's activity between various semiconductor materials [141].

The NaNbO_3 NRs have intrinsic band gap absorption at around 368 nm in the UV region, as revealed by DRS analysis to investigate the exceptional photocatalytic activity of the CdS/Pt/N-NaNbO_3 NRs shown in Fig. 30a–c. The incorporation of N element and the loading of CdS QDs caused a shift in the absorption edge to around 530 nm, which suggests an expansion of the material absorbance into the visible wavelength range. So relatively high CdS content caused the bandgap edge shifted towards longer wavelengths, and the absorbance in the visible spectrum

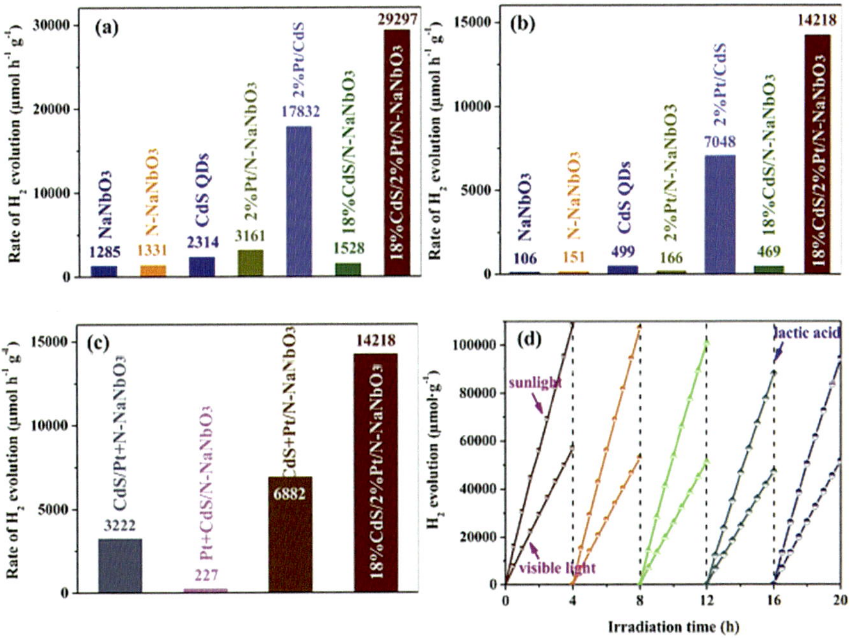

Fig. 2.29 Rate of hydrogen production for NaNbO_3 composite samples under **a** simulated sunlight and **b** and **c** visible light with longer wavelengths (above 420 nm). **d** The 18% CdS/2% Pt/N-NaNbO_3 NRs sample and 10% lactic acid as a sacrificial reagent to carry out the process of H_2 evolution. (Reproduced with permission. Copyright 2023, Elsevier)

(400–600 nm) increased. The maxima absorption wavelength of 18%CdS/N-$NaNO_3$ and 18%CdS/2%Pt/N-$NaNO_3$ is significantly higher than their counterparts without CdS. The development of Z-scheme heterojunction system and the electron magnetic coupling of Pt and CdS had a significant impact on the absorbance, resulting in CdS/Pt/N-$NaNbO_3$ nanocomposites with improved visible-light absorption in the range of 600–800 nm. The band gap energies of CdS QDs and N-$NaNbO_3$ were found to be 2.17 eV and 3.10 eV, respectively, calculated based on Kubelka–Munk methods. The presence of Pt NPs and CdS QDs leads to a reduction in the band gap of materials with a composition of 18%CdS/2%Pt/N-$NaNbO_3$. The DRS results indicate that the CdS/Pt/N-$NaNbO_3$ NRs possess favorable photocatalytic properties, mainly attributed to their ability to absorb light across a wide range of wavelengths. This characteristic enhances the CdS/Pt/N-$NaNbO_3$ NRs light absorption efficiency and promotes the generation of photo-induced charge carriers, thus leading to improved photo-activity [148]. Further characterizations are shown in Fig. 2.31.

As evidenced by Fig. 29d, the 18%CdS/2%Pt/N-$NaNbO_3$ NRs exhibit exceptional stability under visible light irradiation; both simulated sunlight and cycling tests are required to determine the stability of photocatalysts. Figure 31a shows an XRD pattern confirming that the structure and shape of the composite remain the same even after 20 h of irradiation, indicating its outstanding stability during photocatalytic reactions. The traces of heterogeneous components such as CdS and Pt result in a decrease in the PL intensity (Fig. 31b), implying a significant reduction of recombination of electrons and holes. The photocurrent of 18%CdS/2%Pt/N-$NaNbO_3$ is relatively higher compared to that of 18%CdS/N-$NaNbO_3$ and N-$NaNO_3$

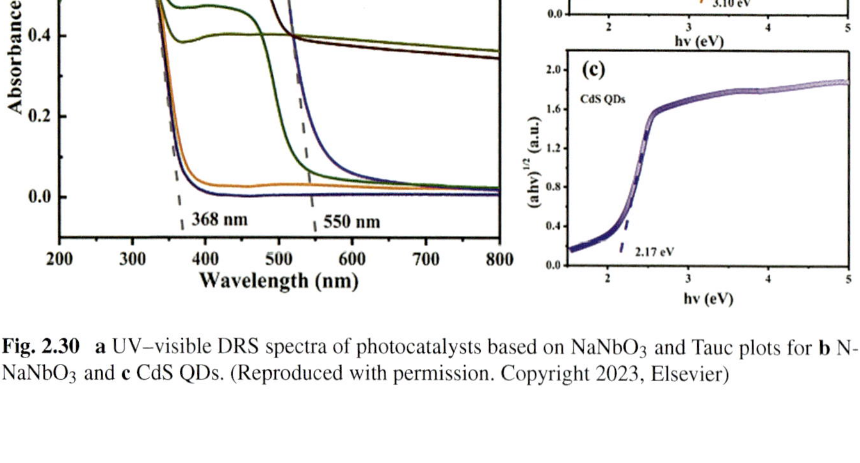

Fig. 2.30 **a** UV–visible DRS spectra of photocatalysts based on $NaNbO_3$ and Tauc plots for **b** N-$NaNbO_3$ and **c** CdS QDs. (Reproduced with permission. Copyright 2023, Elsevier)

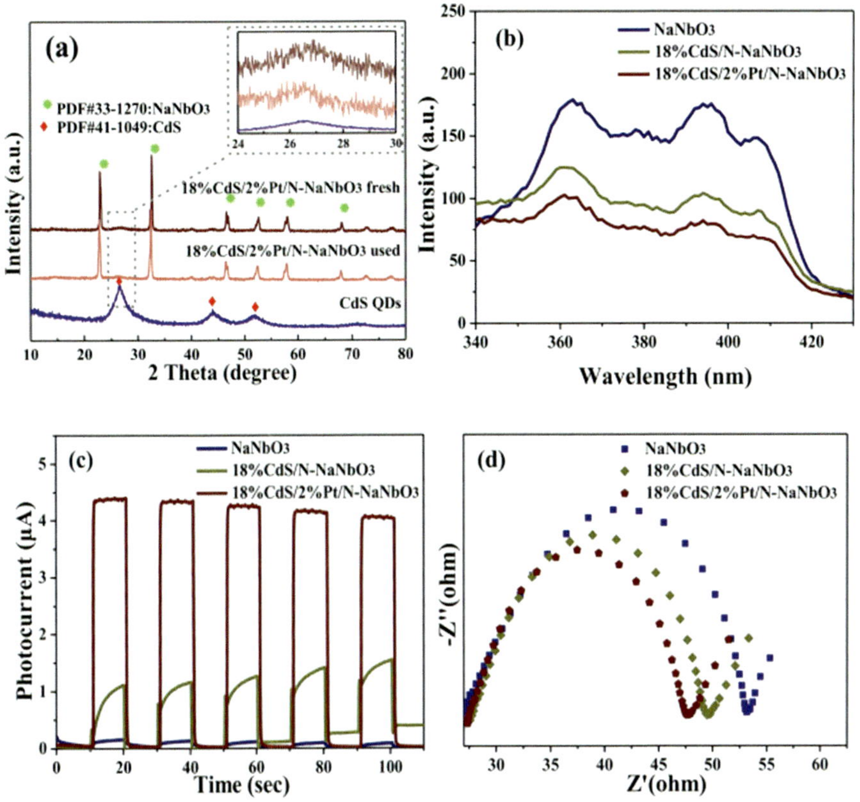

Fig. 2.31 **a** XRD patterns of fresh and spent 18%CdS/2%Pt/N-$NaNbO_3$ NRs, **b** PL spectra, **c** photocurrent as function of time, and **d** Nyquist plot impedance spectra of $NaNbO_3$ composite samples. (Reproduced with permission. Copyright 2023, Elsevier)

(Fig. 31c), elucidating the well-separated of charge carriers with the all-solid-state Z-scheme charge transfer system. The transmission of photoinduced charge carriers of 18%CdS/2%Pt/N-$NaNbO_3$ is much faster than 18%CdS/N-$NaNbO_3$ and N-$NaNO_3$ (Fig. 31d). As presented in Table 2.4, the QE of the 18%CdS/2%Pt/N-$NaNbO_3$ NRs was determined to be approximately 14.1% at 320 nm. A number of band-pass filters were used to acquire certain incoming wavelengths for the purpose of measuring the quantum efficiency of the composite [141].

Table 2.4 Summarized quantum efficiencies (QEs) for the 18%CdS/2%Pt/N-$NaNbO_3$ NRs

Wavelength (nm)	320	350	420	450
QE%	14.1	9.9	2.5	3.6

Reproduced with permission. Copyright 2023, Elsevier

2.4 Application of Photocatalysts

AB, an organic compound extensively employed in the dye and pigment industry, is of significant interest due to its conversion potential to AN and HB through hydrogenation. Nonetheless, conventional hydrogenation approaches pose challenges related to high energy consumption and environmental impact. Therefore, the utilization of photocatalysts as a sustainable and efficient method for AB hydrogenation has attracted growing attention. This chapter focuses on the utilization of four distinct types of photocatalysts, namely, TiO_2, CdS, $NaNbO_3$, and Zn(O, S), in the context of AB hydrogenation. TiO_2 is a common material utilized in the field of photocatalytic, such as water treatment or hydrogen production [149] due to its excellent properties and low cost [150, 151]. For instance, Kaliyamoorthy et al. (2015) [111] used TiO_2-loaded Pt for AB conversion to AN under visible light with a yield of 15% after 4 h. CdS is another commonly used photocatalyst for AB hydrogenation. The narrow bandgap and high electron mobility make it efficient in absorbing visible light. Hairus et al. (2021) [124] reported a new material of Ni-incorporated CdS for cleaving the azo bonding of AB under visible with a yield of 89.7% within 120 min. Besides, $NaNbO_3$ fibers are good photocatalysts with high chemo-selectivity for AB conversion and can reach the yield of 100% for conversion AB to HB due to high activity, stability, and low band gap energy [152].

Zn(O, S) is a novel photocatalyst that combines the advantages of both ZnO and ZnS. Its application in AB hydrogenation has been explored recently. Hardy et al. (2020) [124] synthesized Zn(O, S) nanoparticles for the conversion of AB to AN under visible light with a yield of 99%. The reaction rate was significantly improved under visible light irradiation compared to that under UV or solar light.

The application of photocatalysts in AB hydrogenation has shown great potential as a green and efficient method. However, there are still challenges that need to be addressed, such as the low conversion rate under visible light irradiation, the stability of photocatalysts, and the selectivity of the hydrogenation products. Future research could focus on the design and synthesis of novel photocatalysts with high activity, stability, and selectivity. Additionally, the scale-up of the photocatalytic process for industrial applications needs to be considered. Another promising direction for future research is the development of photocatalytic systems that can utilize solar energy more efficiently. This could be achieved by exploring new photocatalytic materials or modifying the existing ones to enhance their light absorption and charge separation properties. Moreover, the integration of photocatalysts with other renewable energy sources could provide a sustainable and cost-effective approach to AB hydrogenation. The application of photocatalysts, including TiO_2, CdS, $NaNbO_3$, and Zn(O, S), has shown great potential for the green and efficient hydrogenation of AB. However, there is still much research needed to overcome the challenges and optimize the reaction conditions. Ongoing progress in this area opens up the possibility of utilizing photocatalytic AB hydrogenation as a sustainable and eco-friendly alternative to conventional hydrogenation methods.

2.5 Performances of Catalytic Materials

Hu et al. (2016) used a magnetron sputtering method to synthesize Cu-incorporated TiO_2 films under various atmospheres [153]. The photocatalytic activity was tested by using an aqueous methanol solution under UV light illuminant fixed onto the surface of films, as shown in Fig. 2.32**.** Moreover, the Cu-TiO_2 synthesized under an oxygen-rich condition (C-T3) performs relatively high HER up to 2.80 μmol/cm^2.h, which is 55 times higher compared to pristine TiO_2.

Zhu et al. (2019) synthesized new photocatalytic materials containing CdS and carbon dots (CDs) [49]. In this study, they investigate the influences of morphology and CDs concentration on photo-activity in water and seawater. The HER of photocatalysts composites with nanosheet structure (CDs/CdS-S) under UV light is 4.64 and 6.70 mmol/h g for seawater and H_2O, respectively. It is much greater than the ones with nanorods (CDs/CdS-R), nanoparticles (CDs/CdS-N), and irregular (CDs/CdS-I) structures. In detail, the results are demonstrated in Fig. 33a–e.

Hairus et al. (2017) synthesized Zn(O, S) NP aggregates at low temperatures in the presence of various amounts of thioacetamide [100]. The nanocrystallite size of ZnOS NPAs is below 10 nm. The energy band gap of these materials is in the range of 3.2–3.6 eV. It is much smaller compared to ZnS (4.1 eV). The ZnOS-0.5 prepared at 70 °C shows the HER up to 3159 umol/g after 5 h for 10% ethanol solution. For 50% ethanol solution, pure ethanol, or water, the performance is lower compared to 10% ethanol. Interestingly, ZnOS-0.5 prepared at 90 °C performs an excellent HER up to 17,023 umol/g for 5 h. The results are shown in Fig. 34a. Additionally,

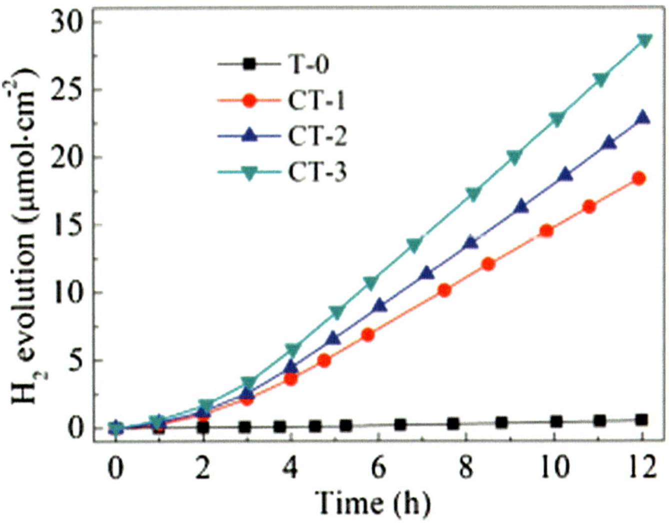

Fig. 2.32 Relationship of amount of evolved hydrogen versus time (Reproduced with permission. Copyright 2023, Elsevier)

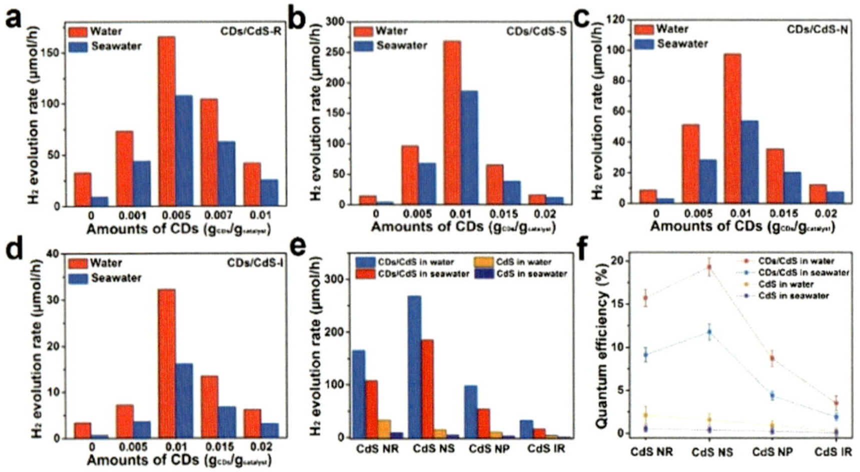

Fig. 2.33 Influences of morphologies and CD concentration on photocatalytic HER of **a** CDs/CdS-R; **b** CDs/CdS-S; **c** CDs/CdS-N; **d** CDs/CdS-I. Comparison of **e** the optimal HER and **f** QE of CdS and CDs/CdS composites under 420 nm irradiation upon different morphologies (Reproduced with permission. Copyright 2023, Elsevier)

ZnOS-0.5 also illustrates good stability. After four recycles, the HER is relatively high (Fig. 34b) without losing the efficiency.

Gu et al. (2015) utilized the solvothermal method to synthesize cubic $NaNbO_3$ nanowires (NN NWs) and cubic $NaNbO_3$ nanoparticles (NN NPs). The band gap energy of NN NWs and NN NPs was 3.54 and 3.37 eV, respectively (Fig. 35a). Figure 35b shows that NN NWs perform better HER compared to NN NPs and relatively greater than the previous report for cubic NN NPs [154].

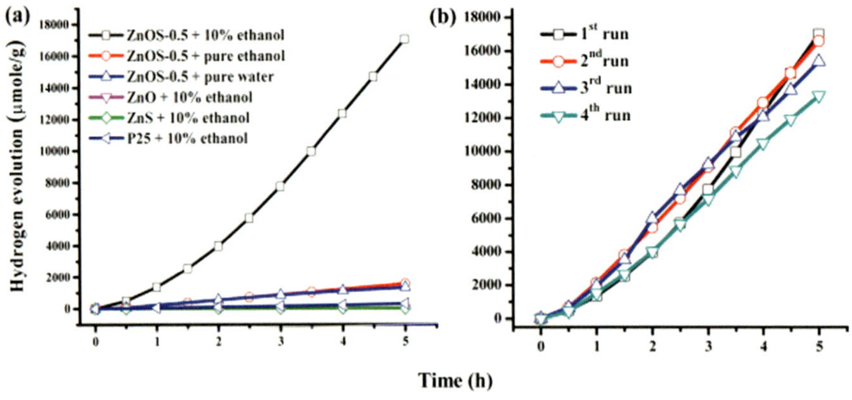

Fig. 2.34 **a** Photo-activity of ZnOS-0.5 with various mediums and **b** the reusability of the material in 10% C_2H_5OH solution (Reproduced with permission. Copyright 2023, Elsevier)

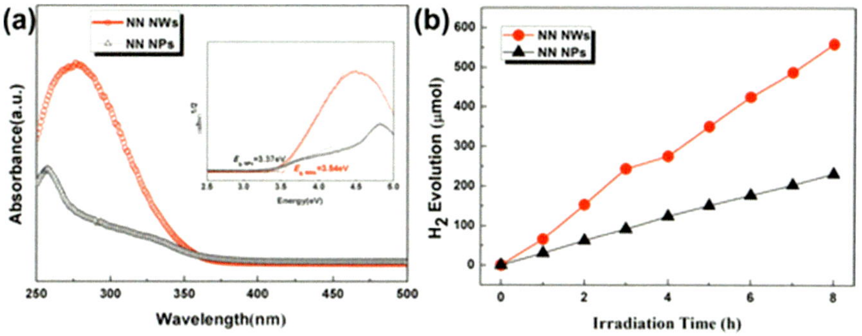

Fig. 2.35 **a** Optical properties and band gap energy of NN NWs and NN NPs by UV–vis spectroscopy. **b** Photocatalytic HER of NN NWs and NN NPs in CH_3OH solution with 0.5 wt.% Pt loading (Reproduced with permission. Copyright 2023, American Chemical Society)

Selvam et al. (2015) used TiO_2-loaded Pd nanoparticles (TiO_2/Pd NPs) for conversion AB [111]. The successful photo-reduction of AB was confirmed by the presence of secondary amines. The results are shown in.

Figure 2.36 and Table 2.5. Accordingly, the TiO_2/Pd NPs are outperformance, such as high selectivity and high activity compared to commercial TiO_2.

Hairus et al. (2021) used Ni-incorporated CdS for breaking the azo bonds of AB with a hydrogenation reaction under visible light [58]. Figure 2.37a–d shows UV–vis spectra of 10 ppm AB using different amounts of Ni concentration incorporated in CdS catalyst with reaction time from 0 to 120 min. Under light illumination, trans-AB is converted relatively to cis-AB, as shown by the absorption peak at about 320 nm. According to pristine CdS, there is no azo-bond cleavage until 120 min.

Fig. 2.36 The influences of reaction times on photo-conversion over Pd_2/TiO_2 (Reproduced with permission. Copyright 2023, Royal Society of Chemistry)

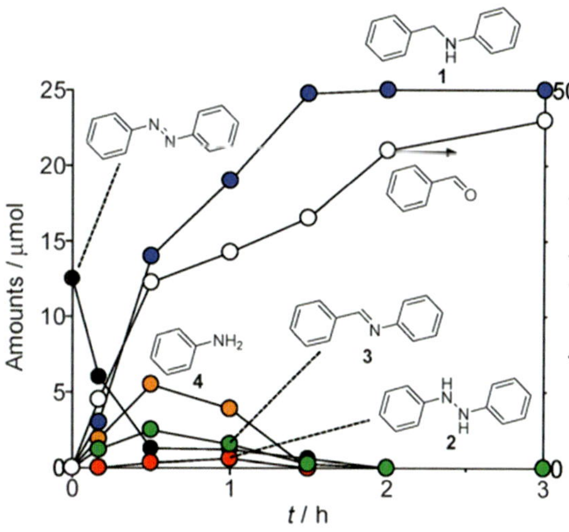

Table 2.5 Summarized photoreaction of AB and benzyl alcohol with different catalysts

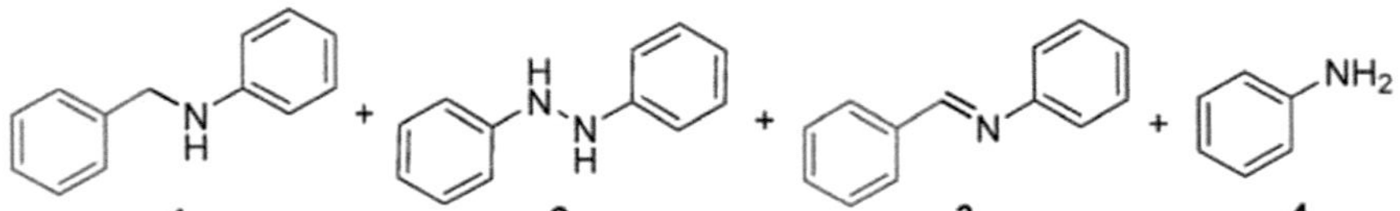

1 + 2 + 3 + 4

Entry	Catalyst	AB conv. (%)	Product yield (%)			
			1	2	3	4
1	None	0	0	0	0	0
2	TiO_2	15	0	7	0	8
3	$Pd_{0.5}/TiO_2$	94	47	27	10	10
4	Pd_1/TiO_2	>99	85	10	5	0
5	Pd_2/TiO_2	>99	98	0	0	0
6	Pd_4/TiO_2	>99	68	0	20	12
7	Pt_2/TiO_2	95	6	30	24	30

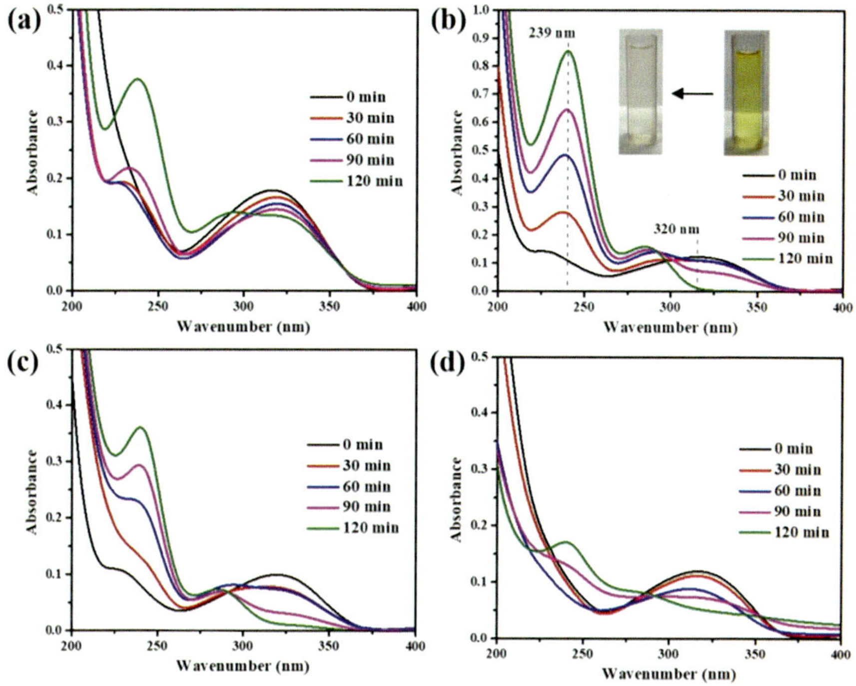

Fig. 2.37 UV–vis spectra of AB solution corresponding to various conversion time with **a** 0%, **b** 2%, **c** 5%, and **d** 10% Ni-CdS catalysts (Reproduced with permission. Copyright 2023, Wiley-VCH GmbH)

Interestingly, when Ni is introduced to CdS, an obvious absorption peak appears at 239 nm, corresponding to AN.

Pei et al. (2020) utilized $NaNbO_3$ fibers for converting AB to HB with a yield of 100% [25]. Figure 2.38 depicts the influences of reaction time on AB conversion. The molar of HB after 12 h is about 95% and reaches 100% after 16 h. $NaNbO_3$ fibers demonstrate an outstanding selectivity for HB conversion compared to P25. The catalytic activity has a close relationship with *cis–trans* photo-isomerization of AB under light illuminance.

Hardy et al. (2021) synthesized different concentrations of Y doped into Zn(O, S) matrix for AB conversion under visible light [155]. The photo-reduction of AB is a close relationship with the cis–trans photo-isomerization of AB. Under light illuminance, trans-AB relatively transforms to cis-AB (Fig. 2.39). The molar of AB gradually decreases with the increase in reaction time. In detail, the absorption peak at 320 nm assigned to AB disappeared after 45 min. Contrarily, an obvious peak appeared at 239 nm, confirming the successful photo-conversion of AB to AN.

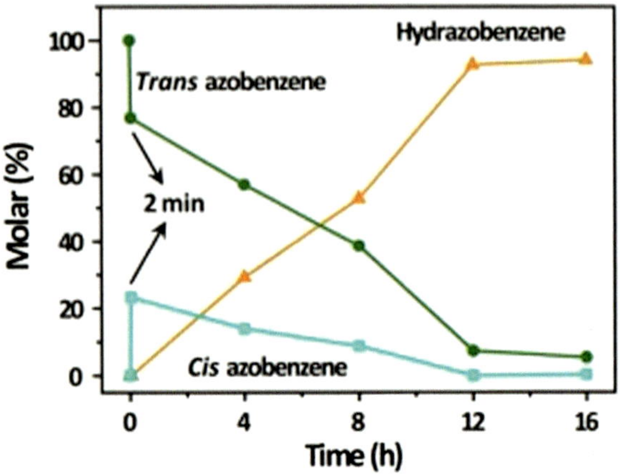

Fig. 2.38 The influences of reaction times on photocatalytic hydrogenation of AB over $NaNbO_3$ fibers. (Reproduced with permission. Copyright 2023, Elsevier)

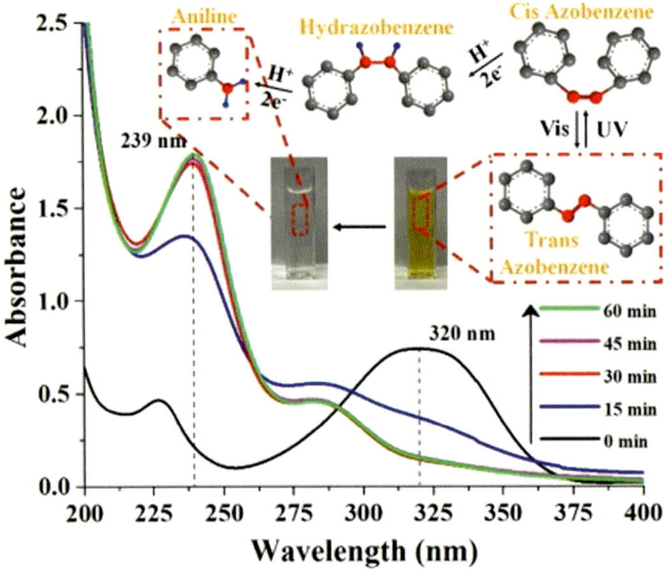

Fig. 2.39 UV–Vis spectra of AB solution during the PHR process. (Reproduced with permission. Copyright 2023, Elsevier)

2.6 Mechanisms of Catalytic Materials

Figure 2.40 demonstrates the general mechanism of the hydrogenation of AB to AN [156]. Firstly, AB uses two protons and two electrons to form HB as an intermediate. Then, HB keeps being reduced to AN. However, depending on each specific kind of catalytic material, the mechanism will occur with some differences. In this section,

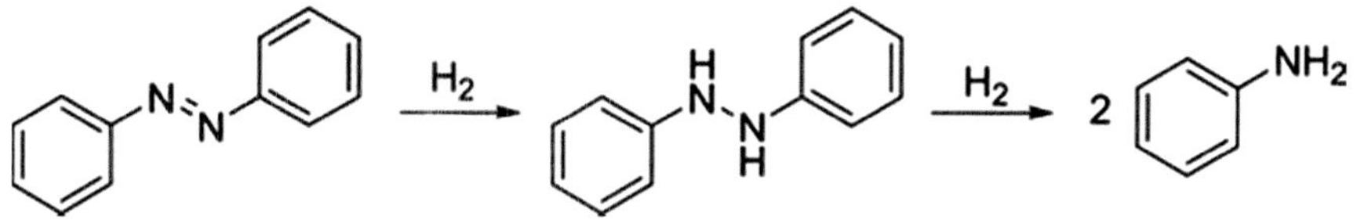

Fig. 2.40 General mechanism of hydrogenation of AB to AN (Reproduced with permission. Copyright 2023, American Chemistry Society)

we will discuss the plausible mechanism of hydrogenation of AB of TiO_2, Zn(O, S), CdS, and $NaNbO_3$ in detail.

2.6.1 *TiO_2-based Photocatalysts*

Figure 2.41 shows the plausible mechanism of secondary amine (**1**) generation from AB and alcohol over TiO_2/Pd photocatalysts under UV light [111]. At first, under a photoexcitation agent, one electron from the ground state (VB) of TiO_2 jumps to an excited state (CB) to form electron (e^-) and hole (h^+) pairs, as follows:

$$TiO_2 + h\nu \rightarrow e^- + h^+ \tag{2.11}$$

The hole (h^+) will carry out oxidizing alcohol to generate aldehyde [157]. This process, as illustrated:

$$R_1{-}CH_2OH + 2h^+ \rightarrow R_1{-}CHO + 2H^+ \tag{2.12}$$

Moreover, the photo-generated electron (e^-) on the VB of TiO_2 migrates to the Pd site. Herein, the reduction of H^+ occurs to produce H atom on the surface of Pd (H–Pd species) [158], as follows:

$$H^+ + e^- + Pd \rightarrow H{-}Pd \tag{2.13}$$

AB is reduced to AN (**4**) on the surface of the Pd-H species. In this process, HB (**2**) is formed as an intermediate [159]. The detailed processes are shown in Eqs. 2.14 and 2.15.

$$R_0{-}NN{-}R_0 + 2H{-}Pd \rightarrow HB(2) + 2Pd \tag{2.14}$$

$$R_0{-}NH{-}NH{-}R_1(2) + 2H{-}Pd \rightarrow 2AN(4) + 2Pd \tag{2.15}$$

In addition, amine (**3**) is also formed by the catalytic condensation between generated aldehyde and AN on the TiO_2 site [160]. This process is explained in detail in Eq. 2.16.

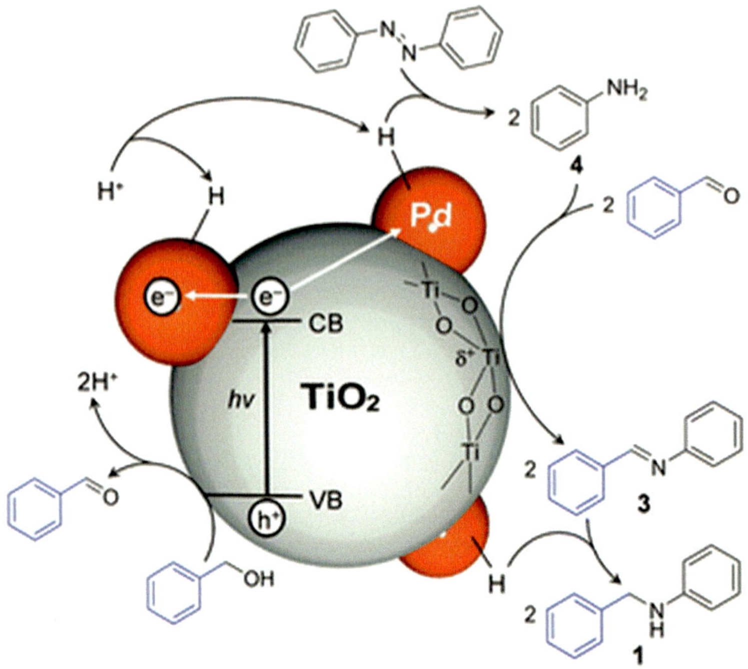

Fig. 2.41 Plausible mechanism of secondary amine generation from AB and alcohol over TiO_2/Pd photocatalysts under UV light. (Reproduced with permission. Copyright 2023, Royal Society of Chemistry)

$$R\text{−CHO} + R\text{−NH}_2 \rightleftarrows R\text{−CHN−}R(3) + \text{H}_2\text{O} \tag{2.16}$$

Finally, the amine **(3)** is reduced to secondary amine **(1)** by the H–Pd species. Equation 2.17 shows the reaction between imine and the H–Pd species.

$$R_0\text{−CHN−}R(3) + 2\text{H−Pd} \rightarrow R_0\text{−CH}_2\text{−NH−}R_1(1) + 2\text{Pd} \tag{2.17}$$

2.6.2 *Zn(O, S)-Based Photocatalysts*

Using the Y_n-Zn(O, S) nanocatalysts as an example, this part would explain the mechanism of PHR of nitrobenzene and AB in detail [155]. The core step of the kinetic mechanism for PHR of NB and AB relates to the formation of oxygen vacancy. Under UV light, e^- and h^+ are generated in CB and CB, respectively (shown in Eq. 2.18).

$$Y_5 - \text{Zn(O, S)} + \text{hv} \rightarrow \text{e}^- + \text{h}^+ \tag{2.18}$$

The holes (h^+) would oxidize H_2O to generate the oxygen vacancy sites (V_o^{2+}) on the surface of photocatalytic materials. This process is explained in Eq. 2.19.

$$H_2O + O^{2-} + 2h^+ \rightarrow 2OH^- + V_o^{2+} \quad (2.19)$$

In the trace of C_2H_5OH, more V_o^{2+} can be generated in the system, detail demonstrated in Eq. 2.20.

$$C_2H_5OH + O^{2-} + 4h^+ \rightarrow CH_3CHO + H_2O + V_o^{2+} \quad (2.20)$$

When the concentration of V_o^{2+} is high, the oxidization of water will keep occurring, causing protonated surfaces on the catalyst (Eq. 2.21).

$$H_2O + V_o^{2+} \rightarrow 2h^+ + O_o^{o+} \quad (2.21)$$

There are two possible pathways for self-generated protons. Firstly, water can be reduced to form proton or hydrogen gas (Eq. 2.22). Secondly, the generated H^+ can hydrogenate 4-nitrophenol (4-NP), NB, and AB to generate smaller molecular-weight compounds such as 4-aminophenol (4-AP) and AN, respectively (Eqs. 2.23–2.25).

$$H^+ + 2e^- \rightarrow H_2 \quad (2.22)$$

$$4 - NP + 6H^+ \rightarrow 4 - \text{aminophenol} + 2H_2O \quad (2.23)$$

$$NB + 6H^+ + 6e^- \rightarrow AN + 2H_2O \quad (2.24)$$

$$AB + 4H^+ + 4e^- \rightarrow AN \quad (2.25)$$

More visualization about the phenomena of molecular adsorption and reduction on the nanocatalyst surface. Figure 2.42 illustrates the scenarios for the mechanism for hydrogenation of 4-NP and AB [155]. Accordingly, 4-NP (Fig. 42a), nitro (–NO_2) groups, and hydroxyl groups (–OH) are adsorbed and trapped at the active site (oxygen vacancies). The reduction reaction would occur when nitro groups interact with oxygen vacancy sites. However, due to the presence of the remaining trapped -OHs of 4-nitrophenol, the yield of HER decreased. Moreover, the gradual release of 4-AP molecules to the solution during the HER causes the re-adsorption of hydroxyl groups on the active sites, resulting in a lower hydrogen evolution rate. In the case of AB (Fig. 42b), the organic molecules are absorbed on the surface of catalysts at the position of azo groups (–N = N–). Then, the –N = N– bonds of AB molecules are cleaved, followed by the release of AN molecules on the surface.

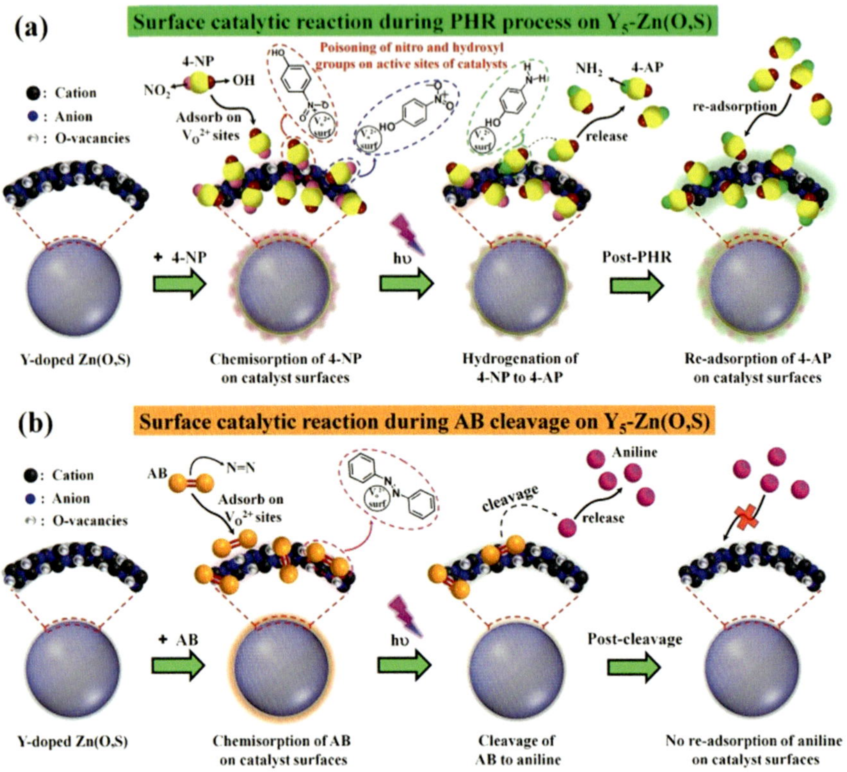

Fig. 2.42 Mechanism for hydrogenation of 4-NP and AB (Reproduced with permission. Copyright 2023, Elsevier)

2.6.3 *CdS-Based Photocatalysts*

The mechanism of AB and methyl orange (MO) hydrogenation of Ni-CdS nanocatalysts in aqueous media in the presence of ethanol is shown in Fig. 2.43 [58]. Under the effects of the photoexcitation agent, electron–hole pairs are photo-generated, as shown in Eq. 2.26.

$$\mathrm{Ni-CdS + hv \rightarrow e^- + h^+} \tag{2.26}$$

The positive charges (h^+) would oxidize H_2O and C_2H_5OH to generate protons (H^+), illustrated in Eqs. 2.27 and 2.28, respectively.

$$\mathrm{H_2O + h^+ \rightarrow H^+ + OH^-} \tag{2.27}$$

$$\mathrm{C_2H_5OH + h^+ \rightarrow H^+ + CH_3COH} \tag{2.28}$$

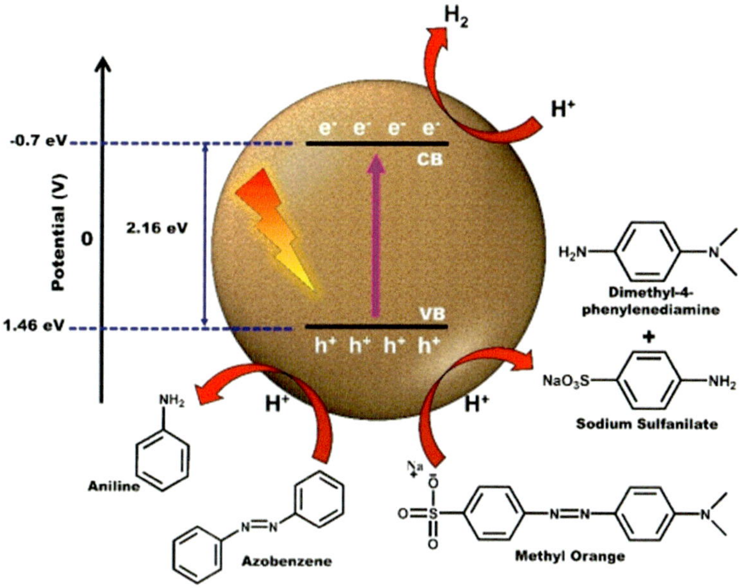

Fig. 2.43 Mechanism of AB and methyl orange (MO) hydrogenation of Ni-CdS nanocatalysts in aqueous media in the presence of ethanol (Reproduced with permission. Copyright 2023, Wiley-VCH GmbH)

The free hydrogens could recombine to form hydrogen gas if there are no any AB compounds in the solution (Eq. 2.29).

$$2H^+ + 2e^- \rightarrow H_2 \tag{2.29}$$

In the presence of AB or MO, the generated protons (H^+) would break N = N bonding and form AN, or sodium sulfanilate and p-aminodimethylamine, respectively. These processes are shown in Eq. 2.30 and Eq. 2.31, respectively.

$$H^+ + AB \rightarrow AN \tag{2.30}$$

$$H^+ + MO \rightarrow C_6H_6NNaO_3S + (CH_3)_2NC_6H_4NH_2 \tag{2.31}$$

2.6.4 *NaNbO$_3$-based Photocatalysts*

The plausible hydrogenation mechanism of AB over $NaNbO_3$ is shown in Fig. 2.44 [25]. Under UV light illumination, electron–hole pairs are photo-generated in $NaNbO_3$. The e^- and h^+ migrate to the surface of photocatalysts, and herein, the

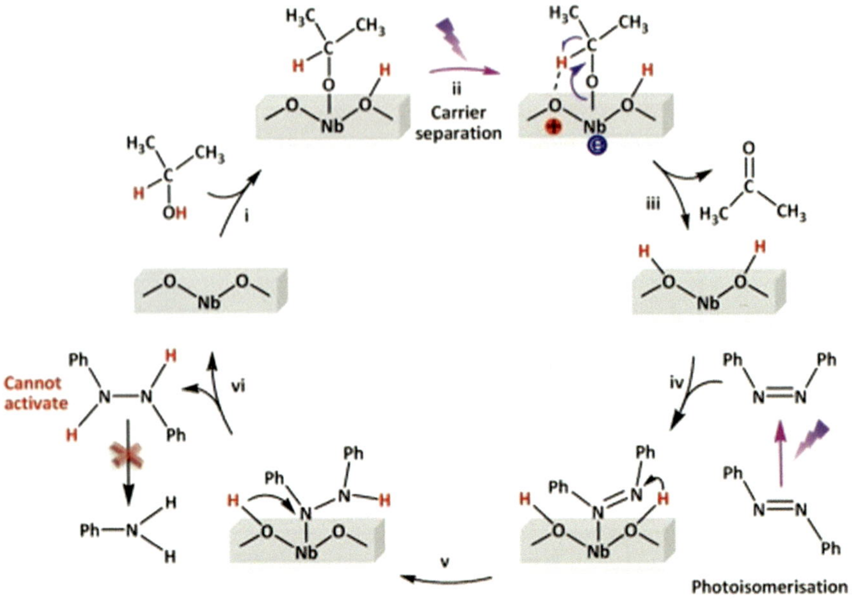

Fig. 2.44 Plausible mechanism of hydrogenation of AB over $NaNbO_3$ (Reproduced with permission. Copyright 2023, Elsevier)

photocatalytic reaction is occurred [161]. Through a photo-induced hole oxidation process, isopropanol protons are dissociated and adsorbed [162] on $NaNbO_3$. Details are described in steps (i), (ii), and (iii). Simultaneously, trans-AB is photo-isomerized to the cis form. Cis AB is adsorbed and activated on the surface of catalyst (iv), and then, hydrogenated selectively to HB (v). Due to weak interaction between HB and $NaNbO_3$, the further hydrogenation of HB to AN is prohibited.

2.7 Perspectives

Given the impact of the greenhouse effect, growing environmental concerns, and the need for energy security, the utilization of metal sulfide and oxide nanocatalysts for green hydrogen production and AB hydrogenation towards valuable chemicals through photoreactions is emerging as a promising applications. As a result, research on this topic is expected to be conducted extensively in the future, focusing on the following areas:

(i) Designing novel photocatalysts: Despite significant progress in the field of photocatalysis, future research will focus on the exploration of new materials possessing excellent selectivity, favorable thermodynamics, high stability, and a large surface area to achieve enhanced efficiency.

(ii) Band energy modification for solar utilization: The utilization of solar energy in the field of photocatalysis, particularly in hydrogen production and the cleavage of azo bonds in AB, will attract significant attention from scientists and experts due to its unlimited and environmentally friendly nature.
(iii) Synthesis of eco-friendly materials: In line with environmental requirements, the development of new materials must prioritize eco-friendliness, ensuring the absence of secondary waste or the production of non-toxic secondary waste.
(iv) Investigation of kinetics and mechanisms: A better understanding of the kinetics and mechanisms involved in photocatalytic processes is crucial for achieving optimal performance and establishing a solid foundation for future research. This exploration can be conducted through experimental conditions or by employing computational science, machine learning, and intelligent artificial intelligence techniques.
(v) Cost considerations: To bridge the gap between laboratory-scale research and industrial applications, it is imperative to focus on developing photocatalytic materials that are not only highly efficient but also cost-effective. This aspect is of great significance in terms of commercialization.

2.8 Conclusion

Metal sulfide and oxide nanocatalysts have emerged as highly promising candidates for photoreactions, offering solutions to numerous challenges related to energy security, environmental sustainability, and quality of life. As a result, research in this field is expected to expand significantly in the future. In the context of energy and environmental concerns, designing a new materials can lead to higher efficiency, improved performance, and reduced environmental impact. Moreover, the exceptional properties of metal sulfide and oxide nanocatalysts will enable their application in diverse areas.

In summary, the utilization of photocatalysts for hydrogen production and AB hydrogenation using metal oxide and sulfide nanocatalysts represents a promising avenue for sustainable energy production and environmental preservation. These technological advancements offer eco-friendly and efficient alternatives to traditional energy sources while tackling critical issues such as climate change and industrial waste pollution. Investing in research and innovation in these areas is vital for ensuring a sustainable and environmentally friendly energy future. Governments, academia, industry, and international organizations all play a crucial role in driving this research forward to meet the growing energy demand while safeguarding our planet.

References

1. Abdullah H, Ginting RT, Ikhtiari R, Gultom NS, Shuwanto H, Kuo DH (2021) One-step synthesis of configurational-entropy in-doped Zn(O, S)/Zn-doped $In(OH)_{3-x}S_x$ composite for visible-light photocatalytic hydrogen evolution reaction. Int J Hydrogen Energy 46:29926–29939
2. Shanmugasundaram S, Abdullah H, Gultom NS, Shuwanto H, Kuo DH (2021) Influence of sulfur amount in Ni-incorporated $ZnIn_2(O, S)_4$ on phase formation and the visible light photocatalytic hydrogen evolution reaction. New J Chem 45:10959–10970
3. Abdullah H, Ginting RT, Shuwanto H, Kuo DH (2022) Banana peel biowaste-derived carbon composited with Zn(O, S) for solar-light photocatalytic hydrogen generation. Int J Hydrogen Energy 47:41021–41033
4. Kishor R, Purchase D, Saratale GD, Saratale RG, Ferreira FR, Bilal M, Chandra R, Bharagava RN (2022) Ecotoxicological and health concerns of persistent coloring pollutants of textile industry wastewater and treatment approaches for environmental safety. J Environ Chem Eng 9:105012
5. Vajnhandl S, Valh JV (2014) The status of water reuse in European textile sector. J Environ Manage 141:29–35
6. Velusamy S, Roy A, Sundaram S, Mallick TK (2021) A review on heavy metal ions and containing dyes removal through graphene oxide-based adsorption strategies for textile wastewater treatment. Chem Rec 21:1570–1610
7. Islam MdS, Tanaka M (2004) Impacts of pollution on coastal and marine ecosystems including coastal and marine fisheries and approach for management: a review and synthesis. Mar Pollut Bull 48:624–649
8. Pinheiro HM, Touraud E, Thomas O (2004) Aromatic amines from azo dye reduction: status review with emphasis on direct UV spectrophotometric detection in textile industry wastewaters. Dyes Pigm 61:121–139
9. Ali SH, de Oliveira JAP (2018) Pollution and economic development: an empirical research review. Environ Res Lett 13:123003
10. Alexander AM, Hargreaves JSJ (2010) Alternative catalytic materials: carbides, nitrides, phosphides and amorphous boron alloys. Chem Soc Rev 39:4388–4401
11. Corma A, Serna P (2006) Chemoselective hydrogenation of nitro compounds with supported gold catalysts. Science 313:332–334
12. Corma A, Concepción P, Serna P (2007) A different reaction pathway for the reduction of aromatic nitro compounds on gold catalysts. AngewandteChemie Int 46:7266–7269
13. Boronat M, Concepción P, Corma A, González S, Illas F, Serna P (2007) A molecular mechanism for the chemoselective hydrogenation of substituted nitroaromatics with nanoparticles of gold on TiO_2 catalysts: a cooperative effect between gold and the support. J Am Chem Soc 129:16230–16237
14. Abdullah H, Ko YR, Kuo DH, Gultom NS (2020) Effects of tin in La–Sn-codoped Zn(O, S) photocatalyst to strongly cleave the azo bond in azobenzene with in situ generated hydrogen. ACS Appl Mater Interfaces 12:16186–16199
15. Shuwanto H, Gultom NS, Abdullah H, Kuo DH (2020) Environmentally benign photoreactions for hydrogen production and cleavage of N=N bond in azobenzene over co-doped Zn(O, S) nanocatalyst: the role of in situ generated $^{H}+$. ACS Appl Mater Interfaces 3:12692–12702
16. Abdullah H, Shuwanto H, Kuo DH (2021) Multifunctional Ni–Mg bimetal-activated Zn(O, S) for hydrogen generation and environmental remediation with simulated solar-light irradiation. Catal Sci Technol 11:7200–7216
17. Abdullah H, Kuo DH, Gultoma NS (2019) N=N bond cleavage of azobenzene via photocatalytic hydrogenation with Dy-doped Zn(O, S): the progress from hydrogen evolution to green chemical conversion. Catal Sci Technol 9:2651–2663
18. Tada H, Kubo M, Inubushic Y, Itob S (2000) N=N bond cleavage of azobenzene through Pt/TiO_2 photocatalytic reduction. Chem Commun 11:977–978

19. Brown C, Rastogi SK, Barrett SL, Anderson HE, Twichell E, Gralinski S, McDonald A, Brittain WJ (2017) Differential azobenzene solubility increases equilibrium cis/trans ratio in water. J Photochem Photobiol, A 336:140–145
20. Chi H, Mya KY, Lin T, He C, Wang F, Chin WS (2013) Thermally stable azobenzene dyes through hybridization with POSS. New J Chem 37:735–742
21. NTP (National Toxicology Program) (2021) Report on carcinogens, 5th edn. U.S. Department of Health and Human Services, Public Health Service, Research Triangle Park, NC
22. Chan SS, Khoo KS, Chew KW, Ling TC, Show PL (2022) Recent advances biodegradation and biosorption of organic compounds from wastewater: microalgae-bacteria consortium—a review. Biores Technol 344:126159
23. Dąbrowski A, Podkościelny P, Hubicki Z, Barczak M (2005) Adsorption of phenolic compounds by activated carbon-a critical review. Chemosphere 58:1049–1070
24. Acuña P, Lin X, Calvo MS, Shao Z, Pérez N, Villafañe F, Rodríguez-Pérez MÁ, Wang DY (2020) Synergistic effect of expandable graphite and phenylphosphonic-aniline salt on flame retardancy of rigid polyurethane foam. PolymDegrad Stab 179:109274
25. Pei L, Wang J, Fan C, Ge H, Waclawik ER, Tan H, Liu M, Gu X, Zheng Z (2020) The key role of photoisomerisation for the highly selective photocatalytic hydrogenation of azobenzene to hydrazobenzene over $NaNbO_3$ fibre photocatalyst. Chemistry 400:112655
26. Nawaz H, Akhter Z, Yameen S, Siddiqi HM, Mirza B, Rifat A (2009) Synthesis and biological evaluations of some Schiff-base esters of ferrocenyl aniline and simple aniline. J Organomet Chem 694:2198–2203
27. Naji SH (2011) Synthesis, pectroscopic and biological studies of some metal complexes with orthoamino hydrazo benzene. Baghdad Sci J 8:375–387
28. Yu S, Wang X, Chen Z, Wang J, Wang S, Hayat T, Wang X (2017) Layered double hydroxide intercalated with aromatic acid anions for the efficient capture of aniline from aqueous solution. J Hazard Mater 321:111–120
29. Wang J, Yuan Z, Nie R, Hou Z, Zheng X (2010) Hydrogenation of nitrobenzene to aniline over silica gel supported nickel catalysts. Ind Eng Chem Res 49:4664–4669
30. Lee SP, Chen YW (2000) Nitrobenzene hydrogenation on Ni–P, Ni–B and Ni–P–B ultrafine materials. J Mol Catal A: Chem 152:213–223
31. Jiang H, Xu J, Sun B (2018) Selective hydrogenation of aromatic compounds using modified iridium nanoparticles. Appl Organomet Chem 32:e4260
32. Li H, Zhao Q, Wan Y, Dai W, Qiao M (20065) Self-assembly of mesoporous Ni–B amorphous alloy catalysts. J Catal244:251–254
33. Daems N, Wouters J, Goethem CV, Baert K, Poleunis C, Delcorte A, Hubin A, Vankelecom IFJ, Pescarmona PP (2018) Selective reduction of nitrobenzene to aniline over electrocatalysts based on nitrogen-doped carbons containing non-noble metals. Appl Catal B: Environ 226:509–522
34. Corma A, González Arellano C, Iglesias M, Sánchez F (2009) Gold complexes as catalysts: chemoselective hydrogenation of nitroarenes. Appl Catal A 356:99–102
35. Reis PM, Royo B (2009) Chemoselective hydrogenation of nitroarenes and deoxygenation of pyridine N-oxides with H_2 catalyzed by MoO_0C_{12}. Tetrahedron Lett 50:949–952
36. Wienhöfer G, Baseda-Krüger M, Ziebart C, Westerhaus FA, Baumann W, Jackstell R, Jungea K, Beller M (2013) Hydrogenation of nitroarenes using defined iron–phosphine catalysts. Chem Commun 49:9089–9091
37. Xiao G, Li P, Zhao Y, Xu S, Su H (2018) Visible-light-driven chemoselective hydrogenation of nitroarenes to anilines in water through graphitic carbon nitride metal-free photocatalysis. Chem Asian J 13:1950–1955
38. Zhao D, Yang CF (2016) Recent advances in the TiO_2/CdS nanocomposite used for photocatalytic hydrogen production and quantum-dot-sensitized solar cells. Renew Sustain Energy Rev 54:1048–1059
39. Gemeda TN, Chang LH, Liang YT, Phan VHK, Fadhilah G, Prasetyo F, Ahmed MT (2023) A recent review on photocatalytic CO_2 reduction in generating sustainable carbon-based fuels. In: Abullah H (ed) Photocatalytic activities for environmental remediation and energy conversion. Green energy and technology. Springer, Singapore

40. Mishra M, Chun DM (2015) α-Fe_2O_3 as a photocatalytic material: a review. Appl Catal A 498:126–141
41. Abe R, Takami H, Murakami N, Ohtani B (2008) Pristine simple oxides as visible light driven photocatalysts: highly efficient decomposition of organic compounds over platinum-loaded tungsten oxide. J Am Chem Soc 130:7780–7781
42. Kudo A, Omori K, Kato H (1999) A novel aqueous process for preparation of crystal form-controlled and highly crystalline $BiVO_4$ powder from layered vanadates at room temperature and its photocatalytic and photophysical properties. J Am Chem Soc 121:11459–11467
43. Hu C, Zhou X, Chen Y, Li L, Li B, Ren H, Zhao S, Zou C (2021) Selective hydrogenation of WO_3 for erasable conducting circuit. J Appl Phys 129:235702
44. Wang G, Mao D, Guo X, Yu J (2018) Enhanced performance of the CuO-ZnO-ZrO2 catalyst for CO_2 hydrogenation to methanol by WO_3 modification. Appl Surf Sci 456:403–409
45. Yi Z, Jinhua Y, Kikugawa N, Kako T, Ouyang S, Williams HS, Yang H, Cao J, Luo W, Li Z, Liu Y, Withers RL (2010) An orthophosphate semiconductor with photooxidation properties under visible-light irradiation. Nat Mater 9:559–564
46. Wang X, Maeda K, Thomas A, Takanabe K, Xin G, Carlsson JM, Domen K, Antonietti M (2009) A metal-free polymeric photocatalyst for hydrogen production from water under visible light. Nat Mater 8:76–80
47. Bortolus P, Monti S (1979) Cis-trans photoisomerization of azobenzene. solvent and triplet donor effects. J Phys Chem 83:648
48. Kima YK, Park H (2011) Light-harvesting multi-walled carbon nanotubes and CdS hybrids: application to photocatalytic hydrogen production from water. Energy Environ Sci 4:685–694
49. Zhu C, Liu C, Fu Y, Gao J, Huang H, Liu Y, Kang Z (2019) Construction of CDs/CdS photocatalysts for stable and efficient hydrogen production in water and seawater. Appl Catal B 242:178–185
50. Zhang B, Yao W, Huang C, Xu Q, Wu Q (2013) Shape effects of CdS photocatalysts on hydrogen production. Int J Hydrogen Energy 38:7224–7231
51. Shiraishi Y, Katayama M, Hashimotoa M, Hiraia T (2018) Photocatalytic hydrogenation of azobenzene to hydrazobenzene on cadmium sulfide under visible light irradiation. Chem Commun 54:452–455
52. Ho W, Jimmy CJ, Lee S (2006) Low-temperature hydrothermal synthesis of S-doped TiO_2 with visible light photocatalytic activity. J Solid State Chem 179:1171–1176
53. Yang J, Zeng JH, Yu SH, Yang L, Zhou G, Qian Y (2000) Formation process of CdS nanorods via solvothermal route. Chem Mater 12:3259–3263
54. Mahdavi R, Talesh SSA (2017) The effect of ultrasonic irradiation on the structure, morphology and photocatalytic performance of ZnO nanoparticles by sol-gel method. Ultrason Sonochem 39:504–510
55. Shu Y, Zhang Q, Fumio S, Tsugio S (2003) Preparation of visible light-activated titania photocatalyst by mechanochemical method. Chem Lett 32:358–359
56. Długosz O, Szostak K (2020) Photocatalytic properties of zirconium oxide–zinc oxide nanoparticles synthesised using microwave irradiation. Appl Nanosci 10:941–954
57. Abdullah H, Kuo HD, Gultom NS (2019) N=N bond cleavage of azobenzene via photocatalytic hydrogenation with dy-doped Zn(O, S): the progress from hydrogen evolution to green chemical conversion. Cataly Sci Technol 9:2651–2663
58. Abdullah H, Siburian R, Pasaribu SP, Panggabean AS (2021) Visible-light driven Ni-incorporated CdS photocatalytic activities for azo-bond cleavages with hydrogenation reaction. Chem Select 6:2041–2050
59. Patzsch J, Berg B, Bloh JZ (2019) Kinetics and optimization of the photocatalytic reduction of nitrobenzene. Front Chem 7:289
60. Xu F, Meng K, Cao S, Jiang C, Chen T, Xu J, Yu J (2022) Step-by-step mechanism insights into the TiO_2/Ce_2S_3 S-scheme photocatalyst for enhanced aniline production with water as a proton source. ASC Cataly 12:164–172
61. Daniels CL, Liu DJ, Adamson MAS, Knobeloch M, Vela J (2021) Azo(xy) vs aniline selectivity in catalytic nitroarene reduction by intermetallics: experiments and simulations. J Phys Chem C 125:24440–24450

62. Wang C, Yang Q, Wang JY, Zhao J, Wan XY, Guo Z, Yang YH (2019) Application of support vector machine on controlling the silanol groups of silica xerogel with the aid of segmented continuous flow reactor. Chem Eng Sci 199:486–495
63. Roy D, Mandal SC, Pathak B (2021) Machine learning-driven high-throughput screening of alloy-based catalysts for selective CO_2 hydrogenation to methanol. ASC Appl Mater Interfaces 13:56151–56163
64. Mamiyev Z, Balayeve NO (2022) Metal sulfide photocatalysts for hydrogen generation: a review of recent advances. Catalysts 12:1316
65. Møller KT, Jensen TR, Akiba E, Li H (2017) Hydrogen-asustainable energy carrier. Progress Nat Sci: Mater Int 27:34–40
66. Stocker T, Plattner GK, Dahe Q, Climate Change IPCC (2013) The physical science basis. Intergovernmental Panel Clim Change 2013:17003
67. Bonde JM, Moses PG, Jaramillo TF, Nørskov JK, Chorkendorff I (2009) Hydrogen evolution on nano-particulate transition metal sulfides. Faraday Discuss 140:219–231
68. Wojtyła S, Baran T (2019) Photocatalytic H_2 production over RuO_2@ZnS and RuO_2@CuS nanostructures. Int J Hydrogen Energy 44:14624–14634
69. Liu Y, Meng X, Wang Z, Qiu J (2022) A Li_2S-based all-solid-state battery with high energy and superior safety. Sci Adv 8:eabl8390
70. Maeda K, Teramura K, Lu D, Takata T, Saito N, Inoue Y, Domen K (2006) Photocatalyst releasing hydrogen from water. Nature 440:295
71. Baran T, Wojtyla S, Dibenedetto A, Aresta M, Macyk W (2015) Zinc sulfide functionalized with ruthenium nanoparticles for photocatalytic reduction of CO_2. Apply Cataly B: Environ 178:170–176
72. Mamiyev ZQ, Balayeva NO (2015) Preparation and optical studies of PbS nanoparticles. Opt Mater 46:522–525
73. Balayeva NO, Mamiyev ZQ (2016) Synthesis and characterization of Ag_2S/PVA-fullerene (C60) nanocomposites. Mater Lett 175:231–235
74. Ganapathy M, Chang CT, Alagan V (2022) Facile preparation of amorphous $SrTiO_3$-crystalline PbS heterojunction for efficient photocatalytic hydrogen production. Int J Hydrogen Energy 47:27555–27565
75. Frame FA, Osterloh FE (2010) CdSe-MoS_2: a quantum size-confined photocatalyst for hydrogen evolution from water under visible light. J Phys Chem C 114:10628–10633
76. Keimer B, Moore JE (2017) The physics of quantum materials. Nat Phys 13:1045–1055
77. Amirav L, Alivisatos AP (2010) Photocatalytic hydrogen production with tunable nanorod heterostructures. J Phys Chem Lett 1:1051–1054
78. Shiga Y, Umezama N, Srinivasan N, Koyasu S, Sakai E, Miyauchi M (2016) A metal sulfide photocatalyst composed of ubiquitous elements for solar hydrogen production. Chem Commun 52:7470–7473
79. Sivula K, Krol RVD (2016) Semiconducting materials for photoelectrochemical energy conversion. Nat Rev Mater 1:15010
80. Flores C, Kubacka A, Berestok T, Zhang T, Llorca J, Arbiol J, Cabot A, García MF (2020) Hydrogen photogeneration using ternary $CuGaS_2$-TiO_2-Pt nanocomposites. Int J Hydrogen Energy 45:1510–1520
81. Hou H, Yuan YF, Cao S, Yang Y, Ye X, Yang W (2020) $CuInS_2$ nanoparticles embedded in mesoporous TiO_2 nanofibers for boosted photocatalytic hydrogen production. J Mater Chem C 8:11001–11007
82. Singh J, Sono RK (2021) Enhanced sunlight driven photocatalytic activity of In_2S_3 nanosheets functionalized MoS_2 nanoflowers heterostructures. Sci Rep 11:15352
83. Balayeva NO, Mamiyev Z (2022) Integrated processes involving adsorption, photolysis, and photocatalysis. In: Hybrid and combined processes for air pollution control. Elsevier, Amsterdam, The Netherlands
84. Lin W, Cheng H, Ming J, Yu Y, Zhao F (2012) Deactivation of Ni/TiO_2 catalyst in the hydrogenation of nitrobenzene in water and improvement in its stability by coating a layer of hydrophobic carbon. J Catal 291:149–154

85. Yang XJ, Chen B, Zheng LQ, Wu LZ, Tung CH (2014) Highly efficient and selective photocatalytic hydrogenation of functionalized nitrobenzenes. Green Chem 16:1082–1086
86. Tada H, Kubo M, Inubushi Y, Ito S (2000) N=N bond cleavage of azobenzene through Pt/ TiO_2 photocatalytic reduction. Chem Commun 977−978
87. Abdullah H, Kuo HD (2019) Utilization of photocatalytic hydrogen evolved (Zn, Sn)(O, S) nanoparticles to reduce 4-nitrophenol to 4-aminophenol. Int J Hydrogen Energy 44:191–201
88. Abdullah H, Gultom NS, Kuo HD (2019) Synthesis and characterization of la-doped Zn(O, S) photocatalyst for green chemical detoxification of 4-nitrophenol. J Hazard Mater 363:109–118
89. Jiang B, Guo Y, Kim J, Whitten AE, Wood K, Kani K, Rowan AE, Henzie J, Yamauchi Y (2018) Mesoporous metallic iridium nanosheets. J Am Chem Soc 140:12434–12441
90. Lee YY, Jung HS, Kim JM, Kang YT (2018) Photocatalytic CO_2 conversion on highly ordered mesoporous materials: comparisons of metal oxides and compound semiconductors. Appl Cataly B 224:594–601
91. Tang J, Durrant JR, Klug DR (2008) Mechanism of photocatalytic water splitting in TiO_2 reaction of water with photoholes, importance of charge carrier dynamics, and evidence for four-hole chemistry. J Am Chem Soc 130:13885–13891
92. Singhal R, Sharma P, Vishnoi R, Avasthi DK (2017) Synthesis and characterizations of Au-C60 nanocomposite. J Alloy Compd 696:9–15
93. Devi SA, Singh KJ, Devi KN (2022) Visible light driven photocatalytic activities of metal sulfides synthesized by simple co-precipitation method. Mater Today: Proc 65:2819–2824
94. Mamiyev ZQ, Balayeva NO (2016) CuS nanoparticles synthesized by a facile chemical route under different pH conditions. Mendeleev Commun 26:235–237
95. Meerbach C, Klemmed B, Spittel D, Bauer C, Park YJ, Hubner R, Jeong HY, Erb D, Shin HS, Lesnyak V, Eychmuller A (2020) General colloidal synthesis of transition-metal disulfide nanomaterials as electrocatalysts for hydrogen evolution reaction. ACS Appl Mater Interfaces 12:13148–13155
96. Tada H, Ishida T, Takao A, Ito S (2004) Drastic enhancement of TiO_2-photocatalyzed reduction of nitrobenzene by loading Ag clusters. Langmuir 20:7898–7900
97. Abdullah H, Kuo HD (2016) Photocatalytic performance of the SiO_2 sphere/N-Type TiO_2/ P-type $CubiS_2$ composite catalysts coated with different contents of Ag nanoparticles under ultraviolet and visible light irradiations. Appl Phys A 122:739
98. Chen X, Kuo HD, Saragih AD, Wu ZY, Abdullah H, Lin J (2019) The effect of the Cu^+/Cu^{2+} ratio on the redox reactions by nanoflower cunios catalysts. Chem Eng Sci 194:105–115
99. Abdullah H, Gultom NS, Kuo HD (2017) Synthesis of Zn(O, S)/Ga_2O_3 nanocomposite photocatalyst for hydrogen production and 4-nitrophenol reduction. New J Chem 41:12397–12406
100. Abdullah H, Kuo HD, Chen X (2017) High efficient noble metal free Zn(O, S) nanoparticles for hydrogen evolution. Int J Hydrogen Energy 42:5638–5648
101. Bekena FT, Abdullah H, Kuo HD, Zeleke MA (2019) Photocatalytic reduction of 4-nitrophenol using effective hole scavenger over novel Mg-doped Zn(O, S) nanoparticles. J Indus Eng Chem 78:116–124
102. Gultom NS, Abdullah H, Kuo HD, Ke WC (2019) Oriented P-N heterojunction Ag_2O/Zn(O, S) nanodiodes on mesoporous SiO_2 for photocatalytic hydrogen production. ACS Appl Energy Mater 2:3228–3236
103. Chang YC, Chen DH (2009) Catalytic reduction of 4-nitrophenol by magnetically recoverable Au nanocatalyst. J Hazard Mater Lett 165:664–669
104. Huang C, Ye W, Liu Q, Qiu X (2014) Dispersed Cu_2O octahedrons on H-Bn nanosheets for P-nitrophenol reduction. ACS Appl Mater Interfaces 6:14469–14476
105. Narayanan RK, Devaki SJ (2015) Brawny silver-hydrogel based nanocatalyst for reduction of nitrophenols: studies on kinetics and mechanism. Indus Eng Chem Res 54:1197–1203
106. Zelekew OA, Kuo HD (2016) A two-oxide nanodiode system made of double-layered P-type Ag_2O@N-type TiO_2 for rapid reduction of 4-nitrophenol. Phys Chem Chem Phys 18:4405–4414
107. Hansen MJ, Lerch MM, Szymanski W, Feringa BL (2016) Direct and versatile synthesis of red-shifted azobenzenes. Angew Chem Int Ed 55:13514–13518

108. Wegener M, Driessen MJ, Szymanski AJM, Feringa BL (2017) Photocontrol of antibacterial activity: shifting from UV to red light activation. J Am Chem Soc 139:17979–17986
109. Berradi M, Hsissou R, Khudhair M, Assouag M, Cherkaoui O, El Bachiri A, El Harfi A (2019) Textile finishing dyes and their impact on aquatic environs. Heliyon 5:e02711
110. Abe R, Takami H, Murakami N, Ohtani B (2008) Pristine simple oxides as visible light driven photocatalysts: highly efficient decomposition of organic compounds over platinum-loaded Tungsten oxide. J Am Chem Soc 130:7780–7781
111. Selvam K, Sakamoto H, Shiraishi Y, Hirai T (2015) Photocatalytic secondary amine synthesis from azobenzenes and alcohols on TiO_2 loaded with Pd nanoparticles. New J Chem 39:2856–2860
112. Shiraishi Y, Tekada Y, Sugano Y, Ichikawa S, Tanaka S, Hirai T (2011) Highly efficient photocatalytic dehalogenation of organic halides on TiO_2 loaded with bimetallic Pd–Pt alloy nanoparticles. Chem Commun 47:7863–7865
113. Gultom NS, Abdullah H, Kuo DH (2017) Enhanced photocatalytic hydrogen production of noble-metal free Ni-doped Zn(O, S) in ethanol solution. Int J Hydrogen Energy 42:25891–25902
114. Firdous A, Singh D, Ahmad MM (2013) Photocatalytic reduction of 4-nitrophenol using effective hole scavenger over novel Mg-doped Zn(O, S) nanoparticles. Appl Nanosci 3:13–18
115. Gultom NS, Abdullah H, Kuo DH (2019) Effects of graphene oxide and sacrificial reagent for highly efficient hydrogen production with the costless Zn(O, S) photocatalyst. Int J Hydrogen Energy 44:29516–29528
116. Lie J, Tanda S, Liu JC (2020) Subcritical water extraction of valuable metals from spent lithium-ion batteries. Molecules 25
117. Abdullah H, Gultom NS, Kuo DH, Saragih AD (2018) Cobaltdoped Zn(O, S)/Ga_2O_3 nanoheterojunction composites for enhanced hydrogen production. New J Chem 42:9626–9634
118. Abdullah H, Gultom NS, Kuo DH (2020) Depletion-Zone size control of p-type NiO/n-type Zn(O, S) nanodiodes on high-surfacearea SiO_2 nanoparticles as a strategy to significantly enhance hydrogen evolution rate. Appl Catal B 261:118223
119. Cheng C, Zhang X, Wei C, Liu Y, Cui C, Zhang Q, Zhang D (2018) Mesoporous hollow $ZnCO_2S_4$ core-shell nanospheres for high performance supercapacitors. Ceram Int 44:17464–17472
120. Gultom NS, Abdullah H, Kuo DH (2020) Phase transformation of bimetal zinc nickel oxide to oxysulfide photocatalyst with its exceptional performance to evolve hydrogen. Appl Catal B 272:118985
121. Silambarasan K, Narendra Kumar AV, Joseph J (2016) $K_4[Fe(CN)_6]$ immobilized anion sensitive protonated amine functionalized polysilsesquioxane films for ultra-low electrochemical detection of dsDNA. Phys Chem Chem Phys 18:7468−7474
122. Luo J, Sam A, Hu B, DeBruler C, Wei X, Wang W, Liu TH (2017) Unraveling pH dependent cycling stability of ferricyanide/ferrocyanide in redox flow batteries. Nano Energy 42:215–221
123. Slater JC (1964) Atomic radii in crystals. J Chem Phys 41:3199–3204
124. Abdullah H, Siburian R, Pasaribu SP, Panggabean AS (2021) Visible-light driven Ni-incorporated CdS photocatalytic activities for azo-bond cleavages with hydrogenation reaction. Chem Select 6:2041–2050
125. Wang Y, Liu L, Xu L, Cao X, Li X, Huang Y, Meng C, Wang Z, Zhu W (2014) $Ag_2O/TiO_2/V_2O_5$ one-dimensional nanoheterostructures for superior solar light photocatalytic activity. Nanoscale 6:6790–6797
126. Chen X, Shangguan W (2013) Hydrogen production from water splitting on CdS-based photocatalysts using solar light. Front Energy 7:111–118
127. Abdullah H, Gultom NS, Kuo DH (2020) Depletion-zone size control of p-type NiO/n-type Zn (O, S) nanodiodes on high-surface-area SiO2 nanoparticles as a strategy to significantly enhance hydrogen evolution rate. Appl Catal B 261:118223
128. Matsui F, Lovering EG, Curran NM, Watson JR (1983) Determination of azobenzene and hydrazobenzene in phenylbutazone and sulfinpyrazone products by high-performance liquid-chromatography. J Pharm Sci 72:1223–1224

129. Akula MR, Kabalka GW (1996) Reduction of azobenzenes to diphenylhydrazines. Synth Commun 26:3821–3825
130. Trost BM, Fleming I (1991) Comprehensive organic synthesis—selectivity, strategy and efficiency. In: Modern organic chemistry. Elsevier, Kidlington
131. Wawzonek S, Fredrickson JD (1955) The polarographic behavior of the azobenzenehydrazobenzene system. J Am Chem Soc 77:3985–3988
132. Grampp G, Mureşanu C, Landgraf S (2005) Solvent influence on the electrochemicalreduction of photochemically generated cis-azobenzene. J Electroanal Chem 582:171–178
133. Schmöger C, Stolle A, Bonrath W, Ondruschka B, Keller T, Jandt KD (2009) A practical approach for ambient-pressure hydrogenations using Pd on porous glass. Chem Sus Chem 1:77–82
134. Prasad HS, Gowda S, Gowda DC (2004) Facile transfer hydrogenation of azo compounds to hydrazo compounds and anilines by using Raney nickel and hydrazinium monoformate. Synth Commun 34:1–10
135. Lu H, Zhao J, Li L, Gong L, Zheng J, Zhang L, Wang Z, Zhang J, Zhu Z (2011) Selective oxidation of sacrificial ethanol over TiO_2-based photocatalysts during water splitting. Energy Environ Sci 4:3384–3388
136. Zhang YT, Pei L, Zheng Z, Yuan Y, Xie T, Yang J, Chen S, Wang J, Waclawik ER, Zhu H (2015) Heterojunctions between amorphous and crystalline niobium oxide with enhanced photoactivity for selective aerobic oxidation of Benzylamine to imine under visible light. J Mater Chem A 3:18045–18052
137. Fan MS, Hu B, Yan X, Song CG, Chen TJ, Feng Y, Shi WD (2015) Excellent visiblelight-driven photocatalytic performance of Cu_2O densitized $NaNbO_3$ heterojunction. New J Chem 39:6171–6177
138. Iwase A, Udagawa Y, Yoshino S, Ng YH, Amal R, Kudo A (2019) Solar water splitting under neutral conditions using Z-scheme systems with Mo-doped $BiVO_4$ as an O_2-evolving photocatalyst. Energ Technol 7:1980351
139. Feng XW, Chen H, Jiang F, Wang X (2019) One-pot fabrication of double Z-scheme $CeCO_3OH/g\text{-}C_3N_4/CeO_2$ photocatalyst for nitrogen fixation under solar irradiation. Catal Sci Technol 9:2849–2857
140. Yang FL, Zhang Q, Zhang L, Cao MT, Liu QQ, Dai WL (2019) Facile synthesis of highly efficient Pt/N-rGO/N-$NaNbO_3$ nanorods toward photocatalytic hydrogen production. Appl Catal B 257:117901
141. Yang FL, Zhang Q, Zhang JH, Zhang L, Cao MT, Dai WL (2020) Embedding Pt nanoparticles at the interface of CdS/$NaNbO_3$ nanorods heterojunction with bridge design for superior Z-Scheme photocatalytic hydrogen evolution. Appl Catal B 278:119290
142. Blasse G, de Heuvel GPMV (1972) Vibrational and electronic spectra and crystal structure of cubic Na_3NbO_4. Mater Res Bull 7:1041–1043
143. Shiratori Y, Magrez A, Kasezawa K, Kato M, Rohrig S, Peter F, Pithan C, Waser R (2007) Noncentrosymmetric phase of submicron $NaNbO_3$ crystallites. J Electroceram 19:273–280
144. Shanker V, Samal SL, Pradhan G, Narayana C, Canguli A (2009) Nanocrystalline $NaNbO_3$ and $NaTaO_3$: rietveld studies, Raman spectroscopy and dielectric properties. Solid State Sci 11:562–569
145. Ge H, Xu F, Cheng B, Yu J, Ho W (2019) S-scheme heterojunction TiO_2/CdS nanocomposite nanofiber as H_2-production photocatalyst. Chem Cataly Cataly Chem 11:1–10
146. Zhong Y, Wu Y, Chang B, Ai Z, Zhang K, Shao Y, Zhang L, Hao X (2019) CoP/CdS/ WS_2 p-n-n tandem heterostructure: a novel photocatalyst for hydrogen evolution without using sacrificial agents. J Mater Chem A 7:14638–14645
147. Yan T, Ding R, Ying D, Huang Y, Huang Y, Tan C, Sun X, Gao P, Liu E (2019) An intercalation pseudocapacitance-driven perovskite $NaNbO_3$ anode with superior kinetics and stability for advanced lithium-based dual-ion batteries. J Mater Chem A 7:22884–22888
148. Zhou H, Pan J, Ding L, Tang Y, Ding J, Guo Q, Fan T, Zhang D (2014) Biomassderived hierarchical porous CdS/M/TiO_2 (M = Au, Ag, Pt, Pd) ternary heterojunctions for photocatalytic hydrogen evolution. Int J Hydrogen Energy 39:16293–16301

149. Balayeva NO, Fleisch M, Bahnemann DW (2018) Surface-grafted WO_3/TiO_2 photocatalysts: enhanced visible-light activity towards indoor air purification. Catal Today 313:63–71
150. Balayeva NO, Mamiyev Z, Dillert R, Zheng N, Bahnemann DW (2020) Rh/TiO_2-photocatalyzed acceptorless dehydrogenation of N-heterocycles upon visible-light illumination. ACS Catal 10:5542–5553
151. Balayeva NO, Zheng N, Dillert R, Bahnemann DW (2019) Visible-light-mediated photocatalytic aerobic dehydrogenation of N-heterocycles by surface-grafted TiO_2 and 4-amino-TEMPO. ACS Catal 9:10694–10704
152. Peia L, Wang J, Fana C, Gea H, Waclawikc ER, Tana H, Liua M, Gua X, Zhenga Z (2020) The key role of photoisomerisation for the highly selective photocatalytic hydrogenation of azobenzene to hydrazobenzene over $NaNbO_3$ fibre photocatalyst. J Photochem Photobiol A 400:112655
153. Hu Q, Huang J, Li G, Jiang Y, Lan H, Guo W, Cao Y (2016) Origin of the improved photocatalytic activity of Cu incorporated TiO_2 for hydrogen generation from water. Appl Surf Sci 328:170–177
154. Gu Q, Zhu K, Zhang N, Sun Q, Liu P, Liu J, Wang J, Li Z (2015) Modified solvothermal strategy for straightforward synthesis of cubic $NaNbO_3$ nanowires with enhanced photocatalytic H_2 evolution. J Phys Chem C 119:25956–25964
155. Shuwanto H, Abdullah H, Kuo DH, Gultom NS (2021) Surface active sites of Y-doped Zn(O, S) for chemisorption and hydrogenation of azobenzene and nitroaromatic compounds under light via self-generated proton. Appl Surf Sci 552:149508
156. Guillamón E, Oliva M, Andrés J, Llusar R, Pedrajas E, Safont VS, Algarra AG, Basallote MG (2021) Catalytic hydrogenation of azobenzene in the presence of a cuboidal MO_3S_4 cluster via an uncommon sulfur-based H_2 activation mechanism. ASC Catal 11:608–614
157. Shiraishi Y, Sugano Y, Tanaka S, Hirai T (2010) One-pot synthesis of benzimidazoles by simultaneous photocatalytic and catalytic reactions on Pt@TiO_2 nanoparticles. Angew Chem Int Ed 49:1656–1660
158. Shiraishi Y, Fujiwara K, Sugano Y, Ichikawa S, Hirai T (2013) N-monoalkylation of amines with alcohols by tandem photocatalytic and catalytic reactions on TiO_2 loaded with Pd nanoparticles. ASC Catal 3:312–320
159. Burns S, Jackson SD, Vass EM (2011) Propyne and 1,3-butadiene adsorption and co-adsorption over palladium catalysts. Kinet Catal 52:723–732
160. Hakki A, Dillert R, Bahnemann DW (2013) Factors affecting the selectivity of the photocatalytic conversion of nitroaromatic compounds over TiO_2 to valuable nitrogen-containing organic compounds. Phys Chem Chem Phys 15:2992–3002
161. Xu ZF, Xu K, Feng HF, Du Y (2018) Wei-Chang Hao s-p orbital hybridization: a strategy for developing efficient photocatalysts with high carrier mobility. Sci Bull 62:465–468
162. Wang XS, Zhou C, Shi R, Liu QQ, Zhang TR (2019) Two-dimensional $Sn_2Ta_2O_7$ nanosheets as efficient visible light-driven photocatalysts for hydrogen evolution. Rare Met 38:397–403

Chapter 3
Photocatalytic Oxygen Reduction Reaction to Generate H_2O_2 Over Carbon-Based Nanosheet Catalysts

Afandi Yusuf, Salva Salshabilla, Bobby Refokry Oeza, Nurul Ika Damayanti, Hairus Abdullah, and Januar Widakdo

Abstract Photocatalytic oxygen reduction reaction (ORR) to generate hydrogen peroxide (H_2O_2) has attracted significant attention as a sustainable and environmentally friendly approach. Carbon-based nanosheet catalysts have emerged as promising materials for this reaction due to their unique structural and electronic properties. In this chapter, we investigate the photocatalytic performance of carbon-based nanosheet catalysts for the selective synthesis of H_2O_2. The carbon-based nanosheets were synthesized via a facile and scalable method, resulting in a high surface area and excellent dispersion of active sites. The efficient charge transfer and oxygen reduction kinetics were attributed to the carbon-based nanosheets' unique electronic structure and abundant active sites. This chapter provides valuable insights into the design and development of efficient carbon-based nanosheet catalysts for photocatalytic ORR to generate H_2O_2, contributing to the advancement of sustainable energy conversion and storage technologies. In addition, we also include the synergistic effect of combining a semiconductor photocatalyst, metal–organic framework (MOF), and carbon nanosheets for efficient H_2O_2 generation. The semiconductor photocatalyst provides light absorption and charge separation capabilities, while the MOF and carbon nanosheets serve as co-catalysts to enhance catalytic activity and

A. Yusuf · S. Salshabilla · B. R. Oeza
Department of Materials Science and Engineering, National Taiwan University of Science and Technology, Taipei 10672, Taiwan, ROC

B. R. Oeza
Department of Material Engineering, Institut Teknologi Sepuluh Nopember Surabaya, Keputih Sukolilo, Surabaya, East Java 60111, Indonesia

N. I. Damayanti
Department of Chemical Engineering, National Taiwan University of Science and Technology, Taipei 10672, Taiwan, ROC

H. Abdullah
Department of Industrial Engineering, Universitas Prima Indonesia, Medan 20118, Indonesia

J. Widakdo (✉)
Department of Physics, Universitas Indonesia, Jakarta 16424, Indonesia
e-mail: hoant1@eaut.edu.vn

H. Abdullah (ed.), *Solar Light-to-Hydrogenated Organic Conversion*,
https://doi.org/10.1007/978-981-99-8114-4_3

selectivity. The carbon nanosheets, synthesized through a scalable method, offer a high surface area and abundant active sites, promoting efficient charge transfer and catalytic reaction kinetics. With its porous structure and metal active sites, the MOF acts as a promoter by facilitating the adsorption and activation of reactants. This chapter explores the potential of graphitic carbon nitride (g-C_3N_4) as a promising photocatalyst for H_2O_2 generation. The unique electronic and structural properties of g-C_3N_4 make it an ideal candidate for efficient charge separation and redox reactions under visible light irradiation. The synthesis of g-C_3N_4 photocatalysts with varying morphologies and compositions can optimize their photocatalytic performance. The effects of modification to g-C_3N_4 enhance the catalytic activity and selectivity toward H_2O_2 production. The optimized g-C_3N_4 photocatalyst exhibits remarkable activity, surpassing conventional metal-based catalysts in efficiency and stability. The insights gained from this chapter contribute to the fundamental understanding of g-C_3N_4-based photocatalysis for H_2O_2 production and pave the way for the development of sustainable energy conversion technologies. The promising performance of g-C_3N_4 in photocatalytic H_2O_2 generation underscores its potential for practical applications in areas such as wastewater treatment, energy storage, and chemical synthesis.

3.1 Introduction

Considering the worldwide energy dilemma and environmental apprehensions, generating clean, friendly energy is crucial. The rising demand for energy and the adverse effects of fossil fuels on the environment have increased the need for green technologies and alternative energy sources. This demand arises from the need to find sustainable energy sources and address the environmental disruption caused by the extensive use of hydrocarbon fuels. The rapid expansion of society has also contributed to this need. As pollution levels escalate and fossil fuel reserves decline, there is a growing demand for renewable energy sources. Fossil fuels currently account for the most significant proportion of global energy production. The environment is adversely affected by fossil fuels as they contaminate the air and contribute to global warming. Moreover, fossil fuels are the predominant energy source for virtually all the world's energy needs (86.4%). Alternative sources of energy can decrease or eradicate the use of fossil fuels. Energy from the sun is a sustainable and renewable source of power that can be transformed into both electricity and heat and is present all around the world. The Earth's atmosphere receives roughly 3.85×10^{24} J of solar power each year [1–3]. To convert solar energy into fuel, photocatalytic materials are necessary. A stable and quantum-efficient photocatalyst is mandatory for hydrogen production, a promising green and renewable fuel for the future. Hydrogen is a promising alternative energy source that could replace fossil fuels. It has a higher energy density than hydrocarbons, with a value of 2.75 times that of hydrocarbon and 2.40 times that of methane. Fuel cells powered by hydrogen have the potential to generate both electricity and water as byproducts. Scientists are developing sustainable and economical techniques and technologies for producing and storing hydrogen energy. Nanotechnology has enhanced hydrogen generation and storage capacity. Several methods for producing hydrogen include autothermal reforming, steam reforming, water electrolysis, and partial oxidation. Hydrogen can also be generated through various processes such as electrochemical, photochemical, and solar-assisted thermochemical systems [4–6].

H_2O_2, called hydrogen peroxide, is one of the most essential compounds in the world [7]. H_2O_2 is one of the most valuable chemicals used for industrial applications and civil applications, including medical equipment sterilization, water filtration, chemical production, and fuel cell technology [8]. Despite that, in treating wastewater, bleaching paper and pulp, and eradicating bacteria and viruses, hydrogen peroxide (H_2O_2) is an eco-friendly popular oxidant and disinfectant [9]. Moreover, H_2O_2 is a promising liquid fuel that can be stored and transported more easily than compressed hydrogen. Because of that, H_2O_2 has high production worldwide, approaching approximately 4 million tons in 2020, and will continue to increase over the year [10]. This inorganic molecule's range of uses encompasses a wide range of diverse applications, making H_2O_2 an extremely adaptable and multifunctional chemical that currently supports an industry worth roughly 19.4 billion dollars in 2022, and it is estimated that this value will increase to approach 15.4 million tons in 2030 [11, 12] (see Fig. 3.1).

Hydrogen peroxide (H_2O_2) is a vital green chemical and a powerful oxidant with numerous applications [13]. H_2O_2 is a safe and environmentally friendly oxidant that produces only water and oxygen as byproducts. It is widely used as a clean oxidant in various industries, such as paper production, wastewater treatment, disinfection, and chemical synthesis. Additionally, H_2O_2 is considered a promising energy carrier in fuel cells [14]. Due to its higher oxidizing power compared to oxygen, H_2O_2

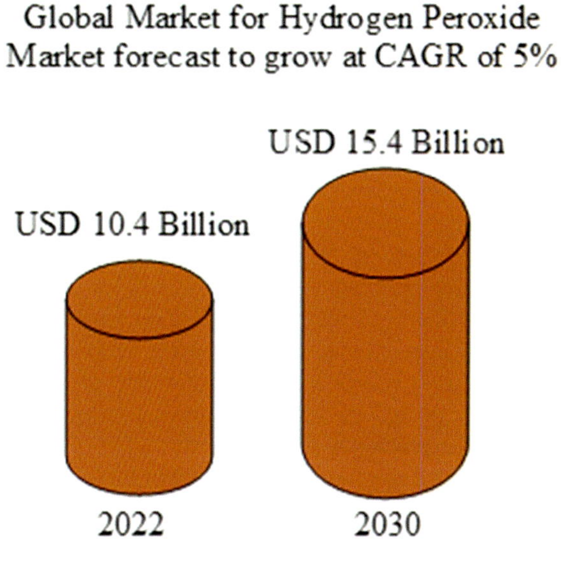

Fig. 3.1 Global market for H_2O_2 in the world, inspired by Refs. [11, 12]

produces more electrons, inhibiting electron–hole pairs from recombining during the photocatalytic process. As an electron acceptor, this compound can generate radicals by reacting with electrons in the conduction band. Using complex pollutant mixtures in photocatalytic treatments can lead to further complications with applying H_2O_2. The presence of H_2O_2 can impede the degradation rate due to its adsorption onto the catalyst surface, inhibition of holes generated in the valence band, and reaction with hydroxyl radicals. These factors all contribute to negative effects on the treatment's efficiency [13]. In addition, H_2O_2 has practical applications as a rocket fuel and in fuel cells, where it can produce electricity, and the only byproducts are water and oxygen. The increasing energy requirements have resulted in hydrogen peroxide being suggested as a potential chemical fuel [14]. Generating H_2O_2 in close proximity to where it will be used can eliminate the potential risks and energy expenses associated with transporting concentrated H_2O_2, particularly given that numerous applications only require diluted H_2O_2. H_2O_2 concentrations of less than 9% by weight are required for most applications, such as pulp bleaching and chemical synthesis. In addition, for various applications such as medical, chemical, and microbial contamination remediation, studies have revealed that concentrations of H_2O_2 as low as 3 and 0.1 wt.% can be adequate for various applications. Many methods are currently being used to produce H_2O_2 on a large scale, including electrochemical synthesis, alcohol oxidation, and anthraquinone autoxidation [14]. The anthraquinone oxidation process is a popular method for producing H_2O_2, but several factors limit it. The process involves multistep oxidation and hydrogenation reactions, requires the use of numerous organic reagents, and generates exhaust gas and solid waste [15]. Currently, the industrial-scale method for producing H_2O_2, which is anthraquinone oxidation, requires a significant amount of energy [16]. The current large-scale production method for H_2O_2, which accounts for over 95% of the global demand, involves a four-step cycle. This process includes hydrogenating AQ; oxidizing the hydrogenated AQ to regenerate it and produce H_2O_2, extracting, purifying, and concentrating the H_2O_2; and recovering the working solutions. However, despite the high H_2O_2 yield achieved by this process, it consumes a considerable amount of energy and produces a significant amount of chemical waste. As a result, dilution of the high-concentration H_2O_2 product obtained through the AQ process is required (up to 70 wt.%), resulting in energy waste. It is now possible to directly synthesize H_2O_2 from H_2 and O_2, and this method shows promise. Additionally, it can reduce the production of toxic byproducts typically generated when using AQ as a reaction carrier.

Hydrogen peroxide is a non-toxic and eco-friendly oxidizing agent extensively used in numerous areas, such as organic synthesis, the chemical industry, environmental applications, and wastewater treatment [17]. H_2O_2 also attracted world attention as a storable and transportable promising liquid fuel [18]. H_2O_2 is mainly produced via the anthraquinone, direct synthesis, electricity-driven, and light-driven processes [19]. However, most of the H_2O_2 is manufactured by the anthraquinone method, resulting in harmful organic byproducts to the environment and requiring high energy [20]. The potential explosion concern also hinders a potential direct synthesis procedure that employs noble metal-based catalysts to catalyze the combining of H_2 and O_2 [21]. Thus, it is indeed desirable to develop eco-friendly H_2O_2 in the most effective, inexpensive, and sustainable ways possible to fulfill the increasing demand for H_2O_2 (see Fig. 3.2).

Commercially, the industrial process to produce H_2O_2 and approximately 95% of all H_2O_2 produced in this country is created using the anthraquinone (AQ) method, which calls for the successive hydrogenation of alkyl-anthraquinone and the subsequent oxidation of alkyl-anthrahydroquinone in organic solvents (such as heavy aromatics) with catalysts made of Ni or Pd [22]. The AQ method is appropriate for large scale and at a single location manufacture of economically concentrated H_2O_2 due to the complicated procedures, including hydrogenation, oxidation, separation, concentration, and purification. Unfortunately, the several hydrogenation and oxidation processes in the anthraquinone oxidation process demand a substantial amount of energy, generate a significant quantity of wastewater, and potential safety problems [7, 23]. However, there are severe problems with energy use, environmental damage, and possible safety concerns with this method. First, the continuous usage of oxygen and hydrogen in the reaction tower increases the possibility of explosive combustion and detonation. Second, there are significant potential safety risks

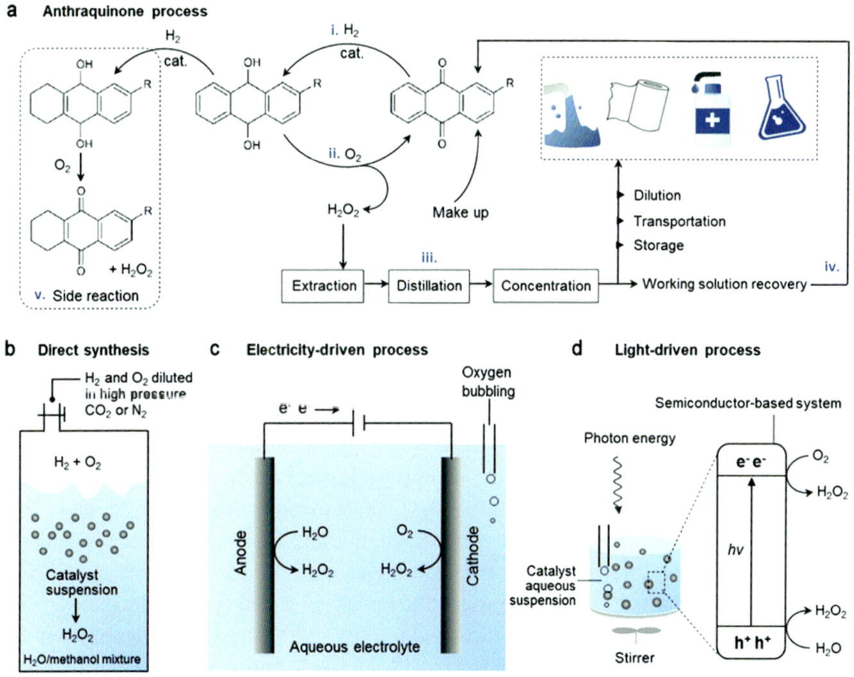

Fig. 3.2 Schematic diagram illustrating the several catalytic processes used to synthesize H_2O_2. **a** Conventional anthraquinone process with the (i–iv) steps cycle and (v) anthraquinone process side reaction. **b** Direct synthesis of H_2 and O_2, which diluted in high-pressure inert gas. **c** Electrocatalyst and **d** Photocatalyst from water oxidation and oxygen reduction [19]. With license ID 1357049-1

associated with the breakdown of hydrogen peroxide during storage and transportation. The purification process of H_2O_2 is repeated to be able to increase the purity and decrease decomposition rates. And this process will increase the energy and resource consumption [24]. Consequently, it is worthwhile to develop economical and environmentally responsible procedures for large-scale development of H_2O_2.

Directly producing hydrogen peroxide from H_2 and O_2 in a liquid medium is one option that enables decentralized production of H_2O_2 and eliminates the need for long-distance transportation, which has gained significant attention in the past few decades. However, safety concerns regarding mixing H_2 and O_2 gases have hindered the implementation of this technology. As a result, recent efforts have focused on safer and more efficient methods [25].

The increasing demand for clean energy sources in recent times has presented various obstacles, such as the exploration of substitutes for conventional and local sources that significantly harm the environment. Hydrogen peroxide (H_2O_2) is an appropriate chemical candidate for fulfilling this role due to its favorable characteristics, which include a high calorific value, strong oxidation capacity, environmentally friendly nature, ease of storage and transportation, and production of clean byproducts, namely, water (H_2O) and oxygen (O_2), without any harmful pollutants. The exceptional qualities of hydrogen peroxide make it a vital ingredient in various fields, including but not limited to wastewater treatment, environmental remediation, chemical synthesis, pulp bleaching, and numerous other industries. As the demand for hydrogen peroxide continues to rise, there is a need for increased production output to meet the growing demand. Currently, the industry's conventional method of hydrogen peroxide production involves anthraquinone oxidation. However, this method has several significant drawbacks, such as excessive energy consumption, liquid and solid waste generation, use of organic solvents, and reliance on costly noble metal catalysts, despite its efficiency. Given the drawbacks of the anthraquinone oxidation process, it is crucial to explore cleaner methods of hydrogen peroxide production that can serve as an alternative to AO [26].

H_2O_2 production using a semiconductor photocatalytic technique has gained much attention since it uses abundant and renewable sunlight as a driving component. Additionally, the photocatalytic process does not involve using H_2 and can be a safe and environmentally beneficial technique [27]. When producing H_2O_2 directly from H_2 and O_2 in a liquid medium, several undesired reactions occur along with the intended response of $H_2 + O_2 \rightarrow H_2O_2$. One of these undesired reactions is $2H_2 + O_2 \rightarrow 2H_2O$, which is even more thermodynamically favored than the production of H_2O_2. The explosive nature of the H_2/O_2 gaseous mixture poses a significant challenge, necessitating the use of inert carrier gases to minimize the danger, but this results in reduced selectivity and productivity. The use of costly noble metal catalysts has been the only way to enhance selectivity and activity in the reaction for H_2O_2 production, which emphasizes the necessity for cost-effective, efficient, environmentally friendly methods for localized H_2O_2 synthesis that are mild. Additionally, hydrogenation of the aromatic ring can produce organic byproducts and AQ consumption, which limits AQ recovery during the cycling process. Transporting and storing unstable H_2O_2 from a centralized factory to end users can raise safety concerns [13]. These approaches have some drawbacks as they require significant amounts of energy and organic solvents, which may not be environmentally friendly. Additionally, the presence of organic impurities can lead to contamination of the produced H_2O_2, making extraction more complex [14]. As a result, significant efforts have been put toward developing novel technologies for manufacturing H_2O_2 that can operate under more moderate reaction settings while producing no waste [23]. Electrochemical reduction of O_2 at the cathode of an electrochemical cell is one of the several processes for producing H_2O_2 that has garnered significant interest in recent times [16].

A feasible alternative to the industrial anthraquinone methods is developed with photocatalytic H_2O_2 using existing semiconductor materials. This method offers sustainable low-density solar energy to a storable chemical energy technique to convert water and oxygen [28]. The process of photocatalytic H_2O_2 production includes multiple steps such as absorption of light, trapping, separation, and transfer of charge carriers, adsorption of reactant surfaces, surface redox reactions, intermediate conversion, and stabilization/desorption of the final product [29]. These crucial processes intrinsically determine the effectiveness of H_2O_2 generation across the entire process. Optical absorption, photoexcited charge separation, surface charge transfer, and light absorption are crucial elements in photocatalytic technology for transforming solar energy into chemical energy. These elements play a crucial role in various stages of the process, such as intermediate conversion, excitation/trapping/separation/transfer of charge carriers, surface redox reaction, adsorption of reactant surfaces, and stabilization/desorption of the product [30]. The process involves using a proton-coupled electron transfer (PCET) technique on semiconductor materials to perform an oxygen reduction reaction (ORR) and produce photocatalytic H_2O_2 [31].

Solar energy-driven hydrogen production with the aid of a catalyst is a promising avenue for sustainable energy production in the future. Photocatalysis offers a more straightforward option, using a light-absorbing catalyst suspended in water to split water molecules through the absorption of photons. This method is cheaper and easier to mass-produce. During photocatalytic water splitting, hydrogen gas is produced through water reduction, while oxygen gas is produced through water oxidation. However, the generation of hydrogen peroxide as a byproduct can be problematic, as it can potentially poison the catalyst. Despite its potential use as a chemical raw material, finding a photocatalyst that can simultaneously produce H_2O_2 and H_2 in a single system remains a challenge of great significance [32].

The best technique for obtaining a future energy source is converting water using a catalyst and solar power to produce H_2 [33]. Photocatalysis is a less complicated choice, in which water is suspended with a powdered light-absorbent catalyst, which absorbs photons and speeds up the process. The process of generating hydrogen peroxide through photocatalytic oxygen reduction reaction (ORR) is considered a more eco-friendly process than the widely used anthraquinone oxidation method. The latter method involves high temperatures, pressures, and organic solvents, which can harm the environment and pose a risk to workers. In contrast, the photocatalytic ORR process is safer and more sustainable, as it doesn't require these harmful elements and uses sunlight as an energy source, reducing energy consumption and production costs.

Overall, the photocatalytic ORR technique shows potential as a substitute for the conventional anthraquinone oxidation method, given its environmental friendliness and cost-effectiveness [15]. Photocatalysis has emerged as a possible alternative since it necessitates an optical semiconductor, oxygen, oxygen, and adequate and renewable light as its driving force. Theoretically, this is less expensive and simpler to industrialize [32]. An eco-friendly substitute for conventional production onsite of H_2O_2 is the photocatalyst ORR, which catalytically converts O_2 (oxygen) to H_2O_2 (hydrogen peroxide). The photocatalyst plays a crucial role in influencing the photocatalyst oxygen reduction reaction process' activity and selectivity.

The process of using the photocatalytic ORR to generate hydrogen peroxide is seen as a more sustainable and environmentally friendly option when compared to the anthraquinone oxidation process [1]. A concentration of 5% wt.% is often necessary for various applications like medical disinfection, wastewater treatment, and pulp bleaching. The use of semiconductor materials for photocatalytic hydrogen peroxide (H_2O_2) generation has shown significant potential as it solely needs oxygen, water, and light. Selective ORR via two-electron transfer can produce H_2O_2 during the photocatalytic process ($O_2 + 2e^- + H^+ \rightarrow H_2O_2$). The lack of stability of H_2O_2 under exposure to heat and UV light is why non-metal catalysts based on visible-light-driven H_2O_2 evolution have shown promise [15]. As a result, there has been an increase in the development of alternative methods for synthesizing H_2O_2 by business and research organizations. These methods include small-scale and cost-effective facilities that can replace anthraquinone technology. Furthermore, the novel technique might lower the risks and expenses associated with transit and handling. The same photocatalytic guidelines describe the photocatalytic production of H_2O_2. The production of H_2O_2 is believed to occur either through a direct one-step two-electron reduction or an indirect sequential two-step single-electron reduction process. This uphill reaction involves the conversion of H_2O and O_2 into H_2O_2 and has a typical Gibbs free energy change (ΔG°) of 117 kJ mol^{-1}. Metal-free visible light photocatalytic H_2O_2 generation requires a covalent organic framework based on (diarylamino)benzene. Hydrogen peroxide (H_2O_2) can be produced through electrocatalysis or photocatalysis in the presence of photoabsorbers by converting water (H_2O) and oxygen (O_2). The two-electron ($2e^-$) pathway is the main route for H_2O_2 production, although it competes with the four-electron transfer that leads to the formation of H_2O and O_2. The electron transfer pathways in the water/oxygen system are complex, and oxygen can undergo reduction through $1e^-$, $2e^-$, and $4e^-$ pathways, producing $O_2^{\bullet-}$ (a superoxide anion), H_2O_2, and H_2O [19].

A catalyst is necessary for hydrogen peroxide fuel cells to facilitate the oxygen reduction reaction. Usually, precious metals-based photocatalysts have been employed to produce H_2O_2. However, the amount of metal used in generating new catalytic materials must be minimized due to the limited availability of metal-based catalysts and achieve sustainable H_2O_2 production with minimum adverse environmental impacts [34]. The advancement of catalysts that do not contain precious metals for bifunctional purposes and/or multifunctional catalysis has garnered significant attention to decrease or eliminate the necessity of using catalysts that contain precious metals [35]. There has been considerable progress in the development of cost-effective non-precious metal (NPM) oxygen reduction reaction (ORR) catalysts, and several NPM catalysts that have recently been reported have shown ORR catalytic activity that is equivalent to or even superior in comparison to the catalyst that contains metals. The carbon-based catalyst is one of the non-precious metal ORR catalysts that exhibit good performance. The performance of carbon-based catalysts is being improved through significant efforts, and to obtain precise data regarding metal-free carbon-based catalysts, high-purity carbon materials are needed [36]. In the most widely used fuel cell technology, which is the acidic polymer electrolyte membrane fuel cell (PEMFC), even carbon-based metal-free oxygen reduction reaction (ORR) catalysts have demonstrated excellent operational stability and high energy efficiency. The potential of these novel carbon-based catalysts, which do not contain any metals, is to decrease expenses while also preserving high effectiveness and financial feasibility since they use plentiful carbon resources. They can be employed in fuel cells and other energy devices [37].

Solar energy usage to produce hydrogen peroxide (H_2O_2) through photocatalysis is an attractive approach that relies on the generation of photogenerated electrons to promote both oxygen reduction reaction (ORR) and water oxidation. The direct reaction pathway for water oxidation ($2H_2O + 2H^* \rightarrow H_2O_2 + 2H^+$) by photogenerated holes through the two-electron pathway is not thermodynamically favored, as it has a higher oxidation potential compared to the four-electron pathway for water oxidation to form oxygen. Thus, the creation of $2e^-$ ORR catalysts that are exceptionally efficient and selective and are composed of elements that are abundant on earth is vital for this pathway that enables a direct reaction. Composite catalysts based on polymeric carbon nitride (C_3N_4), particularly graphitic carbon nitride (g-C_3N_4), have shown promise as $2e^-$ ORR catalysts for H_2O_2 synthesis because of their ability to respond to visible light, appropriate band configuration, uncomplicated synthesis, and thermal stability, C_3N_4 materials are advantageous. Nevertheless, the activation of C_3N_4 for ORR is still constrained by swift charge recombination and inadequate oxygen adsorption selectivity, which negatively affects the kinetics of the reaction [38].

The development of diverse catalysts for a variety of applications has significantly benefited from advances in nanoscience and nanotechnology [39]. Typically, semiconductor materials are used for photocatalysis. Although a large variety of metal-based semiconductors, including SnO_2, Fe_2O_3, ZnO, TiO_2, CdS, NiS, Ag_3O_4, etc., have been used as photocatalysts, these materials' high bandgaps have made it challenging to harness solar energy effectively for photocatalytic purposes [40]. Yet, because of the low spectral utilization rate caused by the large bandwidth, many strategies have been used to absorb additional solar radiation, thereby boosting photocatalytic activity. Other semiconductors discovered as possible photocatalysts include ZnO, WO_3, SnO_2, and others. Numerous research papers have been published on enhancing photocatalysts to enhance their performance, including element doping, integrating them with heterogeneous semiconductors, and managing defects. However, the progress of photocatalytic materials toward improved catalytic activity still needs to be improved because of their limited ability to absorb sunlight and their weak capability to separate charges generated by light [41]. Semiconductors, such as metal phosphides, nitrides, and sulfides, which have narrow bandgaps were employed for photocatalytic reactions. Nonetheless, their practical applicability was restricted due to their volatility and susceptibility to photocorrosion [42].

Photocatalytic processes utilizing semiconductor-based materials for the production of H_2O_2 are considered eco-friendly due to their consumption of only water, O_2, and light. Conventional photocatalysts, such as TiO_2, $BiVO_4$, and ZnO, could be more efficient in absorbing and utilizing visible light, there is a need to create photocatalysts that respond to visible light. The most commonly used photocatalysts are surface-modified or precious metal-supported TiO_2 and g-C_3N_4-based materials. However, they necessitate external sacrificial agents to achieve efficient charge separation. It isn't easy to employ non-precious metal photocatalysts for producing H_2O_2 in pure water. Nonetheless, in 2019, researchers discovered that resorcinol–formaldehyde resins can serve as metal-free photocatalysts for H_2O_2 production without the need for sacrificial agents in pure water. Nevertheless, the photocatalytic performance could be better when the resin is synthesized at low temperatures. To form C–O groups, high-temperature hydrothermal synthesis with auto-pressurization is necessary in this study. As a result, there is a need to develop H_2O_2 photocatalysts that have milder synthesis conditions and higher solar-to-chemical conversion efficiency [43].

A variety of materials can be used to produce hydrogen peroxide through catalysis and there has been development in modified carbon-based catalysts to enhance the effectiveness and specificity of this procedure. While noble metals like platinum, palladium, and gold have been the most commonly used catalysts for selective H_2O_2 production, their scarcity and high cost make them impractical for large-scale use. As a result, carbon-based materials and other non-noble catalysts have been explored as low-cost alternatives due to their chemical stability, physical strength, and high electrical conductivity. Recent advances have shown that carbon-based materials can be effective at selectively producing H_2O_2 through a two-electron reduction pathway while inhibiting the competitive four-electron pathway [16]. Scientists are exploring the use of carbon-based nanomaterials (CNMs), such as graphene, carbon nanotubes (CNTs), graphitic carbon nitride (g-C_3N_4), fullerene (C60), carbon dots (CDs), and carbon nanofibers (CNFs), to enhance photocatalysis. These substances exhibit remarkable physicochemical traits, including significant chemical and thermal endurance; an extensive surface area; and upgraded electrical, mechanical, and optical attributes. Typically, carbon-based nanomaterials (CNMs) are produced via various techniques such as ball-milling, laser ablation, arc discharge, and chemical vapor deposition (CVD) [19, 44]. Carbon-based nanomaterials (CNMs), such as graphene, carbon nanotubes (CNTs), graphitic carbon nitride (g-C_3N_4), fullerene (C60), carbon dots (CDs), and carbon nanofibers (CNFs), possess outstanding physicochemical features such as elevated chemical and thermal endurance; vast surface area; and improved electrical, mechanical, and optical attributes. These materials have unique morphologies and precisely regulated structures, with narrow bandgaps, making them interesting for developing photocatalysts. Even though CNMs themselves have low photocatalytic activity, they can enhance the effectiveness of photocatalysts by serving as electron carriers. By adding CNMs to photocatalysts, the active center's adsorption site in the reagent can be increased, and the electron–hole pairs present in CNMs can function as electron acceptors or transmission channels, reducing photoexcitation. In addition, CNMs' photothermal effects or photosensitivity can be enhanced in the presence of light, and they can decrease particle agglomeration and increase particle distribution, thereby improving photocatalytic activity. Due to their low cost and vast availability, CNMs have emerged as an attractive alternative for photocatalysis, particularly for the production of H_2 through light photocatalysis, which researchers have thoroughly investigated. Carbon dots, for example, are frequently utilized as photocatalysts for H_2 production. CNM-photocatalyzed H_2 synthesis has made progress [4] (see Fig. 3.3).

To produce H_2O_2, several photocatalysts have been used, including bismuth vanadate, graphitic carbon nitride (g-C_3N_4), and titanium dioxide. Among them, g-C_3N_4 is particularly intriguing since it is a substance-free of metals, non-toxic, and has

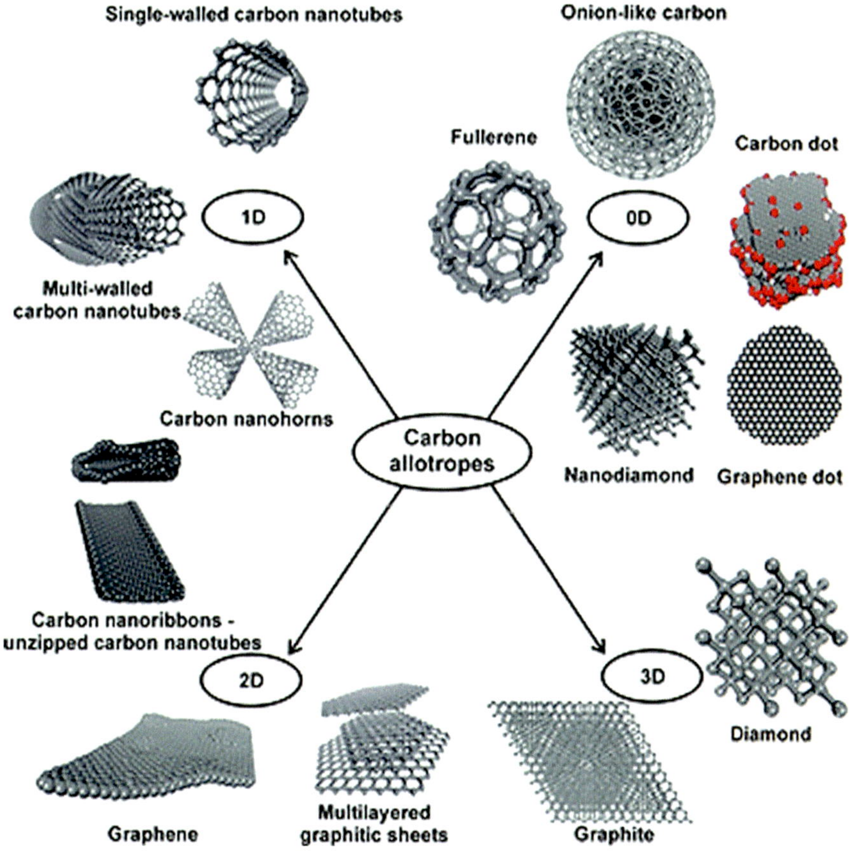

Fig. 3.3 Classification of carbon allotropes for catalysts based on dimensionality [45]. With license ID 1357052-1

chemical instability that has demonstrated remarkable potential not only to produce H_2O_2 but also to produce hydrogen and wastewater treatment, among other things [10]. Researchers have recently become interested in graphite-like carbon nitride (g-C_3N_4) for photocatalytic applications because of its ideal bandwidth. These materials have other beneficial features, such as being easy to prepare, having a low cost, being highly tolerant to extreme temperatures, acidic or alkaline solutions, and having a narrow bandgap with a suitable band position (ca. -1.3 eV and ca. $+1.4$ eV) [41], strong chemical stability, and thermal stability [46]. This has a broad scope of uses, including disinfection, CO_2 reduction, dye material degradation, and water splitting. Moreover, precursors, such as CH_4N_2S (thiourea), CH_4N_2O (urea), and $C_3H_6N_6$ (melamine), which are abundant and high in N_2, can be used to synthesize g-C_3N_4 easily [47].

Recently, there has been increasing interest in utilizing graphene-based nanomaterials for photocatalysis. One strategy to create these photocatalysts is to combine graphene materials with semiconducting materials, like TiO_2, to develop nanocomposite photocatalysts. The combination of these materials is thought to enhance photocatalytic efficiency by promoting electron transfer from semiconducting photocatalysts to graphene-based materials upon photoexcitation, which prevents electron–hole pair recombination. However, in such a system, the hydroxyl radical generated during photocatalysis may interact with certain graphene materials, causing them to quickly degrade. Photocatalysis for H_2O_2 production has become popular due to its ability to occur in water without organic solvents and its potential for utilizing solar energy. Photocatalysts commonly used for producing H_2O_2 include metal oxides, metal–organic complexes, and polymeric materials. Typically, these systems require organic electron donors to achieve high concentrations of H_2O_2. Nevertheless, more sustainable photocatalysts are being studied that do not require organic electron donors. For example, rare earth metal–organic complex photocatalysts have been found to produce millimolar levels of H_2O_2 without the need for organic electron donors [3].

Carbon nitride catalysts that are used for the production of H_2O_2 are typically two-dimensional structures composed of heptazine rings that are repeated on basal planes, known as graphitic carbon nitride (g-C_3N_4). These catalysts have van der Waals forces existing between each layer. The electronic bandgap of g-C_3N_4 is 2.7 eV and responds to visible light wavelengths less than 460 nm, making it distinct from semi-metallic graphene and graphite carbon, which have higher conductivity. Due to its semiconductor properties and unique structure, g-C_3N_4 has been extensively researched as a light absorber and catalyst in heterogeneous photo- and electrocatalysis. However, there has been minimal research on the electrocatalytic production of H_2O_2 using g-C_3N_4 [48, 49]. One potential impediment to its future use could be the difficulty in creating a balance between selectivity toward the $2e^-$ ORR and appropriate overpotential to allow for increasing activity before achieving a breakthrough at this time. In the absence of alcohol, the efficiency of pure g-C_3N_4 was relatively low (an organic scavenger). This is owing to two flaws in pure g-C_3N_4: (1) the rapid recombination of photogenerated electrons and holes and (2) the limited chemical adsorption capability of O_2 on its surface. A photogenerated electron and hole-trapping vacancy defect have been observed. Meanwhile, vacancy defects can increase gas molecule adsorption and activation due to their abundance of localized electrons [50]. Innovative research utilized g-C_3N_4 for the photosynthesis of H_2O_2, achieving superior results compared to the previously studied TiO_2 system with visible-light excitation (>420 nm) and 90% selectivity. Further study by the same group revealed that the intermediate 1,4-endoperoxide species on the surface of g-C_3N_4 promote the two-electron ORR route with higher selectivity than peroxo species coordinated with Ti^{4+} on the surface of TiO_2. These findings were primarily based on the photogenerated h+ side oxidation half-reaction of alcohols, which provided protons and electrons for O_2 reduction to H_2O_2. The researchers then demonstrated the direct oxidation of water to produce O_2 and H^+ by modifying g-C_3N_4 with aromatic diamide and graphene for band structure tuning and electron trapping. This solar-to-chemical reaction, which utilizes only water, oxygen, and metal-free catalysts, is a promising method for the economical and sustainable onsite synthesis of H_2O_2. The study opens up possibilities for further research on g-C_3N_4-based catalysts, aiming to improve their activity and selectivity [19]. The presence of carbon vacancies and amino group termination in g-C_3N_4 has significantly improved the non-sacrificial reduction of oxygen to H_2O_2. Additionally, this modification has shifted from the sequential one-electron reduction reaction (ORR) pathway to a direct two-electron ORR pathway.

g-C_3N_4 is an organic semiconductor that is widely used in photocatalytic H_2O_2 production and organic pollutant degradation because of its appropriate energy band configuration and high stability, ease of modification, and specific photoresponse ability. However, the thick lamella stacking in the g-C_3N_4 framework can limit the exposure of catalytic active sites, making it challenging for unmodified g-C_3N_4 to achieve high photocatalytic activity. Various methods have been developed to improve the photocatalytic activity of g-C_3N_4. One of these methods involves altering the structure of g-C_3N_4 by adding dopants, including metals or nonmetals, to increase the active sites and adjust their electronic structure. Another method combines g-C_3N_4 with other materials, such as carbon-based nanomaterials, to enhance its efficiency, metal oxides, or noble metals, forming heterostructures synergizing photocatalytic activity. In addition, morphology engineering, such as controlling the size, shape, and surface area of g-C_3N_4, can also improve its photocatalytic performance. These strategies have shown promising results in enhancing the photocatalytic activity of g-C_3N_4 for H_2O_2 production and other photocatalytic applications [51].

Polymeric carbon nitride (PCN) is being considered a potential metal-free photocatalyst for producing H_2O_2 through visible light exposure. However, its natural photocatalytic activity is restricted by a wide bandgap, a low number of charge carriers, and a brief lifespan of charge carriers. One proposed solution is introducing nitrogen defects or vacancies into the PCN structure, which can modify the electronic structure, narrow the bandgap, and act as active sites. However, existing methods for introducing these defects require multiple steps and generate significant inorganic waste, highlighting the need for a more direct and environmentally friendly approach. Further research is needed to understand better the influence of nitrogen defects on the visible light-induced reduction of O_2 to H_2O_2 [52].

3.2 Requirements of in situ Photocatalytic Organic Chemical Conversion

The worsening condition of the environment has gotten more severe as industrialization has proliferated. The semiconductor-based photocatalytic technology has drawn much attention for its possible use in environmental protection as a solution to this issue. The photocatalytic activity has been increased using the heterojunction construction, which is an efficient and promising method. The chemical transformation method refers to creating new materials by dissolving the source materials in a solution containing the required ions and an etching agent. This results in the formation of a new phase and an in situ heterojunction. This technique is beneficial for creating a heterojunction with strong interactions [53].

Photocatalysts have different ways of initiating chemical reactions, including visible light photoredox catalysis. This process is known as energy transfer or sensitization, which can broaden the range of reactions the photocatalyst can achieve. For example, the energy transfer from a photosensitizer to a substrate can produce singlet oxygen, a reactive species that can be used in various organic transformations. Overall, these processes illustrate the complex mechanisms involved in photocatalysis and highlight the importance of understanding the fundamental principles of excited state chemistry to design efficient and selective photocatalytic systems [54] (see Fig. 3.4).

In general, selecting appropriate photocatalysts for reduction and/or oxidation reactions is based on specific criteria. Semiconductors must possess band potentials suitable for the corresponding oxidation or reduction reactions to facilitate photocatalytic responses. For instance, to facilitate the H_2 reduction half-reaction, the electrons in the conduction band must have a more negative potential than the H^+/H_2 reduction potential (0 eV vs. NHE). In contrast, the water oxidation reaction can happen when photoholes in the valence band have a more positive potential than the oxidation potential of O_2/H_2O (+1.23 eV vs. NHE). Additionally, it is essential to consider the inherent characteristics of n- or p-type conductors, especially when a p-type and n-type semiconductor combines to form a p-n heterojunction. This can result in better separation of charges and increased effectiveness in photocatalysis. Additionally, selecting the appropriate amount of co-catalyst on the semiconductor is crucial for optimizing charge separation, controlling the reaction pathway, and determining which products are produced. Consequently, this article will comprehensively examine various enhancement methods to assist readers in creating effective photocatalysts for small molecule photocatalytic conversions [55] (see Fig. 3.5).

1. **Light harvesting**

Photocatalytic technology holds great promise for developing and utilizing solar energy, providing a potential solution to environmental pollution and energy crises. It is essential to fully use the solar spectrum to maximize the efficiency of photocatalysts. However, most semiconductor photocatalysts are only active under UV and some visible light, leaving out the near-infrared (NIR) light that accounts for almost

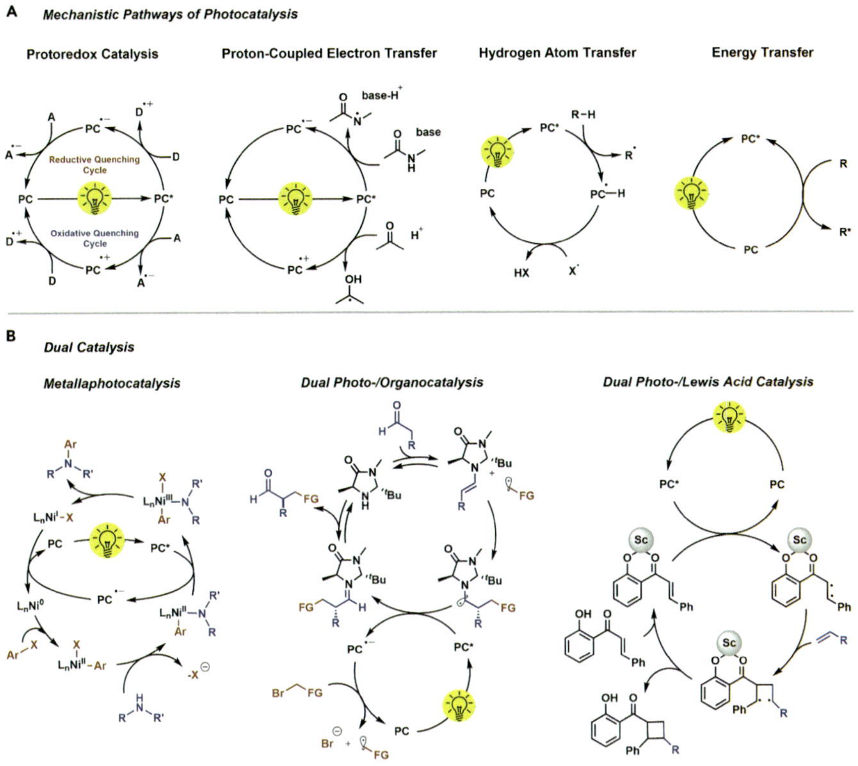

Fig. 3.4 In photocatalytic synthesis, there are different modes of operation (**a**) and some examples of dual catalysis (**b**) that involve combining different catalysts to achieve a desired chemical reaction [54] with license number 5523510634785

half of the solar spectrum. This limits the practical application of semiconductor photocatalysts. Moreover, the solid penetrating capacity of NIR light is advantageous for effective contact with substrates and solid–liquid photocatalytic processes compared to UV and visible light [56]

Photocatalysis is a process that involves the activation of semiconductors by light, resulting in the production of high-energy electrons and holes that can catalyze oxidation and reduction reactions. When light with energy more significant than the bandgap of the semiconductor photocatalyst is absorbed, it generates both electrons and holes. The electrons are excited to the conduction band (CB), while the photoactivated holes are left in the valence band (VB). These electrons and holes can either recombine or move to the catalyst's surface to initiate various photoredox reactions [57].

Polymeric graphitic carbon nitride (g-C_3N_4) is a type of semiconductor photocatalyst with a planar 2D structure and distinct characteristics. However, its ability to catalyze photochemical reactions is restricted due to its poor absorption of visible light and rapid charge recombination. To address this, coupling g-C_3N_4 with other

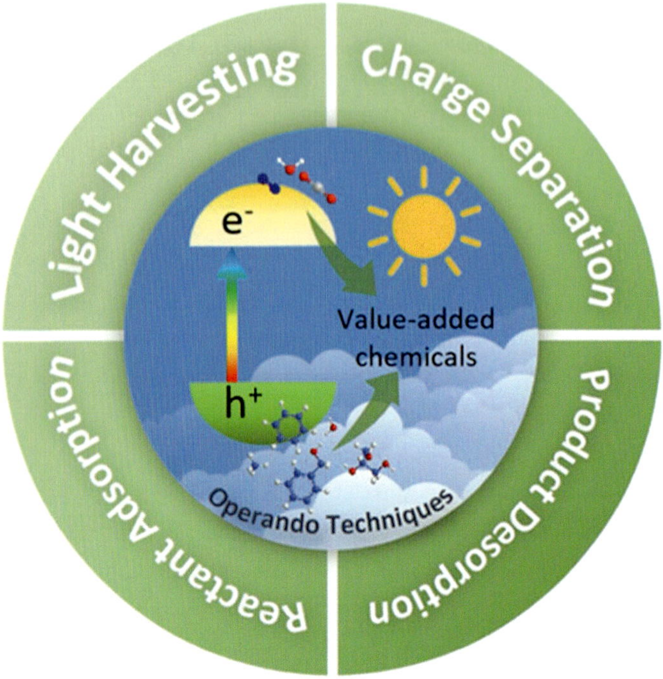

Fig. 3.5 General process in photocatalytic process, under a Creative Commons license Ref. [55]

semiconductors to form heterojunctions is a promising modification strategy that broadens the absorption range and accelerates charge separation. One commonly used method to enhance the photocatalytic activity of g-C_3N_4 is to combine it with titanium dioxide (TiO_2), which has appropriate energy levels, tunable morphology, and stable chemical properties. Two types of heterojunctions have been established with similar energy band structures but different charge transfer mechanisms, namely, type-II and step-scheme (S-scheme) systems, which depend on the direction of energy band bending. The focus of the flow of electrons across the heterojunction interface is from semiconductors with higher Fermi levels (more negative vs. NHE) to semiconductors with lower Fermi levels (less damaging vs. NHE) due to Fermi level alignment [58].

2. **Charge separation**

Photo-induced charge separation is a critical process in natural photosynthesis and in many chemical research fields such as solar fuels, artificial photosynthesis, photoredox catalytic organic reactions, and solar cells [59]. The effectiveness of photocatalysts is often restricted by inefficient charge separation resulting from the small size of the particles and multiple redox reactions co-occurring on the surface sites. To address this issue, various strategies have been employed to enhance charge

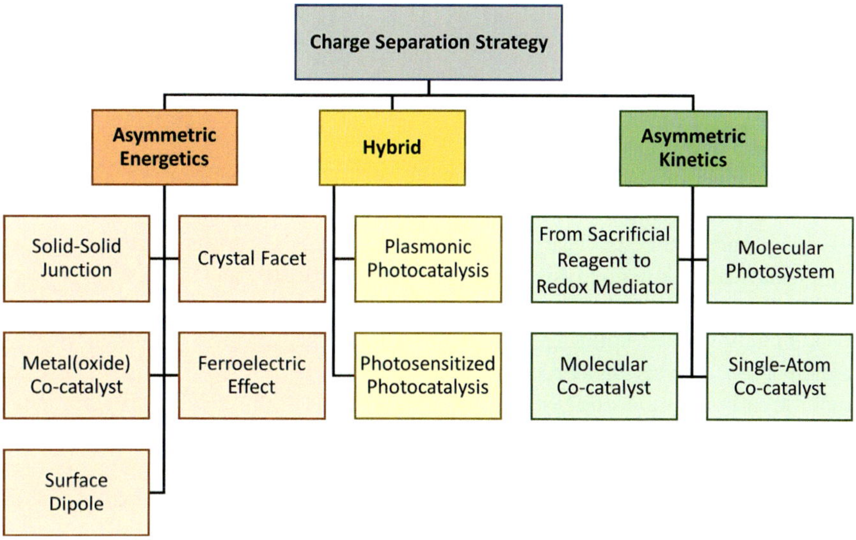

Fig. 3.6 Summary of charge separation strategy. Reprinted (adapted) with permission from Yanagi [60]. Copyright 2022 American Chemical Society

separation in photocatalyst systems, which are summarized and organized in Fig. 3.6 [60].

3. **Reactant adsorption**

In order to enhance the adsorption of reactant molecules, it is an effective approach to load a reactant adsorbent onto the photocatalyst. Additionally, the absorbance of reactants can be improved through the structure engineering of semiconductor photocatalysts [55].

4. **Product desorption**

One of the primary benefits of photocatalysis is that it utilizes photonic energy instead of thermal energy to drive surface reactions, which typically allows the reaction to occur under ambient conditions. However, a potential drawback is that, at lower temperatures, reactants and products may be more strongly bound to the surface, leading to desorption limitations. Therefore, understanding surface coverage is essential in optimizing process conditions for a given photocatalytic reaction [61].

3.3 Carbon-Based Photocatalysts for Oxygen Reduction Reaction (ORR)

3.3.1 g-C_3N_4

H_2O_2 is a potent oxidizer with multiple uses in chemical synthesis, industrial bleaching, and fuel cells. The process of producing H_2O_2 from water using solar energy, known as photocatalysis, is environmentally friendly and promising for practical applications. It is also an excellent metal-free photocatalyst for other applications, including water splitting, CO_2 reduction, and degradation of pollutants [62]. Several photocatalysts have been used for generating H_2O_2, such as titanium dioxide, graphitic carbon nitride, and bismuth vanadate. GCN is a non-metallic polymer semiconductor that stands out among other photocatalysts due to its high effectiveness in promoting the two-electron reduction process of O_2, resulting in a selectivity of over 90% for the production of H_2O_2 under visible light. The atomic and electronic structure of the photocatalyst's surface plays a crucial role in determining the molecular oxygen reduction pathway on the surface of the photocatalyst. GCN has emerged as a highly promising photocatalyst for H_2O_2 production owing to its favorable band structure, chemical stability, eco-friendliness, abundance in the Earth's crust, and a suitable bandgap of 2.7 eV. Moreover, its conductive band (CB) position at −1.3 V versus NHE makes it well suited for the reduction of O_2, which has garnered global interest. GCN has emerged as a promising candidate for photocatalytic H_2O_2 production due to its suitable band structure, chemical stability, earth abundance, and low cost. It can be easily prepared by polymerizing cheap raw materials like urea, dicyandiamide, and melamine. In photocatalysis, the electrons generated by light are found at the C1 and N4 locations of the melem unit in GCN. These electrons then reduce oxygen and create a peroxy group that is swiftly reduced by electrons from another location to produce 1,4-endoperoxide. This reaction leads to the formation of H_2O_2 through a two-electron reduction process. The rapid formation of 1,4-endoperoxide helps to prevent the production of superoxide radicals and supports the formation of H_2O_2 [63]. Under light illumination, GCN can produce both superoxide radicals (O2-) and 1,4-endoperoxides, which can then be used to generate H_2O_2.

Compared to other photocatalysts like ZnO and $BiVO_4$, TiO_2 has a tendency to form a complex with H_2O_2. This characteristic impedes the sustained production of H_2O_2 because the H_2O_2 molecules desorb from the surface of TiO_2. Furthermore, a majority of these semiconducting materials have a restricted light absorption range and exhibit strong absorption in the ultraviolet region. This UV light absorption can lead to the degradation of the produced H_2O_2, which ultimately reduces the efficiency of H_2O_2 production [64]. GCN has the ability to adjust its band structure to enable the conversion of water to H_2O_2 under visible light through photocatalysis. It has been demonstrated that GCN exhibits a high level of selectivity for the oxygen reduction reaction (ORR), leading to the production of H_2O_2, which is attributed to the effective creation of 1,4-endoperoxide species on its surface. Recently, ORR has emerged as a promising technique for H_2O_2 production, as it offers efficient energy conversion and storage [10]. GCN has shown remarkable improvement in the realm of photocatalytic H_2O_2 synthesis [65]. Despite its significant performance in photocatalytic H_2O_2 synthesis, bare GCN still faces some limitations, such as inadequate response to visible light, quick recombination of photoexcited carriers, and insufficient surface area. Researchers have proposed several measures to optimize GCN catalyst to address these catalysts, such as adjusting the specific surface area, reducing manufacturing defects, doping organic polymers, and creating heterojunction catalytic systems. These measures aim to enhance the efficiency of H_2O_2 production by GCN-based photocatalysis. The morphology and nanostructure of GCN play a crucial role in its photocatalytic performance. According to a recent study by Guo et al., GCN nanoplates possess a characteristic of absorbing oxygen to stabilize their states, which enhances the formation of 1,4-endoperoxide. This property improves the mobility of two electrons and enables oxygen activation from one-electron reduction to two-electron reduction by peeling off the bulk GCN [66].

In addition, converting bulk GCN into two-dimensional (2D) nanoplates can enhance its physiochemical characteristics. This includes an increase in surface area, a decrease in charge diffusion distance, and an improvement in redox ability [66]. The 2D framework of GCN has gained much attention in terms of its activity toward photosynthesizing H_2O_2. This is because of its suitable energy levels and its ability to respond to visible light [67]. Doping aromatic rings into the GCN backbone via phi-conjugated bonds is a simple and effective method to improve its photocatalytic performance significantly. This technique extends the phi-conjugated system and delocalizes the phi-electron structure of GCN, leading to enhanced visible light response and efficient separation and transfer of photoexcited charges, resulting in higher photocatalytic activity. Another effective strategy is constructing a porous structure, which provides an enlarged specific surface area with more active sites and channels for mass adsorption and transfer in photocatalysis reactions. Various approaches, including complex templates, soft templates, gaseous templates, and freeze-drying treatments, have been employed to develop high-performance porous GCN-based photocatalysts [68].

Various morphologies of g-C_3N_4, including nanosheets, nanospheres, nanorods, nanofibers, and nanotubes, have been synthesized. Of these materials, nanotubes with a one-dimensional hollow structure have been identified as having exceptionally large surface areas, high light absorption, and fast electron transport, all of which can significantly improve photocatalytic performance. Additionally, appropriate doping of g-C_3N_4 is crucial for adjusting its band structure and charge carrier concentration, ultimately leading to enhanced light absorption and promoted charge injection. The addition of oxygen and phosphorus dopants to GCN can adjust the valence band structure and improve the separation efficiency of photogenerated electrons and holes, ultimately enhancing the selectivity of the OER toward the two-electron pathway. To further strengthen charge separation ability and promote hydrogen evolution performance, co-catalysts such as Pt, Pd, and Au can be applied to the surface of GCN. Given the high cost of noble metals and the moderate H_2O_2 production rates associated with their use, it is crucial to investigate alternative co-catalysts that are based on abundant and low-cost elements, such as MoS_2, Ni, NiP, or CoP. By using these co-catalysts, it may be possible to achieve efficient and high-rate photocatalytic H_2O_2 production at a lower cost [26].

Using a heterojunction photocatalytic system leads to an improved ORR process by producing more photogenerated electrons through the use of ethanol as a hole-scavenger, resulting in the formation of H_2O_2. Heterojunction catalytic systems have gained popularity in the development and application of composite catalysts for diverse purposes, such as pollutant degradation, disinfection, and water splitting. These systems feature a unique Z-scheme charge transfer pathway that facilitates the spatial separation of electron–hole pairs, resulting in their full utilization within the composite catalyst. This feature is particularly advantageous for H_2O_2 production and photocatalysis. Recent work by Fattahimoghaddam et al. has demonstrated increased H_2O_2 production through the use of nitrogen-deficient GCN nanosheets within a heterojunction catalytic system [10]. Yunxiao et al. showed that GCN-GCN (M-CN) isotype junctions with N vacancies and type-II junction structures have remarkable photocatalytic H_2O_2 production performance. They prepared a range of ZnO/GCN heterojunction catalysts through thermal polycondensation and utilized natural water as a case study to establish a disinfection process by in situ H_2O_2 generation with sunlight as the energy source [69]. An example of successful modification of GCN for improved H_2O_2 production is the use of pyromellitic diimide (PDI) units, which interact with GCN via phi–phi interactions. This modification leads to a positive shift in the conduction band of GCN, resulting in enhanced selectivity for H_2O_2 formation through the two-electron reduction of dioxygen [67].

To simplify the modification process and reduce the preparation cost of the catalyst, a more straightforward method is needed to enhance the photocatalytic activity of GCN for H_2O_2 production while preserving its basic structure. Such a method would be highly significant for improving the cost-effectiveness of the photocatalytic H_2O_2 production reaction [63]. In order to improve the photocatalytic performance of pristine-GCN, previous reports have suggested various strategies due to the challenges of fast recombination and low surface area, which lead to poor photocatalytic H_2O_2 generation rates [65]. The properties of a catalyst can be improved by treating it in a plasma atmosphere, which can alter its electronic structure, hydrophilicity, acid strength, and active component dispersion. Plasma treatment can also induce changes in the catalyst's morphology, resulting in a larger active surface area. PT-GCN has been reported to have improved photocatalytic activity for H_2O_2 production compared to pristine-GCN due to the introduction of various functional groups and the increased surface area resulting from plasma treatment. The improved photocatalytic activity of PT-GCN can be attributed to the increased concentration of surface-active sites and enhanced separation of photogenerated electron–hole pairs. The plasma treatment can also modify the surface charge of the photocatalyst, leading to increased adsorption of reactants and improved photocatalytic performance. Therefore, DBD plasma treatment is a promising method for improving the performance of GCN-based photocatalysts for H_2O_2 production [63].

3.3.2 *Modified g-C_3N_4*

Although g-C_3N_4 has potential as a semiconductor material, it has drawbacks that limit its effectiveness for photocatalysis, such as low light absorption, higher recombination of electron–hole charges, poor recovery, and low quantum yield. However, several methods have been suggested to overcome these limitations and improve its photocatalytic capabilities. These methods include manipulating the structure, modifying the surface, incorporating catalysts with grafted single atoms, introducing K or Eu doping, and forming heterostructure interfaces. Through the combination of g-C_3N_4 with a suitable material, it is feasible to enhance the material's photocatalytic effectiveness and accuracy while maintaining low costs and high stability. The adaptable bandgaps, adjustable crystal structure, chemical stability, optical properties, long charge carrier lifetimes, and ample oxygen vacancies in g-C_3N_4 make this possible [6].

The conventional method of producing g-C_3N_4 involves heating nitrogen-rich precursors such as melamine, urea, cyanamide, dicyandiamide, and thiourea through high-temperature calcination. Although the chemical reaction process for high-temperature calcination is relatively complex, it has several benefits, including cost-effectiveness, mild reaction conditions, and easy control. The characteristics of the resulting g-C_3N_4 are strongly dependent on the type of precursor and reaction conditions used. Recent studies have found that calcining g-C_3N_4 at 600 and 650 °C, instead of melamine calcination at 450–550 °C in nitrogen, enhances visible light absorption and narrows the bandgap. Urea is a commonly used precursor that yields optimal g-C_3N_4 when calcined at 550 °C, leading to significant increases in specific surface area and reductions in optical bandgap. In contrast, g-C_3N_4 produced from thiourea as the precursor has a smaller bandgap and higher catalytic performance than g-C_3N_4 produced from urea. Furthermore, it has been demonstrated that the calcination atmosphere can cause structural irregularities and flaws, which can also impact the structure and characteristics of g-C_3N_4 [70] (see Fig. 3.7).

The modification of g-C_3N_4 can be accomplished through a variety of methods, including morphology control, elemental doping, and the creation of heterojunctions with other materials. These modifications are intended to improve the photocatalytic performance of g-C_3N_4 through multiple approaches.

1. Morphology controlling

Photocatalytic reactions take place on the surface of the photocatalyst, involving processes such as adsorption–desorption, redox reactions, and mass transfer. The morphology of the photocatalyst plays a critical role in determining the distance traveled by the photogenerated charge to reach the reaction site, the number of reaction sites available, and the extent of light utilization. Various techniques, including soft/hard template methods, template-free methods, exfoliation techniques, and molecular self-assembly, can be employed to control the morphology of g-C_3N_4. By adjusting the reaction conditions, g-C_3N_4 can be produced with different morphologies, such as zero-dimensional (0D) g-C_3N_4 quantum dots (g-C_3N_4 QDs), one-dimensional (1D)

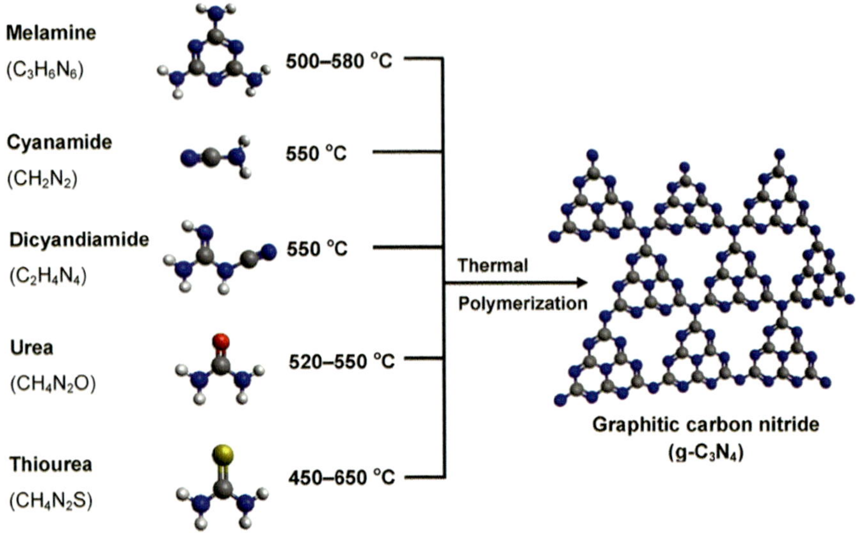

Fig. 3.7 A diagrammatic illustration that represents synthesis process for g-C_3N_4 using precursors such as melamine, cyanamide, dicyanamide, urea, and thiourea [70]. With license number 5541350791874

g-C_3N_4 nanorods/nanotubes, two-dimensional (2D) g-C_3N_4 nanosheets, and various three-dimensional (3D) g-C_3N_4 structures, depending on their dimensionality [71].

- One-dimensional g-C_3N_4

g-C_3N_4 QDs are small-sized nanomaterials with dimensions of less than 10 nm in three dimensions. These nanomaterials exhibit remarkable optical properties that enable the transformation of near-infrared light to visible light, which is crucial for harnessing solar energy. Furthermore, g-C_3N_4 QDs possess superior electron transport properties, as well as more active sites, which enhances the potential to improve photocatalytic efficiency. Bulk g-C_3N_4 is typically produced by heating melamine in air from room temperature to 550 °C and maintaining it for 4 h. To convert bulk g-C_3N_4 into g-C_3N_4 QDs, the bulk g-C_3N_4 powder is ground and then heated at 500 °C for 2 h to produce g-C3N4 nanosheets. Subsequently, the g-C_3N_4 nanosheets are treated with concentrated H_2SO_4 and HNO_3 for 12 h, followed by washing with deionized H_2O to remove the acids. The g-C_3N4 nanoribbon samples are then dispersed in deionized H_2O in a Teflon-lined autoclave and heated at 200 °C for 10 h. The solution of g-C_3N_4 QDs is obtained once it has been cooled down to the temperature of the surrounding environment [72].

The one-dimensional structure of 1D g-C_3N_4 allows the effective movement of electrons in the axial direction without any lateral migration, which can enhance the separation of photogenerated carriers during photocatalytic reactions. Additionally, by altering precursor types, experimental conditions, and other variables, it is possible to modify the physicochemical characteristics of 1D g-C_3N_4. At present, the family of 1D g-C_3N_4 photocatalysts encompasses nanofibers, nanowires, nanorods, and nanotubes [70].

- Two-dimensional g-C_3N_4

The 2D layered configurations in photocatalysis are widely investigated due to their outstanding electrical conductivities and chemical stability. These structures possess a broad surface area, shorter vertical transport distances, and abundant active sites, resulting in improved charge separation. Nevertheless, interlayer covalent bonds and π-conjugated structures hinder horizontal carrier migration and interlayer exciton dissociation. Kang et al. discovered a potential barrier of 33.2 eV between the g-C_3N_4 layers, which confirms this restriction. To overcome this problem, a possible solution is to exfoliate g-C_3N_4 into ultra-thin lamellar structures that are only nanometers thick. Various techniques can be utilized to obtain 2D g-C_3N_4 with atomic thickness, such as thermal etching, liquid stripping, chemical stripping, KCl/NH_4Cl-assisted thermal polymerization, and phosphoric-acid-assisted stripping [72].

- Three-dimensional g-C_3N_4

Despite the advantages of the previously mentioned structures, they are susceptible to agglomeration, and the quantum confinement effect may restrict their light absorption range. Furthermore, producing ultra-thin or single-layer g-C_3N_4 nanosheets often involves acid treatment or chemical exfoliation, leaving room for further enhancement. 3D porous g-C_3N_4 structures, such as nanospheres, nanoflowers, and aerogels, offer benefits such as increased light absorption, abundant active sites, shortened diffusion pathways, and additional pathways for transfer and adsorption reactions. These features can aid in addressing the previously mentioned issues [72].

2. Elemental doping

Doping is the commonly used strategy that can effectively reduce the bandgap of g-C_3N_4 by introducing additional energy levels within the bandgap, leading to enhanced light absorption and improved photocatalytic performance. Rare earth metals, such as La, Ce, Pr, and Eu, have been found to be effective dopants for g-C_3N_4 due to their unique electronic structures and chemical properties. Non-metal doping is also a widely used strategy, with P and S being the most commonly used non-metal dopants. P-doped g-C_3N_4 can significantly improve its photocatalytic performance due to the increased number of active sites, while S-doped g-C_3N_4 can effectively reduce the bandgap and enhance light absorption. Co-doping combines doping techniques to improve the photocatalytic efficiency of g-C_3N_4 by introducing additional energy levels and active sites. For example, the co-doping of N and S has been discovered to be a successful approach for improving the photocatalytic efficacy of g-C_3N_4 by promoting electron–hole pair separation [70].

Incorporating oxygen atoms into g-C_3N_4 can aid in transferring charges from the N_2 aromatic ring to the O atom by decreasing the bond length between the C and N atoms in g-C_3N_4. By substituting the O atom for the most stable tertiary or amino site on the g-C_3N_4 ring, the electronic polarization and charge redistribution of the g-C_3N_4 ring were amplified. This facilitated the separation and transfer of photoinduced electron–hole (e^-–h^+) pairs. Upon S atom doping, it was observed that the Fermi level moved toward the conduction band (CB), indicating that the S atom acts as an n-type co-catalyst for g-C_3N_4. The interaction between the S 3p and N 2p states resulted in a reduction of the bandgap and an expansion of the valence band (VB) width of g-C_3N_4. However, experimental findings demonstrated that the bandgap increased after S atom doping. Additionally, the S atom was primarily responsible for the LUMO, which interacts with CB and causes an increase in the number of electrons [73].

3. Defect engineering

Creating vacancy defects in g-C_3N_4 is an additional approach for altering its photocatalytic activity. These defects have been discovered to have a notable impact on the thermodynamics, kinetics, and mechanism of photocatalytic reactions. In particular, vacancy defects can affect the photocatalytic efficacy of g-C_3N_4 in the following ways:

(a) Vacancy defects can trap and store electrons, facilitating charge separation and boosting the photocatalytic performance of g-C_3N_4.
(b) By generating defect states, introducing vacancy defects in g-C_3N_4 can increase the material's ability to capture light and expand its light absorption range to include the near-infrared region. This, in turn, can enhance its photocatalytic capacity.
(c) The presence of vacancy defects can modify the surface characteristics of g-C_3N_4, resulting in variations in its adsorption behavior toward diverse pollutants. This enables g-C_3N_4 to preferentially adsorb certain pollutants, enhancing its photocatalytic effectiveness for specific applications.
(d) The number, configuration, and position of vacancy defects play a crucial role in determining the characteristics of g-C_3N_4. Therefore, the investigation of defects like nitrogen and carbon vacancies is vital.

When C and N atoms are missing in g-C_3N_4, the bandgap (Eg) becomes smaller due to impurity states. In particular, the conduction band (CB) is connected to the C 2p orbital, and when a C atom is missing, the CB shifts positively, creating a middle gap below the CB. Conversely, the valence band (VB) is associated with the N 2p orbital, and when an N atom is missing, the VB shifts negatively, creating a middle gap above the VB. Although deficient vacancies have a minimal effect on electronic structure, excessive vacancies can act as recombination centers for carriers and negatively impact the performance of g-C_3N_4 [71].

4. Heterojunction building

Several studies have shown that combining g-C_3N_4 with other semiconductors with compatible energy band structures can create heterojunctions. These heterojunctions can leverage the strengths of each component and generate an electric field between the materials by exploiting the differences in energy band structures. This electric field accelerates the transfer of photogenerated carriers and inhibits the recombination of photogenerated electrons and holes, ultimately improving photocatalytic performance. Various types of g-C_3N_4-based heterojunctions have been proposed, including Type-II, p-n, Z-scheme, new S-scheme, and Schottky heterojunctions. These heterojunctions facilitate the separation and transfer of photogenerated electron–hole pairs through different mechanisms. A schematic illustration of the charge transfer mechanism of a Type-II heterojunction is depicted in 0a.

When light falls on a semiconductor, it creates electron–hole pairs. In Type-II heterojunctions, an electric field drives the electrons from the first semiconductor (SC I) to the CB of the second semiconductor (SC II). At the same time, the holes move from the VB of SC II to the VB of SC I. This efficient separation of photogenerated carriers results in the CB of SC II participating in reduction reactions, and the VB of SC I in oxidation reactions, thereby improving photocatalytic activity. However, Type-II heterojunctions can decrease the redox capability of compound semiconductors and hinder the continuous transfer of electrons and holes due to electrostatic repulsion. Z-scheme heterojunctions, on the other hand, mimic natural photosynthesis and address these issues. In Z-scheme heterojunctions, electrons on SC II with a lower CB position recombine with the holes on the VB of SC I through the heterojunction or an intermediate conducting medium under visible light irradiation, maintaining the strong redox ability of the photocatalyst [70] (see Fig. 3.8).

5. Another approach

There are alternative techniques to improve the photocatalytic properties of g-C_3N_4, including the addition of functional groups like hydroxyl (–OH), carboxyl (–COOH), and amino (–NH_2) groups on the surface of g-C_3N_4. These functional groups can augment the surface activity, assist in the adsorption of specific pollutants, and advance the separation and movement of photogenerated carriers.

Another method to enhance the photocatalytic activity of g-C_3N_4 is through the formation of composites with noble metal nanoparticles such as platinum (Pt), palladium (Pd), or gold (Au). These noble metals act as co-catalysts that improve the charge transfer and overall photocatalytic performance of g-C_3N_4.

Moreover, combinations of g-C_3N_4 with other carbon materials, such as graphene and carbon nanotubes, have been created. The exceptional characteristics of these carbon materials, such as their significant surface area and high electrical conductivity, can facilitate the separation and transfer of photogenerated carriers, leading to an improved photocatalytic performance of g-C_3N_4.

3.3.3 Metal–Organic Framework/Covalent Organic Framework

Metal–organic frameworks (MOFs) are coordination polymers of clusters or metal ions connected by organic ligands and coordination bonds. Due to their large surface areas and high porosity, the resulting structures can be used for gas storage and separation purposes. MOFs have drawn a lot of interest recently because of their

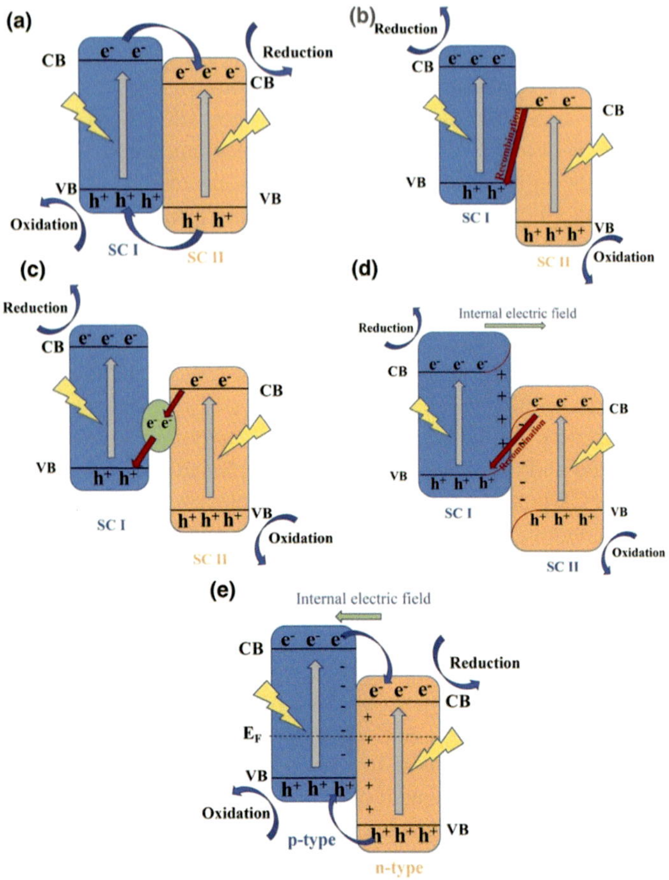

Fig. 3.8 The diagram shows how photoinduced charge carriers are transferred in different types of heterojunction nanocomposites: **a** type-II heterojunction; **b** semiconductor–semiconductor Z-scheme heterojunction; **c** semiconductor–conductor-semiconductor Z-scheme heterojunction; **d** S-scheme heterojunction; **e** p-n heterojunction [70] with license number 55412803073333

adaptable qualities and possible uses in industries including catalysis, sensing, and drug delivery [74] (see Fig. 3.9).

COFs are porous materials made of covalently bonded organic molecules that create extended structures with well-defined porosity and a large surface area. COFs are interesting prospects for use in the storage, separation, and catalysis of gases because of their distinctive features, including excellent thermal and chemical stabilities, variable pore diameters, and chemical capabilities [75]. Both MOFs and COFs have unique properties that make them useful for a variety of applications. For example, MOFs have tunable properties, high surface area, and high selectivity for gas adsorption, while COFs have high thermal and chemical stabilities and tunable pore sizes and chemical functionalities.

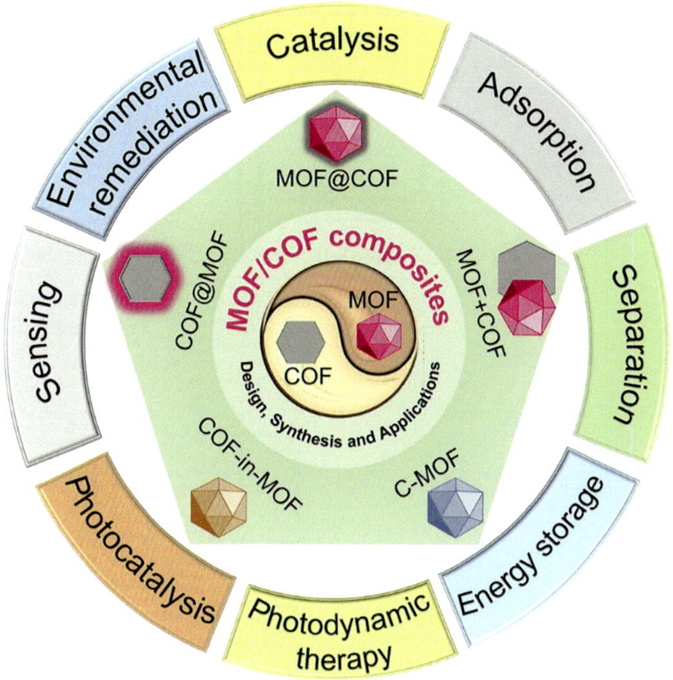

Fig. 3.9 Metal–organic framework and covalent organic framework [74]. It was adapted with permission with license number 5525260933436

One of the most used applications of COFs is highly suitable materials for electrochemical sensors; however, the application is still in its early stages, and the weak conductivity's inherent constraint would affect the sensor's sensitivity. Therefore, creating metal-covalent organic frameworks (MCOFs) with a unique structure through the incorporation of active metal ions into COFs backbones may be beneficial to resolve these limitations [76]. MCOFs, which show an optimal combination of crystallinity, porosity, stability, and tunability, can be considered as a bridge between MOFs and COFs, which results in self-complementary properties between these two types of materials [77] (see Fig. 3.10).

MOFs produced typically include guest molecules, which must be removed through activation methods before the MOFs can be used in their applications. Although pre-treatment of the MOFs by solvent exchange and vacuum application is necessary, thermal activation is commonly used for conventional porous materials. However, some MOFs have low stability and require highly effective thermal and chemical stabilities to undergo direct activation via conventional thermal activation [78]. For the synthesis of COFs, the reaction characteristics and topology demand the utilization of reversible covalent bond-forming processes. The majority of COFs have been produced via solvothermal methods, where the circumstances of the reaction are greatly influenced by the solubility, reactivity, and reversibility of the building

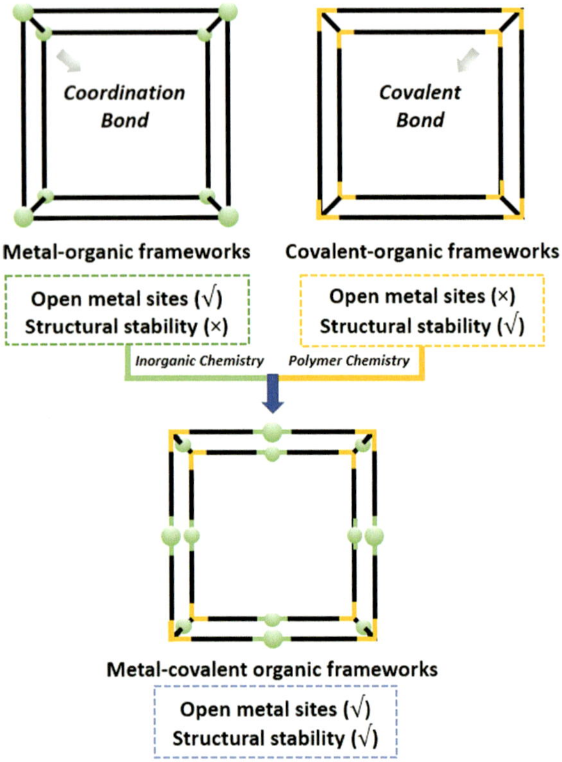

Fig. 3.10 Illustration of MCOFs acts as a link between MOFs and COFs [77]

blocks. Furthermore, the most critical variables to take into account while creating crystalline porous COFs by employing the solvothermal method are reaction time, temperature, solvent conditions, and catalyst concentration [79] (see Fig. 3.11).

These days, batteries are an essential target of research as an energy storage and conversion technology. MOFs have been useful in the development of lithium–ion batteries as they can be used as precursors and templates for carbon-based anodes. Additionally, in solid-state lithium batteries, MOFs may also serve as electrolytes to enhance the transport kinetics of Li^+ within the solid-state electrolyte and the electrode materials [81]. MOF-derived functional materials have shown great promise in the development of advanced electrodes and efficient electrocatalysts for sustainable and secure Electrochemical Energy Storage and Conversion (EECS) systems. Derivatives' large inner spaces and high specific surface area enable the electrochemical reaction process's volumetric expansion while also facilitating the exposure of the active sites and mass transfer. As a result, MOF development is a suitable match for energy storage and conversion technologies [82].

Due to their built-in pores, COFs show possibilities as gas storage materials. Modernization and industrialization have led to an accelerated release of CO_2, which is a significant greenhouse gas that contributes to climate change and global warming. COFs are lightweight, porous materials with customizable frameworks. One reliable

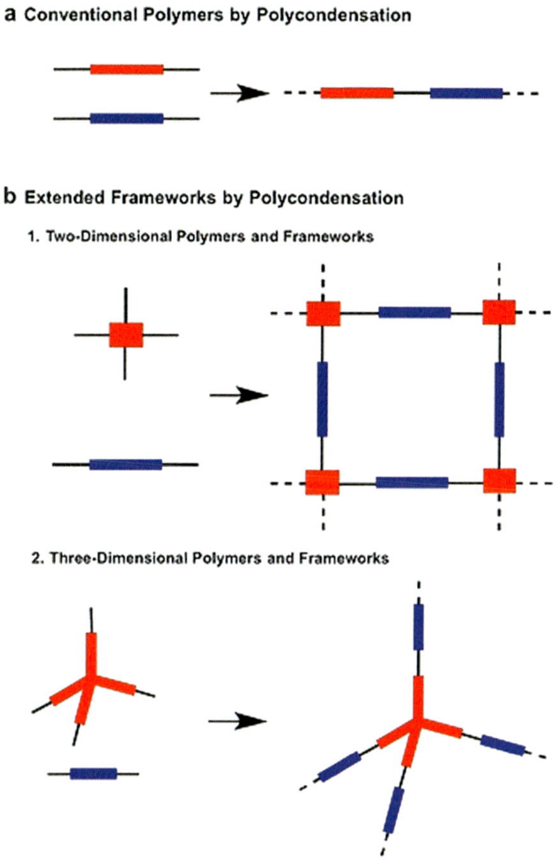

Fig. 3.11 Topology diagram of polymers and frameworks. **a** Conventional polymers produced via polycondensation, **b** Extended frameworks by polycondensation 1. Two dimensionally, and 2. Three dimensionally [80]. Adapted with permission with license number 5525260887923

method for creating custom pores with well-ordered interfaces is to engineer the pore surfaces. The ability of the structure to be designed is crucial for CO_2 adsorption [80].

3.3.4 *Graphene Oxide*

Graphene is a flat layer made up of sp2 carbon atoms arranged in a honeycomb pattern, possessing various advantageous characteristics such as high mechanical strength, electrical conductivity, large surface area, and other impressive features. GO, graphene oxide, a material that has been discovered relatively recently, refers to graphene that has undergone an oxidation process and consists of solution-dispersible polyaromatic two-dimensional carbon sheets derived from the acid exfoliation of graphite [83]. The edges of graphene oxide consist of carboxylic acid groups, while its basal planes are composed of epoxy and hydroxyl groups. Additionally, graphene oxide exhibits other oxygen-based chemical reactivity capabilities throughout its structure [84]. Graphene oxide shares a hexagonal carbon structure with graphene but additionally contains numerous oxygen-based functional groups, such as hydroxyl (–OH), alkoxy (C–O–C), carbonyl (C=O), and carboxylic acid (–COOH). These functional groups offer certain benefits over graphene, including enhanced solubility and the ability to alter the surface of graphene oxide with different functional groups, thereby broadening the scope of its possible uses in nanocomposite materials. Additionally, the ease of synthesis is another advantage [85] (see Fig. 3.12).

Some synthesis method for graphene oxide from graphite is summarized in Table 3.1. The advantages and disadvantages are discussed too. The most standard synthesis used today is Hummers. This method is safer and more efficient than the other synthesis methods, such as Brodie and Staudenmaier methods. As this method has disadvantages of emitting toxic gases during the oxidation process, some modified method is still being developed [87].

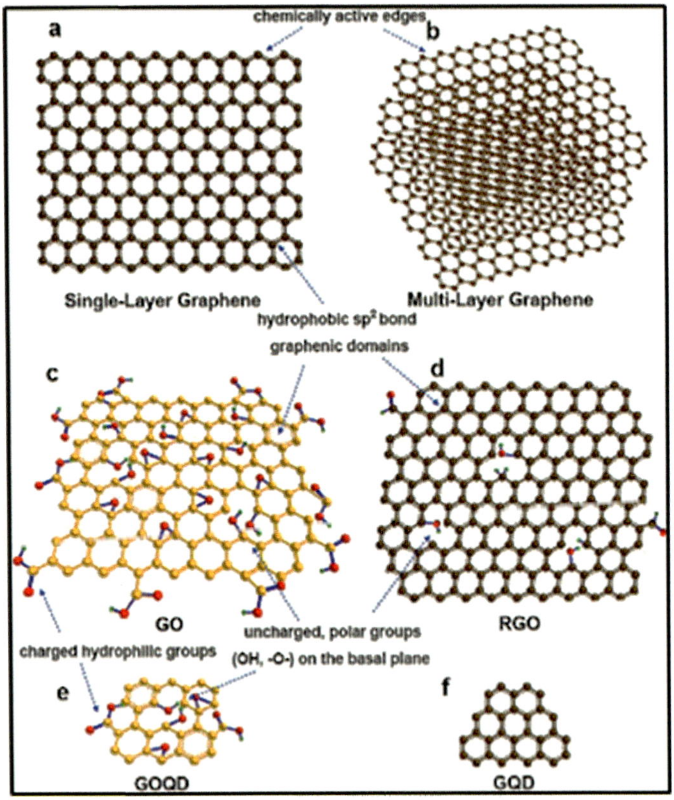

Fig. 3.12 The figure from the reference shows various forms of graphene and its derivatives, including **a** single-layer graphene, **b** multiple-layer graphene, **c** graphene oxide (GO), **d** reduced graphene oxide (rGO), **e** graphene oxide quantum dots (GOQD), and **f** reduced graphene oxide quantum dots (rGQD), with the permission from Ref. [86]

Table 3.1 Synthesis method for graphene oxide from graphite oxide

Name of method and year	Oxidizing agents/steps	Positive aspect	Negative aspect	References
Brodie (1859)	$KClO_3 + HNO_3$, 60 °C, 4 days	The initial or primary method of synthesis	Requesting that the four phases of oxidation be repeated	[88]
Staudenmaier (1898)	$KClO_3 + HNO_3 + H_2SO_4$	Less complicated and efficient than the Brodie approach	The creation of toxic gases	[87]
Hofmann (1937)	Equivalent to Staudenmaier process, but with HNO	Avoid using HNO_3 since it is corrosive	Poisonous gases (NO_x) production	[87]
Hummers and Offeman (1958)	$NaNO_3 + KMnO_4 + H_2SO_4$, 45 °C, 2 h	Avoid using HNO_3 to prevent corrosion	Poisonous gases (NO_x) production	[89]
Kovtyukhova (1999)	$H_2SO_4 + K_2S_2O_8$, P_2O_5, 80 °C, 6 h $H_2SO_4 + KMnO_4$, 35 °C, 2 h 30% H_2O_2	Significant oxidation	The number of stages and the generation of poisonous fumes (NO_x)	[90]
Modified Hummers method by Hirata (2004)	$H_2SO_4 + KMnO_4 + NaNO_3$, 20 °C, 5 days 30% H_2O_2	High yield	Long oxidation reaction	[91]
Ang and Loh (2009)	$NaNO_3 + H_2SO_4 + KMnO_4$, 90 °C, 0.5 h Redisperse in DMF + tetrabutyl ammonium hydroxide + H_2O, 90 °C, 2 days	More than 90% of the monolayer at high GO content	Lengthy process, numerous steps	[92]
Marcano and Tour (2010)	$KMnO_4 + H_2SO_4 + H_3PO_4$, 50 °C, 12 h 30% H_2O_2	No generation of harmful gases	Making explosive intermediates for Mn_2O_7	[93]
Eigler (2013)	$NaNO_3 + H_2SO_4 + KMnO_4$, 30 °C, 21 h 30% H_2O_2	Minimum output of CO_2	Produce toxic gases (NO_x)	[94]
Peng-Gao (2015)	$K_2FeO_4 + H_2SO_4$, room temperature, 1 h	High yield, avoidable Mn ion contamination, and efficiency	Addressing challenging pollutants	[95]

Graphene oxide (GO) is a material that was first proposed about 150 years ago. A model of the graphene oxide structure was proposed by Hofmann and Rudolf in 1939 (0) that exhibited numerous randomly distributed epoxy functional groups on the graphene nanosheets. Ruess introduced some simplifications to this structure in 1946, which involved the addition of –OH groups and alterations to the basal plane model (sp2) with a sp3-hybridized carbon structure. On the other hand, Scholz–Boehm presented a less structured model in 1969, which featured C=C double bonds and sporadically broken C–C bonds within the rigid carbon sheets. Additionally, –OH and $–C=O$ groups were present in various locations without any ether oxygen. The majority of the suggested structural models involve a two-dimensional (2D) sheet of hexagonal carbon atoms that include imperfections and oxygen-based functional groups. GO is generally characterized by a combination of sp2 carbon with islands of sp3 carbons that feature oxygen-containing functionalities. It is believed that carboxylic or carbonyl groups are located at the periphery of the graphene sheets. An important characteristic of graphene oxide is that it is hydrophilic and can easily absorb water molecules between its layers [96] (see Fig. 3.13).

GO resembles an insulating material with a wide bandgap, it exhibits a bandgap ranging from 2.4 to 4.3 eV [97], but its electronic structure is dependent on the carbon-to-oxygen atomic ratio. The extent of oxidation can be used as a straightforward method to tune the bandgap of graphene oxide. Fully oxidized GO functions as an electrical insulator and partially oxidized. GO has the ability to behave like a semiconductor. Its desirable characteristics including its ability to disperse evenly in water at a molecular level, its biocompatibility, and its adjustable bandgap have prompted researchers to investigate its potential as a photocatalyst [98].

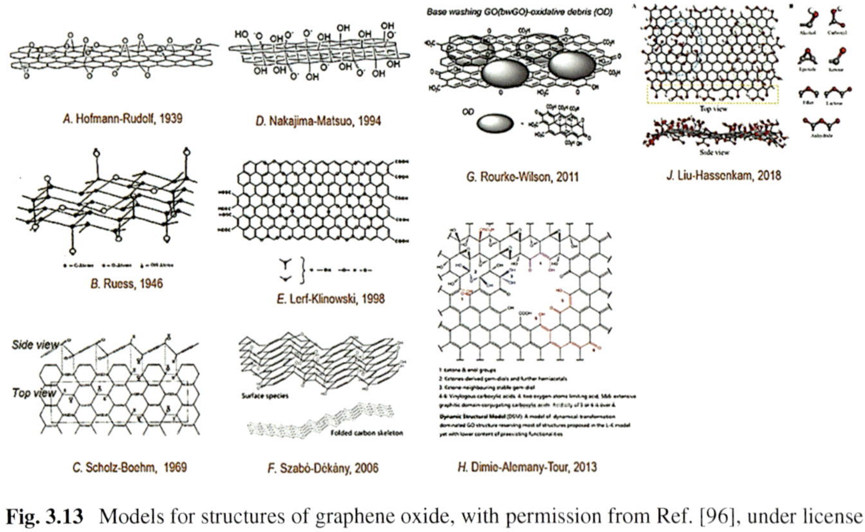

Fig. 3.13 Models for structures of graphene oxide, with permission from Ref. [96], under license number 5520570198829

Recently, there has been a surge of interest in utilizing graphene-based semiconductor photocatalysts due to their superior photocatalytic characteristics, including high charge carrier mobility, large surface area, and exceptional chemical and thermal stabilities. Graphene has become a prospective electron acceptor in these photocatalysts, which helps to enhance charge transfer and reduce electron–hole recombination in the composite, thereby resulting in better photocatalytic performance [99]. Photocatalysts based on GO have been created to perform photocatalytic tasks such as hydrogen production and breaking down organic pollutants. Additionally, they can function as a photocatalyst to produce hydrogen and oxygen from water as a result of their negative conduction band position [100]. Moreover, the broad bandgap of graphene oxide enables its application as a photocatalyst for processes, such as converting CO_2 to methanol. This process not only captures solar energy but also helps in reducing CO_2.

3.4 Application of Photocatalysts in Oxygen Reduction Reaction

Photocatalysts have found widespread application in different fields, including energy conversion and storage. Among the potential uses of photocatalysts, one of the most promising is the oxygen reduction reaction (ORR), which involves the transformation of oxygen molecules into water, typically at the cathode of an electrochemical cell. Photocatalysts can be used to enhance the ORR by accelerating the rate of oxygen reduction and reducing the overpotential required for the reaction to occur. They can also improve the durability and stability of the ORR catalysts. The process of ORR using photocatalysts operates by transferring photoactivated electrons from the photocatalyst to an adsorbed oxygen molecule, which produces superoxide radicals. These radicals can react with protons to create hydrogen peroxide or with electrons to form water. The application of photocatalysts in ORR has significant potential for enhancing the efficiency and endurance of energy conversion and storage mechanisms [101–104].

Photocatalysts are important in various fields, including energy conversion and storage. One such application is in the oxygen reduction reaction (ORR), which converts oxygen molecules into water. Several types of photocatalysts can be used in ORR, including metal oxides, metal sulfides, metal nitrides, and carbon-based materials like g-C_3N_4. g-C_3N_4 has many advantages that make it a promising catalyst material in various electrochemical applications such as fuel cells, metal–air batteries, and electrochemical sensors. It has a high surface area, good electrical conductivity, and chemical stability. In comparison to traditional ORR catalysts, g-C_3N_4 is relatively inexpensive, easy to synthesize, and environmentally friendly since traditional catalysts often use rare or toxic metals. ORR is crucial in fuel cells since it determines their efficiency and performance. During the operation of a fuel cell, ORR occurs at the cathode, where oxygen molecules are reduced to form water

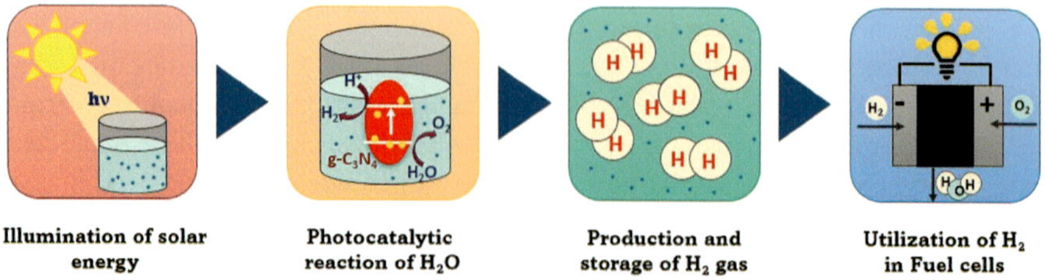

Fig. 3.14 Schematic illustration of photocatalytic H_2 evolution and synergistic photocatalytic (PC) and electrocatalytic (EC) influence of the Pt/g-C_3N_4-modified electrode for methanol oxidation reaction [101], with license number 5554641251675

and electrons. The electrons then travel through an external circuit to the anode, where they react with the fuel to generate more water and electrical energy. Thus, the reaction in the electrochemical cell is essential for producing the electrical current in the fuel cell [105–107] (see Figs. 3.14 and 3.15).

To generate electricity by converting chemical energy into electrical energy, ORR is also utilized in metal–air batteries, which are considered a promising technology for energy storage and conversion due to their high energy density and lightweight properties. A metal–air battery is composed of a metal anode, an air cathode, and an electrolyte that acts as a separator between the anode and cathode. These batteries generate electricity by utilizing a metal anode and an air cathode. During the discharge process of a metal–air battery, the metal anode oxidizes and releases electrons, which flow through an external circuit to power a load or charge a device. At the same time, oxygen from the air cathode reacts with the metal ions at the anode, generating hydroxide ions (OH^-) and completing the electrochemical reaction. The overall reaction can be expressed as:

$$\text{Metal} + \text{Oxygen} + \text{Water} \rightarrow \text{Metal hydroxide}$$

The ORR occurs at the air cathode, where oxygen molecules are reduced to form hydroxide ion (OH^-) and electrons. The hydroxide ions then react with the metal ions at the anode to form metal hydroxides. Improving the efficiency and kinetics of the ORR is crucial for enhancing the performance of metal–air batteries as it is the step that limits the rate of discharge in these batteries. One approach to improving the ORR in metal–air batteries is to use catalysts, such as Pt/g-C_3N_4, to increase the speed of the reaction and decrease the excess voltage required (the voltage required to drive the reaction) [108] (see Fig. 3.16).

Carbon-based materials have the potential to function as structural constituents in the air cathode, besides acting as catalysts, in order to enhance the performance of metal–air batteries. For example, nanostructured carbon materials can be used as conductive fillers to enhance the electron transport properties of the air cathode. In contrast, nanowires and nanotubes can be used to improve the performance of the air cathode in metal–air batteries, carbon-based materials can be utilized as structural

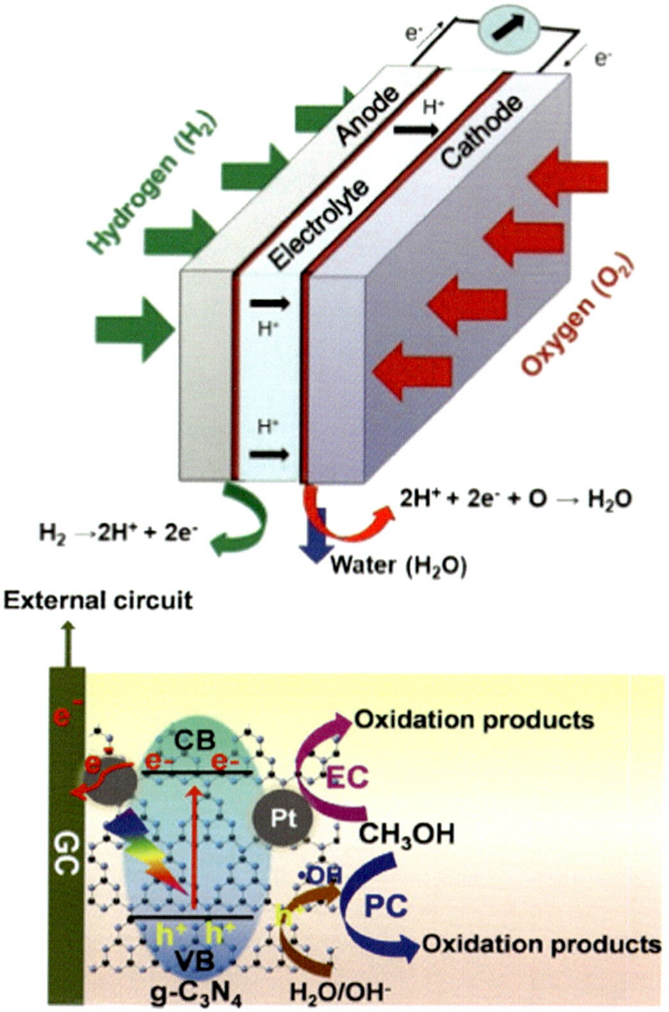

Fig. 3.15 Schematic illustration of the fuel cell operation at the cathode and oxygen molecules are reduced to form water and electrons, with license number 5554650002802

components to augment the surface area and enhance the diffusion properties of oxygen [109, 110] (see Fig. 3.17).

In addition, ORR is also used in electrochemical sensors to detect the presence of various gases, including oxygen, carbon dioxide, and nitrogen oxide. In electrochemical sensing, ORR can be used as a way to detect the presence of oxygen gas in a sample. This technique is commonly used in various applications, such as biomedical devices, environmental monitoring, and industrial process control. The ORR-based electrochemical sensor typically consists of an electrode coated with a catalyst that promotes the ORR. When oxygen gas is introduced to the electrode, it

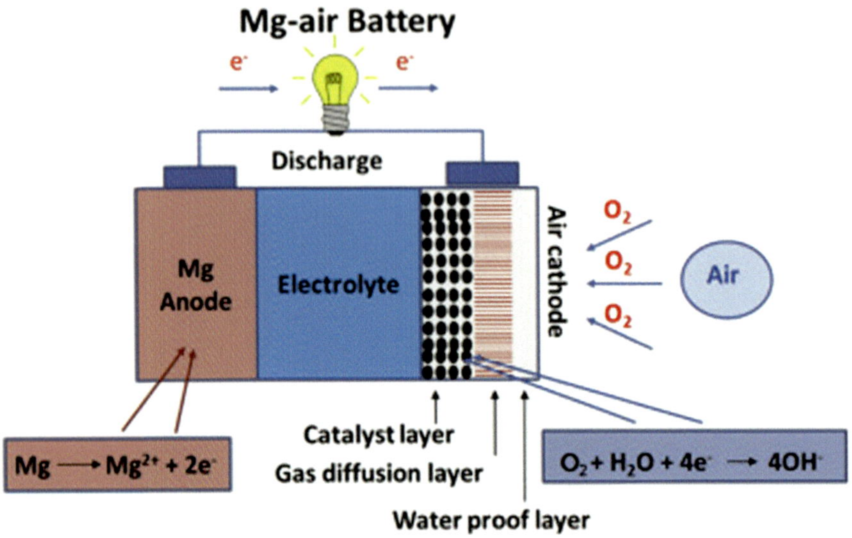

Fig. 3.16 Schematic illustration of discharging/charging process of a Mg–air battery with catalyst layer [109], under a Creative Commons license

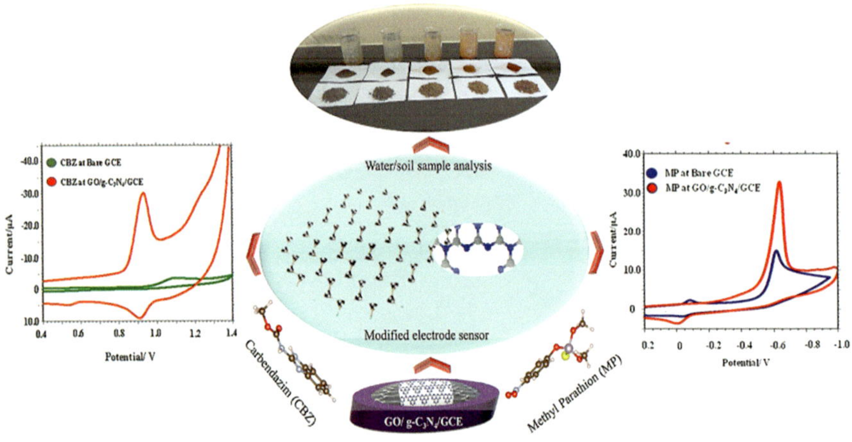

Fig. 3.17 Schematic illustration of the synthetic procedures of various catalysts in electrochemical sensor application [110]. With license number 5554641056214

is reduced by the catalyst, and this reduction generates an electrical current that is proportional to the concentration of oxygen in the sample. By measuring the electrical current, the sensor can determine the concentration of oxygen in the sample. Electrochemical sensors often require a catalyst to facilitate the reaction between the target analyte and the electrode surface. A catalyst is a substance that enhances the rate of a chemical reaction without undergoing any permanent change itself. In electrochemical sensing, a catalyst is typically used to promote the oxidation or reduction of the target analyte at the electrode surface. For example, in an ORR-based electrochemical sensor, a catalyst such as Pt/g-C_3N_4 or other carbon-based materials can be used to promote the reduction of oxygen gas to hydroxide ions at the electrode surface. In other types of electrochemical sensors, different catalysts may be used depending on the target analyte and the electrode material. Catalysts are important in electrochemical sensing because they can significantly enhance the sensitivity, selectivity, and response time of the sensor. Without a catalyst, the reaction between the analyte and the electrode surface may occur too slowly to generate a measurable electrical signal. They are also relatively inexpensive and can be miniaturized for portable and point-of-care applications. However, ORR-based sensors may require calibration to account for changes in the electrode performance over time, and they may also be affected by interfering substances in the sample. Overall, ORR-based electrochemical sensors are a powerful and versatile tool for detecting the presence of oxygen in various applications. The reaction can be used to generate a signal that is proportional to the concentration of the gas being detected, making it a valuable tool in environmental monitoring, industrial process control, and medical diagnostics [108, 110, 111].

In summary, ORR has important applications in fuel cells, metal–air batteries, and electrochemical sensors. The development of more efficient and durable ORR catalysts is critical for improving the performance and expanding the range of applications for these electrochemical devices.

3.5 Performances and Mechanisms of Catalytic Materials

It can be quite complicated to generate high-performance yet affordable catalysts for the electrochemical production of H_2O_2. The photochemical metal–organic deposition approach can be used to create the amorphous nickel oxide NiO_x based on carbon nanosheets. The mesoporous structure of the carbon nanosheets is strongly related to the amorphous NiO_x's high efficiency and selectivity toward $2e^-$ ORR. After the stability issues have been properly solved, the current amorphous NiO_x loaded on carbon nanosheets may become providing electrocatalyst for the synthesis of H_2O_2. Non-noble metal (e.g., Ni) oxides' catalytic performance can be significantly enhanced and even made on par with that of noble metal catalysts with adequate modification [112].

The key to enhancing selectivity and yield was found by maintaining the prompt desorption of H_2O_2 at the active center location. To increase the 2e ORR on carbon-based electrodes, interface engineering techniques and reaction kinetics were used. A significant part of the product's desorption was also played by the carboxyl groups. It was also established that minor carbon flaws played a tiny but significant role in the high H_2O_2 selectivity. The use of the nano-carbon/surfactant composite electrode system provides several advantages beyond its superior catalytic activity in the electrosynthesis of H_2O_2. These advantages include reduced energy consumption, eco-friendliness, sustainability, and outstanding stability [113].

Metal–organic frameworks (MOFs) are a family of adaptable materials that can be used to create functional heterogeneous catalysts, although the most common way is to carbonize them to create porous carbon materials, which eventually results in the loss of their original MOF properties. The development of evenly dispersed NiO nanoparticles within NiFeMOF nanosheets (MOF NSs) using a partially controlled pyrolysis technique would allow for the electrochemical synthesis of hydrogen peroxide (H_2O_2) via the two-electron oxygen reduction reaction (ORR). The NiO NPs created from the MOF NSs are crucial in boosting activity and stability, which results in the optimum binding energy of OOH^* for the superior two-electron ORR to H_2O_2 performance. After catalysis, the NiOOH is transformed and can support high selectivity performance and oxygen adsorption [114].

To fully understand the advantages of the Ni MOF NSs-6 catalyst, it is important to evaluate not only its catalytic performance in terms of activity, selectivity, and stability compared to standard catalysts, but also the methods used to synthesize it and its material properties. The results indicate that this catalyst can significantly enhance catalytic activity and exhibit higher selectivity than most previously reported catalysts, particularly at lower potentials, with an improvement of up to 98%. Moreover, the synthesis process and material characterizations of the catalyst provide additional benefits. A practical and effective liquid–liquid interfacial reaction technique was used to make the Ni MOF NSs at room temperature. When utilized in an alkaline electrolyte, the improved partly unsaturated Ni MOF NSs-6 with a greater amount of Ni^{2+} permits more –NiOOH transformation, which contributes to good catalytic activity [115].

Oxygen-doped carbon nanosheets (OCNS) are another catalyst that can be used to produce H_2O_2. Although oxygen-doped carbon materials have been considered promising catalysts for their high ORR activity and H_2O_2 selectivity, there is a need for direct experimental evidence to identify the actual active sites on the complex carbon surfaces. In order to understand the oxygen-doped carbon nanosheet ($OCNS_{900}$) catalyst for $2e^{-1}$ORR, a chemical titration technique was therefore presented. This catalyst demonstrates exceptional 2e ORR performance potential, mass activity, and H_2O_2 generation rate.

The chemical titration technique can be utilized to distinguish the activity contributions of different oxygen functional groups, specifically C=O, COH, and COOH groups, and the C=O species turn out to be the most active sites for $2e^-$ ORR [116].

In place of the anthraquinone process, the electrochemical reduction reaction of oxygen with two electrons ($2e^{-1}$ ORR) is a potential method for producing renewable H_2O_2. The traditional four-electron oxygen reduction reaction ($4e^{-1}$ ORR) hinders the development of effective and highly selective catalytic materials; therefore, new electrocatalysts such as bismuth titanate ($Bi_4Ti_3O_{12}$) nanosheets have been developed specifically for $2e^{-1}$ ORR H_2O_2 production. The results of the electrochemical analysis indicate that the bismuth titanate ($Bi_4Ti_3O_{12}$) that was produced, particularly the neodymium (Nd)-substituted $Bi_4Ti_3O_{12}$, has a high catalytic activity and selectivity for O_2 to H_2O_2. The exceptional ability of the nanosheets to catalyze the conversion of O_2 to H_2O_2 can be attributed to their extensive surface area and abundance of active sites. The $Bi_4Ti_3O_{12}$ nanosheets doped with Nd display outstanding electrocatalytic performance in the electrochemical production of H_2O_2 from O_2 reduction, with high activity and selectivity of up to 95% in alkaline conditions, thanks to the large surface area and exposed facets of the BNTO nanosheets. Therefore, catalysts based on bismuth titanate are highly suitable as non-precious metal options for generating H_2O_2 electrochemically [117].

Metal-free carbonaceous materials have a lot of potential as heterogeneous catalysts for the destruction of organic contaminants in Fenton-like reactions. This is because they are highly stable and do not result in any secondary contamination from leaching metals. However, the limited scope of carbonaceous materials' applications results from their low catalytic activity toward H_2O_2 activation. The creation of a type of nitrogen-doped carbon nanosheet (NCN) that effectively degrades bisphenol A (BPA) by H_2O_2 activation under neutral circumstances was accomplished using a simple two-step thermal annealing technique. In the presence of 12 mmol/L H_2O_2 and neutral pH levels, NCN catalysts removed nearly all of the BPA in 30 min. When used repeatedly, this NCN catalyst exhibits good stability with barely any activity decline. The creation of catalysts made from carbon-based materials that are capable of breaking down organic pollutants in water using H_2O_2 has expanded the potential for degrading BPA [118].

Finding alternatives to platinum as catalysts for reducing O_2 in fuel cells has been a global goal for a long time. Recently, carbon-based electrocatalysts that do not contain any metals have been rapidly developing. These catalysts have inherent advantages, such as their durability in acid due to the absence of metal leaching, metal–ion contamination, and Fenton reaction-related degradation. To improve their ORR (oxygen reduction reaction) performance, new approaches and concepts have been developed, such as regulating doping/co-doping configurations, building microarchitectures for sp^2 carbon, and controlling porosity. However, it is crucial to increase their ORR activity in acidic electrolytes. Multi-doping, which involves controlling doping configurations to produce sites that carry a positive charge and high spin density simultaneously, is a promising method to achieve this goal [119].

Over the past few decades, photocatalysis has garnered significant interest. It has been employed in a range of research fields, including but not limited to water splitting for H_2 and O_2 production, CO_2 reduction, pollutant degradation, organic synthesis, and H_2O_2 production due to the renewable and sustainable nature of solar energy [120]. Semiconductors are appealing for light-gathering purposes due to their energy bandgap, which is a range of energy where no electronic states are usually present, and this range of energy is comparable to the energy of visible light [121]. In general, the reactions that occur on photocatalysts go through three main stages. Firstly, a photocatalyst from a semiconductor absorbs photons. If the photon's energy exceeds the photocatalyst's bandgap, the electrons in the valence band (VB) become excited and move to the conduction band (CB), leaving holes in the VB. As a result, this mechanism produces sets of electron-negative (–) and hole-positive (+) pairs. In the subsequent stage, these photoactivated electron–hole pairs separate and migrate toward the photocatalyst's surface. During the third stage, the charge carriers interact with the chemical species on the photocatalyst's surface. At the same time, photoactivated electron–hole pairs also combine, but they do not participate in any chemical reactions [122] (see Fig. 3.19).

The fundamental concepts of photocatalysis also apply to the production of H_2O_2 through photocatalysis. It is currently acknowledged that H_2O_2 can be generated through two different pathways. These processes may involve a two-step reduction pathway that occurs indirectly and involves the transfer of a single electron ($O_2^- \rightarrow H_2O_2$), or just in a direct one-step process involving two-electron reduction ($O_2 \rightarrow H_2O_2$). Figures 3.18 and 3.19 illustrate a comprehensive diagram of a photocatalytic setup designed to generate H_2O_2. The positively charged holes (h^+) within the valence band (VB) are responsible for oxidizing H_2O and forming O_2 and H^+ through Eq. (3.1), while electrons (e^-) located in the conduction band (CB) combine with adsorbed O_2 to create H_2O_2. The production of H_2O_2 through an indirect sequential two-step single-electron reduction pathway is demonstrated by Eqs. (3.2)–(3.5). In the first step, O_2 undergoes a one-electron reduction to create the superoxide radical [O_2^- Eq. (3.2)], which then reacts with H^+ to form HO_2^- radical (Eq. (3.3)). Subsequently, HO_2^- radical undergoes another one-electron reduction (Eq. (3.4)), leading to the creation of HO_2^- anions. The final step involves the reaction of HO_2^- with H^+ as demonstrated by Eq. (3.5), which ultimately results in the formation of H_2O_2. Additionally, Eq. (3.6) depicts the direct one-step two-electron reduction pathway to generate H_2O_2. In this process, O_2 undergoes a direct reaction with two H^+ ions, leading to the production of H_2O_2. These two pathways can be summarized through an overall photocatalytic response as demonstrated by Eq. (3.7). It is worth noting that the production of H_2O_2 from H_2O and O_2 through photocatalysis is an energetically uphill process, with standard Gibbs free energy change ($\Delta G^0 = 117$ kJ/mol).

$$2H_2O + 4h^+ \rightarrow O_2 + 4H^+ \tag{3.1}$$

$$O_2 + e- \rightarrow O_2^- \tag{3.2}$$

$$O_2^- + H^+ \rightarrow HO_2^- \tag{3.3}$$

$$HO_2^- + e^- \rightarrow HO_2^- \tag{3.4}$$

$$HO_2^- + H^+ \rightarrow H_2O_2 \tag{3.5}$$

$$O_2 + 2H^+ + 2e^- \rightarrow H_2O_2 \tag{3.6}$$

$$2H_2O + O_2 \rightarrow 2H_2O_2 \Delta G^0 \text{ of } 117 \text{ kJ/mol} \tag{3.7}$$

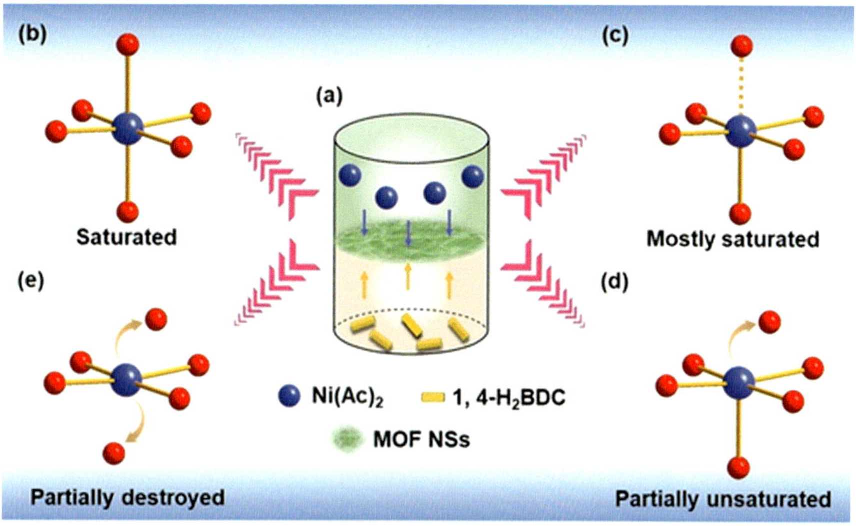

Fig. 3.18 Schematic of the interfacial reaction preparation process for Ni MOF NSs [115]. Adapted with permission with license number 5538270601319

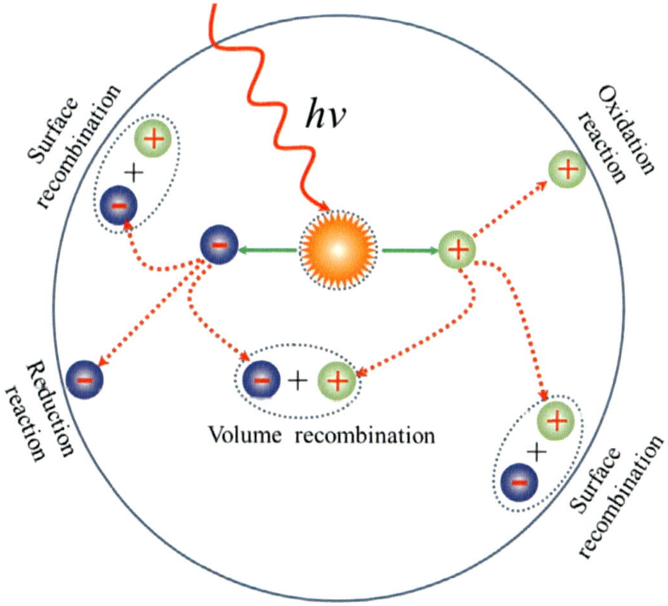

Fig. 3.19 The process of photoactivation and decay of charges in a photocatalyst, licensed under a Creative Common Attribution 4.0 International License Ref. [123]

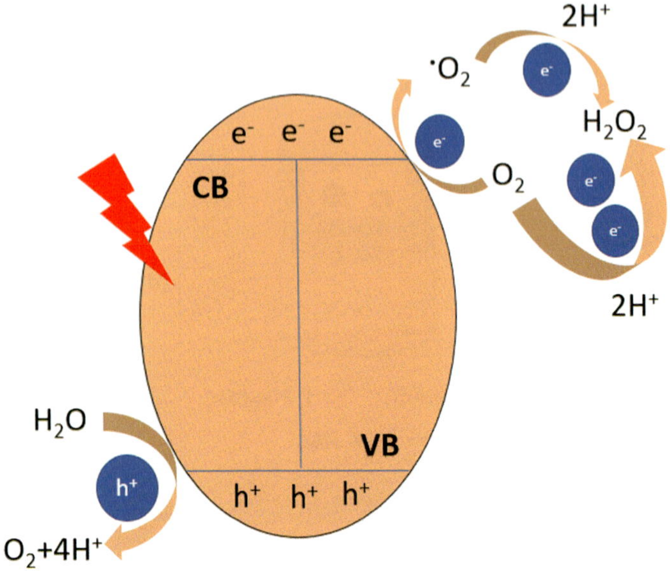

Fig. 3.20 A visual depiction of the photocatalytic setup for generating H_2O_2, inspired by Ref. [122]

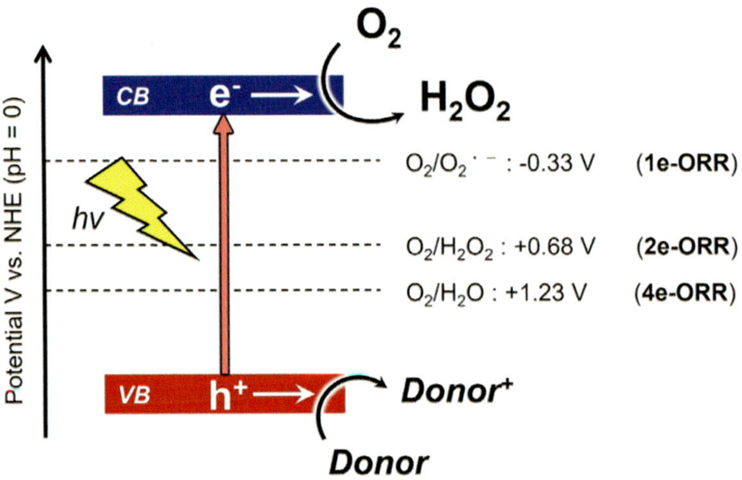

Fig. 3.21 Energy diagram for generation of H_2O_2 via ORR on photocatalytic semiconductors, with permission Ref. [127], under license number 5535951407315

The generation of H_2O_2 using oxygen reduction reaction (ORR) on semiconductors involves a series of three sequential processes, as depicted in Fig. 3.20. The first step consists of the absorption of photons with energy values more significant than the bandgap energy of the semiconductor. This process excites electrons from the valence band (VB) to the conduction band (CB), thereby creating electron–hole pairs. The photo-generated carriers when move toward the surface reaction sites and accumulate there. At these sites, the redox reaction involving these charges leads to the production of H_2O_2. In the ORR process, the photoexcited electrons are utilized directly for H_2O_2 production through a proton-coupled electron transfer process. On the other hand, the photoexcited holes react with the electron donors, supplying protons for the production of H_2O_2. The ORR can occur through two different mechanisms: either in a direct two-electron mechanism that is shown in Eq. (3.8) or an indirect two-electron mechanism involving the formation of a superoxide radical (O_2^-) as an intermediate (Eqs. (3.9, 3.10)). In the latter case, the O_2^- intermediate is involved in the formation of H_2O_2 via a disproportionation reaction (Eq. (3.10)). Figure 3.20 demonstrated that the direct two-electron oxygen reduction reaction pathway (+0.68 V vs. normal hydrogen electrode [NHE]) is more thermodynamically favorable than the indirect two-electron pathway (0.33 V vs. NHE) [124]. However, the indirect oxygen reduction reaction process is favored kinetically because it requires only one electron for each step. The shortcoming of the indirect ORR pathway is that H_2O_2 production is less efficient compared to the direct two-electron ORR process due to the high reactivity of O_2^-. This reactivity can lead to the decomposition of H_2O_2 through over-reduction and back electron transfer to the photocatalyst [125]. Therefore, it would be advantageous to reduce the reactivity of O_2^- to ensure effective H_2O_2 production via the indirect ORR pathway [126] (see Fig. 3.21).

$$O_2 + 2H^+ + 2e^- \rightarrow H_2O_2\ (+0.68\ \text{V versus NHE}) \tag{3.8}$$

$$O_2 + e^- \rightarrow O_2^-\ (-0.33\ \text{V versus NHE}) \tag{3.9}$$

$$O_2^- + 2H^+ + e^- \rightarrow H_2O_2\ (+1.44\ \text{V versus NHE}) \tag{3.10}$$

Photocatalysts used for H_2O_2 production face several challenges, including improving light absorption. In photocatalysis, common challenges include extending the lifespan of excited electrons and holes, as well as facilitating electron transfer to reaction sites. Furthermore, preventing the decomposition of H_2O_2 during the reaction is challenging, resulting in lower H_2O_2 yields. For example, TiO_2 has a maximum selectivity of 33% for H_2O_2, with most of the product decomposing to water. Factors that are important for establishing an efficient photocatalytic system for H_2O_2 production include the morphology and electronic band structure of the photocatalyst, as well as the reaction solution. Examples of the factors that can impact photocatalysis include sacrificial electron donors and the pH level of the solution [122]. Thus, it is imperative to develop new photocatalytic materials that not only exhibit enhanced activity but also mitigate H_2O_2 decomposition. Table 3.2 shows

Table 3.2 Comparison H_2O_2 production rate in different materials

No.	Materials	H_2O_2 production rate $\left(\frac{\mu mol}{hg_{cat}}\right)$	Irradiation conditions (λ, nm)	References
1	g-C_3N_4	63	420–500	[128]
2	g-C_3N_4/PDI-BN0.2-$rGO_{0.05}$	30.80	>420	[129]
3	g-C_3N_4/rGO (10 wt.%)	74.30	>400	[130]
4	C–N–g-C_3N_4	0.98	$420 < \lambda < 700$	[131]
5	OPA/$Zr_{92.5}Ti_{7.5}$-MOF	9700 $\frac{\mu mol}{hg_{cat}}$	>420	[132]
6	TAPD-$(Me)_2$ COF (Solvent: H_2O:EtOH = 9:1)	97	420–700	[133]
7	TAPD-$(Me)_2$ COF (Solvent: H_2O:EtOH = 1:9)	234.52	420–700	[133]

the comparison between some materials used in photocatalytic production of H_2O_2. Consequently, an expanding community of scientists is exploring novel materials and techniques for photocatalytic H_2O_2 production [127].

3.6 Conclusion and Prospective Application

Hydrogen peroxide (H_2O_2) is a versatile oxidant used in various industries and applications, including organic synthesis, wastewater treatment, and as a potential liquid fuel. The conventional production methods for H_2O_2, such as the anthraquinone process, are environmentally harmful and energy-intensive. To address these issues, researchers have developed a photocatalytic method using semiconductor materials, which utilizes solar energy to convert water and oxygen into H_2O_2. The generation of H_2O_2 using oxygen reduction reaction involves three sequential processes. The first step involves the absorption of photons with energy values greater than the bandgap energy of the semiconductor. This process excites electrons from the valence band (VB) to the conduction band (CB), thereby creating electron–hole pairs. The redox reaction involving these charges leads to the production of H_2O_2. In the ORR process, the photoexcited electrons are utilized directly for H_2O_2 production through a proton-coupled electron transfer process. On the other hand, the photoexcited holes react with the electron donors, supplying protons for the production of H_2O_2. The ORR can occur through two different mechanisms: either in a direct two-electron mechanism or in indirect two-electron mechanism involving the formation of a superoxide radical (O_2^-) as an intermediate. In the latter case, the O_2^- intermediate is involved in the formation of H_2O_2 via a disproportionation reaction. The direct two-electron oxygen reduction reaction pathway (+0.68 V vs. standard hydrogen electrode [NHE]) is more thermodynamically favorable than the indirect two-electron pathway (0.33 V vs. NHE). However, the indirect oxygen reduction reaction process is favored kinetically because it requires only one electron for each step. The shortcoming of the indirect ORR pathway is that H_2O_2 production is less efficient than the direct two-electron ORR process due to the high reactivity of O_2^-. This reactivity can lead to the decomposition of H_2O_2 through over-reduction and back electron transfer to the photocatalyst. Therefore, it would be advantageous to reduce the reactivity of O_2^- to ensure effective H_2O_2 production via the indirect ORR pathway.

Traditional precious metal-based catalysts have been commonly used for the oxygen reduction reaction (ORR) in H_2O_2 fuel cells. Still, the limited availability of these metals and environmental concerns have led to the exploration of non-precious metal (NPM) catalysts. Carbon-based catalysts have shown promising performance as NPM ORR catalysts, especially in acidic polymer electrolyte membrane fuel cells (PEMFCs). These carbon-based catalysts offer operational stability, energy efficiency, and cost-effectiveness due to abundant carbon resources. Developing high-purity carbon materials is crucial for optimizing the performance of metal-free carbon catalysts. Overall, these advancements aim to achieve eco-friendly and sustainable H_2O_2 production and improve the efficiency and viability of fuel cells and other energy devices. High-performance and cost-effective catalysts for electrochemical production of H_2O_2 are in demand. Amorphous nickel oxide (NiO_x) based on carbon nanosheets can be created using the photochemical metal–organic deposition approach, which shows high efficiency and selectivity toward the oxygen reduction reaction (ORR). Interface engineering techniques and reaction kinetics are employed to enhance the selectivity and yield of H_2O_2 production. The development of evenly dispersed nickel oxide nanoparticles within nickel–iron metal–organic framework (MOF) nanosheets allows for efficient electrochemical synthesis of H2O2. NiO nanoparticles enhance activity, stability, and selectivity. Oxygen-doped carbon nanosheets (OCNS) and bismuth titanate ($Bi_4Ti_3O_{12}$) nanosheets are also effective catalysts for $2e^-$ ORR H_2O_2 generation.

Metal–organic frameworks (MOFs) and covalent organic frameworks (COFs) are porous materials with unique properties that make them useful in various applications, such as gas storage, separation, and catalysis. COFs, in particular, have shown potential as electrochemical sensors, but their weak conductivity limits their sensitivity. Metal-covalent organic frameworks (MCOFs) have been developed to address this limitation by incorporating active metal ions into COF structures. MCOFs combine the desirable properties of MOFs and COFs, resulting in self-complementary characteristics. Activation methods, including thermal activation, are commonly used to remove guest molecules from as-synthesized MOFs, while solvothermal methods are employed in COF synthesis, with reaction time, temperature, solvent conditions, and catalyst concentration being critical variables. MOFs have also contributed to advancements in lithium–ion batteries as precursors and templates for carbon-based anodes, and their functional derivatives have shown potential as advanced electrodes and electrocatalysts in energy storage and conversion systems. COFs, with their customizable frameworks, have potential for gas storage applications, particularly for CO_2 capture and mitigation, as the ability to engineer pore surfaces is crucial for effective CO_2 adsorption.

The photocatalyst plays a crucial role in influencing the photocatalyst oxygen reduction reaction process' activity and selectivity. Typically, semiconductor materials are used for photocatalysis. Although a large variety of metal-based semiconductors, including SnO_2, Fe_2O_3, ZnO, TiO_2, CdS, NiS, Ag_3O_4, etc., have been used as photocatalysts, these materials' high bandgaps have made it challenging to harness solar energy effectively for photocatalytic purposes. Semiconductor-based photocatalytic technology has gained attention as a potential solution to the environmental issues caused by rapid industrialization. The heterojunction construction, achieved through a chemical transformation method, has enhanced the photocatalytic activity and is considered efficient and promising.

To produce H_2O_2, several carbon-based photocatalysts have been used as alternative metal-based photocatalysts to improve their photocatalytic performance. Some materials have been discussed, such as g-C_3N_4, modified g-C_3N_4, metal–organic framework/covalent organic framework, and graphene oxide. Metal-free carbonaceous materials, such as nitrogen-doped carbon nanosheets, demonstrate catalytic activity for H_2O_2 activation and degradation of organic contaminants. Carbon-based metal-free catalysts with controlled doping configurations show promise as alternatives to platinum in fuel cells, particularly by enhancing oxygen reduction reaction (ORR) activity in acidic electrolytes.

Graphene oxide refers to graphene that has undergone an oxidation process and consists of solution-dispersible polyaromatic two-dimensional carbon sheets derived from the acid exfoliation of graphite. Graphene oxide shares a hexagonal carbon structure with graphene but additionally contains numerous oxygen-based functional groups, such as hydroxyl (–OH), alkoxy (C–O–C), carbonyl (C=O), and carboxylic acid (–COOH). These functional groups offer certain benefits over graphene, including enhanced solubility and the ability to alter the surface of graphene oxide with different functional groups, thereby broadening the scope of its possible uses in nanocomposite materials. GO resembles an insulating material with a wide bandgap; it exhibits a bandgap ranging from 2.4 to 4.3 eV, but its electronic structure depends on the carbon-to-oxygen atomic ratio. GO can behave like a semiconductor. Graphene has become a prospective electron acceptor in these photocatalysts, which helps enhance charge transfer and reduce electron–hole recombination in the composite, resulting in better photocatalytic performance. Photocatalysts based on GO have been created to perform photocatalytic tasks such as hydrogen production and breaking down organic pollutants. Additionally, they can function as a photocatalyst to produce hydrogen and oxygen from water due to their negative conduction band position.

g-C_3N_4 nanosheets possess exceptional catalytic activity, high stability, and efficient charge transfer capabilities, making them well suited for photocatalytic oxygen reduction reaction (ORR) to generate hydrogen peroxide. The unique structural properties of g-C_3N_4 nanosheets, including their high surface area, abundant active sites, and excellent electron transport capability, contribute to their remarkable performance as catalysts in this application. The utilization of light energy as an environmentally friendly and sustainable driving force for H_2O_2 production showcases the potential of g-C_3N_4 nanosheet catalysts in addressing the growing demand for H_2O_2 in various fields.

The use of g-C_3N_4 nanosheet catalysts offers several advantages, including low cost, abundant availability, and environmental sustainability, making them attractive for industrial-scale applications. The photocatalytic production of H_2O_2 using g-C_3N_4 nanosheet catalysts holds promise for various fields, such as wastewater treatment, disinfection, and chemical synthesis, where H_2O_2 plays a vital role. However, further research is needed to optimize the catalyst design, explore the reaction mechanism, and scale up the production process to realize the full potential of g-C_3N_4 nanosheets in H_2O_2 generation.

Prospective application

The prospective application of photocatalytic oxygen reduction reaction (ORR) to generate hydrogen peroxide (H_2O_2) using carbon-based nanosheet catalyst holds significant promise. Carbon-based nanosheets, with their two-dimensional morphology and high electrical conductivity, provide an ideal platform for efficient conversion of oxygen molecules into H_2O_2 under light irradiation. By harnessing the power of light energy, these nanosheets can activate the ORR, converting oxygen molecules into H_2O_2. The photocatalytic approach provides a sustainable and environmentally friendly method for H_2O_2 production and offers advantages such as mild reaction conditions, tunable catalytic properties, and high selectivity. Using carbon-based nanosheets as catalysts allows for many active sites and enhanced charge transfer, leading to improved catalytic performance and overall efficiency in H_2O_2 generation. This emerging application shows excellent potential for addressing the growing demand for H_2O_2 in various fields, including wastewater treatment, disinfection, and chemical synthesis, while minimizing the reliance on traditional energy-intensive and environmentally detrimental methods of H_2O_2 production. For instance, in wastewater treatment, generating H_2O_2 through photocatalytic ORR can facilitate advanced oxidation processes, effectively removing organic pollutants. In disinfection, the produced H_2O_2 can act as a powerful and selective antimicrobial agent. Additionally, carbon-based nanosheets can play a crucial role in chemical synthesis, enabling the green and efficient production of valuable chemicals. Overall, the prospective application section underscores the immense potential of carbon-based nanosheet catalysts in harnessing the photocatalytic ORR for H_2O_2 generation and its multifaceted applications across different sectors, thereby offering novel and sustainable solutions to pressing environmental and societal challenges. Additionally, the prospective application extends to chemical synthesis processes, where H_2O_2 serves as a valuable oxidizing agent. The photocatalytic production of H_2O_2 using carbon-based nanosheet catalysts opens up new possibilities for selective and environmentally benign synthesis routes. These potential applications highlight the significant impact and relevance of utilizing carbon-based nanosheets in the photocatalytic ORR for H_2O_2 generation, paving the way for sustainable and efficient solutions in various fields.

47. Zhurenok AV et al (2021) Synthesis of graphitic carbon nitride-based photocatalysts for hydrogen evolution under visible light. Mendeleev Commun 31(2):157–159
48. Chellammal Gayathri R, Elakkiya V, Sumathi S (2021) Effect of method of preparation on the photocatalytic activity of $NiAl_2O_4$. Inorg Chem Commun 129, 108634
49. Zhang L et al (2020) Performance of WO_3/g-C_3N_4 heterojunction composite boosting with NiS for photocatalytic hydrogen evolution. Appl Surf Sci 499:143862
50. Thakur S et al (2017) Sunlight-driven sustainable production of hydrogen peroxide using a CdS–graphene hybrid photocatalyst. J Catal 345:78–86
51. Jing M et al (2023) Coral-like B-doped g-C(3)N(4) with enhanced molecular dipole to boost photocatalysis-self-Fenton removal of persistent organic pollutants. J Hazard Mater 449:131017
52. Tian J et al (2023) One-step nitrogen defect engineering of polymeric carbon nitride for visible light-driven photocatalytic O(2) reduction to H(2)O(2). J Colloid Interface Sci 634:138–147
53. Zhang D et al (2020) Facile in situ chemical transformation synthesis, boosted charge separation, and increased photocatalytic activity of $BiPO_4$/BiOCl p-n heterojunction photocatalysts under simulated sunlight irradiation. J Phys Chem Solids 147:109630
54. Reischauer S, Pieber B (2021) Emerging concepts in photocatalytic organic synthesis. iScience 24(3):102209
55. Jiao H et al (2022) Insights on carbon neutrality by photocatalytic conversion of small molecules into value-added chemicals or fuels. Acc Mater Res 3(12):1206–1219
56. Li Y et al (2023) Enhancement of charge separation and NIR light harvesting through construction of 2D–2D $Bi_4O_5I_2$/BiOBr:Yb^{3+}, Er^{3+} Z-scheme heterojunctions for improved full-spectrum photocatalytic performance. Adv Sci, 2207514
57. Xiong L, Tang J (2021) Strategies and challenges on selectivity of photocatalytic oxidation of organic substances. Adv Energy Mater 11(8):2003216
58. Li Z et al (2023) Dual step-scheme heterojunction with full-visible-light-harvesting towards synergistic persulfate activation for enhanced photodegradation. J Colloid Interface Sci 640:456–471
59. Hou Y et al (2019) Charge separation, charge recombination, long-lived charge transfer state formation and intersystem crossing in organic electron donor/acceptor dyads. J Mater Chem C 7(39):12048–12074
60. Yanagi R et al (2022) Charge separation in photocatalysts: mechanisms, physical parameters, and design principles. ACS Energy Lett 7(1):432–452
61. Renckens TJA et al (2010) Product desorption limitations in selective photocatalytic oxidation. Catal Today 155(3):302–310
62. Chen Q et al (2023) A hydroxyl-induced carbon nitride homojunction with functional surface for efficient photocatalytic production of H_2O_2. Appl Catal B: Environ 324:122216
63. Chu C et al (2022) Highly efficient photocatalytic H_2O_2 production with cyano and SnO_2 co-modified g-C_3N_4. Chem Eng J 428:132531
64. Geng X et al (2021) H_2O_2 production and in situ sterilization over a ZnO/g-C_3N_4 heterojunction photocatalyst. Chem Eng J 420:129722
65. Lu N et al (2020) Characterization of highly effective plasma-treated g-C_3N_4 and application to the photocatalytic H_2O_2 production. Chemosphere 241:124927
66. Liu W et al (2021) Fabrication of ultra-thin g-C_3N_4 nanoplates for efficient visible-light photocatalytic H_2O_2 production via two-electron oxygen reduction. Chem Eng J 425:130615
67. Farzin F et al (2022) A novel Z-scheme oxygen-doped g-C_3N_4 nanosheet/NaBiS2 nanoribbon for efficient photocatalytic H_2O_2 production and organic pollutants degradation. J Phys Chem Solids 163:110588
68. Wang R et al (2018) Energy-level dependent H_2O_2 production on metal-free, carbon-content tunable carbon nitride photocatalysts. J Energy Chem 27(2):343–350
69. Zhang M et al (2022) Facile synthesis of porous 1,3,5-Trihydroxybenzene substituted g-C_3N_4 for boosted photocatalytic Rhodamine B degradation and H_2O_2 production. J Alloy Compd 925:166604

70. Xing J et al (2022) Synthesis and modifications of g-C_3N_4-based materials and their applications in wastewater pollutants removal. J Environ Chem Eng 10(6):108782
71. Zhang Y et al (2021) Research progress on g–C_3N_4–based photocatalysts for organic pollutants degradation in wastewater: from exciton and carrier perspectives. Ceram Int 47(22):31005–31030
72. Li C et al (2019) Facile fabrication of g-C_3N_4 QDs/BiVO4 Z-scheme heterojunction towards enhancing photodegradation activity under visible light. J Taiwan Inst Chem Eng 95:669–681
73. Mehtab A, Ahmad T (2023) Investigating the spatial charge density flow and molecular structure of g-C_3N_4 photocatalyst from a computational perspective. Appl Catal A: Gen 659:119190
74. Li Y et al (2021) Integration of metal-organic frameworks and covalent organic frameworks: design, synthesis, and applications. Matter 4(7):2230–2265
75. Cusin L et al (2021) Chemical conversion and locking of the imine linkage: enhancing the functionality of covalent organic frameworks. Angew Chem Int Ed 60(26):14236–14250
76. Wang M et al (2021) A novel electrochemical sensor based on MWCNTs-COOH/metal-covalent organic frameworks (MCOFs)/Co NPs for highly sensitive determination of DNA base. Microchem J 167:106336
77. Dong J et al (2020) Metal-covalent organic frameworks (MCOFs): a bridge between metal-organic frameworks and covalent organic frameworks. Angew Chem Int Ed 59(33):13722–13733
78. Zhang X et al (2020) A historical overview of the activation and porosity of metal–organic frameworks. Chem Soc Rev 49(20):7406–7427
79. Wang D-G et al (2021) Covalent organic framework-based materials for energy applications. Energy Environ Sci 14(2):688–728
80. Chen X et al (2020) Covalent organic frameworks: chemical approaches to designer structures and built-in functions. Angew Chem Int Ed 59(13):5050–5091
81. Li X et al (2020) Metal–organic frameworks as a platform for clean energy applications. EnergyChem 2(2):100027
82. Lu XF et al (2021) Metal-organic frameworks derived functional materials for electrochemical energy storage and conversion: a mini review. Nano Lett 21(4):1555–1565
83. Hsu HC et al (2013) Graphene oxide as a promising photocatalyst for CO_2 to methanol conversion. R Soc Chem 2013(5):6
84. Dreyer DR et al (2010) Graphene oxide has carboxylic acid groups at its edges, epoxy and hydroxyl groups at its basal planes, and other chemically reactive oxygen capabilities throughout. The Royal Society of Chemistry 2010. Chem Soc Rev 39:228–240, 12
85. Smith AT et al (2019) Synthesis, properties, and applications of graphene oxide/reduced graphene oxide and their nanocomposites. Nano Mater Sci 1(2019):31–47
86. Cheng C et al (2017) Functional graphene nanomaterials based architectures: biointeractions, fabrications, and emerging biological applications. Chem Rev 117(3):1826–1914
87. Lu KQ et al (2021) Roles of graphene oxide in heterogeneous photocatalysis. ACS Mater Au 2021(1):37–54
88. Brodie CB (1859) XIII. On the atomic weight of graphite. J Philos Trans R Soc Lond (149):249–259
89. Hummers J, William S, Offeman RE (1958) Preparation of graphitic oxide. J Am Chem Soc 80(6):1339–1339
90. Kovtyukhova NI, Ollivier PJ, Martin BR, Mallouk TE, Chizhik SA, Buzaneva EV, Gorchinskiy AD (1999) Layer-by-layer assembly of ultrathin composite films from micron-sized graphite oxide sheets and polycations. Chem Mater 11(3):771–778
91. Hirata M, Gotou T, Horiuchi S, Fujiwara M, Ohba M (2004) Thin-film particles of graphite oxide 1: high-yield synthesis and flexibility of the particles. Carbon 42(14):2929–2937
92. Ang PK, Wang S, Bao Q, Thong JTL, Loh KP (2009) High-throughput synthesis of graphene by intercalation—exfoliation of graphite oxide and study of ionic screening in graphene transistor. ACS Nano 3(11):3587–3594

93. Marcano DC, Kosynkin DV, Berlin JM, Sinitskii A, Sun Z, Slesarev AS, Alemany LB, Lu W, Tour JM (2018) Correction to improved synthesis of graphene oxide. ACS Nano 12(2):2078–2078
94. Eigler S, Enzelberger-Heim M, Grimm S, Hofmann P, Kroener W, Geworski A, Dotzer C, Röckert M, Xiao J, Papp C (2013) Wet chemical synthesis of graphene. Adv Mater 25(26):3583–3587
95. Peng L, Xu Z, Liu Z, Wei Y, Sun H, Li Z, Zhao X, Gao C (2015) An Iron-based green approach to 1-h production of single-layer graphene oxide. Nat Commun 6(1):5716
96. Prasad C et al (2020) An overview of graphene oxide supported semiconductors based photocatalysts: properties, synthesis and photocatalytic applications. J Mol Liq 297:111826. 2019(0167–7322/© 2019 Published by Elsevier B.V.)
97. Yeh TF et al (2010) Graphite oxide as a photocatalyst for hydrogenproduction from water. Adv Funct Mater 20:2255–2262
98. Krishnamoorthy K, Mohan R, Kim SJ (2011) Graphene oxide as a photocatalytic material. Appl Phys Lett 98:244101. 2011(American Institute of Physics)
99. Li Q et al (2011) Highly efficient visible-light-driven photocatalytic hydrogen production of CdS-cluster-decorated graphene nanosheets. J Am Chem Soc 133:10878–10884, 6
100. Yeh TF et al (2011) Graphite oxide with different oxygenated levels for hydrogen and oxygen production from water under illumination: the band positions of graphite oxide. J Phys Chem C 115:22587–22597. 2011(American Chemical Society Publications)
101. Raaja Rajeshwari M, Kokilavani S, Sudheer Khan S (2022) Recent developments in architecturing the g-C_3N_4 based nanostructured photocatalysts: synthesis, modifications and applications in water treatment. Chemosphere 291:132735
102. Zhu Q, Zhang J (2022) Is g-C_3N_4 more suitable for photocatalytic reduction or oxidation in environmental applications? Environ Funct Mater 1(1):121–125
103. Zheng Y et al (2011) Nanoporous graphitic-C_3N_4@Carbon metal-free electrocatalysts for highly efficient oxygen reduction. J Am Chem Soc 133(50):20116–20119
104. Roy P, Pramanik A, Sarkar P (2021) Graphitic carbon nitride sheet supported single-atom metal-free photocatalyst for oxygen reduction reaction: a first-principles analysis. J Phys Chem Lett 12(11):2788–2795
105. Yuda A, Kumar A (2022) A review of g-C_3N_4 based catalysts for direct methanol fuel cells. Int J Hydrogen Energy 47(5):3371–3395
106. Holade Y et al (2015) Recent advances in carbon supported metal nanoparticles preparation for oxygen reduction reaction in low temperature fuel cells. Catalysts 5:310–348. https://doi.org/10.3390/catal5010310
107. Shafi A et al (2022) Exploring multifunctional behaviour of g-C_3N_4 decorated $BiVO_4$/Ag_2CO_3 hierarchical nanocomposite for simultaneous electrochemical detection of two nitroaromatic compounds and water splitting applications. Talanta 241:123257
108. Arafat Y et al (2021) Metal-free carbon based air electrodes for Zn-air batteries: recent advances and perspective. Mater Res Bull 140:111315
109. Zhang L, Shao Q, Zhang J (2021) An overview of non-noble metal electrocatalysts and their associated air cathodes for Mg-air batteries. Mater Rep: Energy 1(1):100002
110. Ilager D et al (2022) Graphene/g-carbon nitride (GO/g-C_3N_4) nanohybrids as a sensor material for the detection of methyl parathion and carbendazim. Chemosphere 292:133450
111. Liu Y et al (2020) Construction of electrochemical sensing interface towards Cd(II) based on activated g-C_3N_4 nanosheets: considering the effects of exfoliation and protonation treatment. Anal Bioanal Chem 412(2):343–353
112. Wu Z et al (2022) Amorphous nickel oxides supported on carbon nanosheets as high-performance catalysts for electrochemical synthesis of hydrogen peroxide. ACS Catal 12(10):5911–5920
113. Bu Y et al (2021) Carbon-based electrocatalysts for efficient hydrogen peroxide production. Adv Mater 33(49):2103266
114. Wang M et al (2020) Partially pyrolyzed binary metal-organic framework nanosheets for efficient electrochemical hydrogen peroxide synthesis. Angew Chem Int Ed 59(34):14373–14377

115. Wang M et al (2021) An efficient interfacial synthesis of two-dimensional metal-organic framework nanosheets for electrochemical hydrogen peroxide production. Angew Chem Int Ed 60(20):11190–11195
116. Chen S et al (2021) Chemical identification of catalytically active sites on oxygen-doped carbon nanosheet to decipher the high activity for electro-synthesis hydrogen peroxide. Angew Chem Int Ed 60(30):16607–16614
117. Zhang Z et al (2021) Highly catalytic selectivity for hydrogen peroxide generation from oxygen reduction on Nd-doped $Bi_4Ti_3O_{12}$ nanosheets. J Phys Chem C 125(45):24814–24822
118. Li L et al (2023) Nitrogen-doped carbon nanosheets for efficient degradation of bisphenol A by H_2O_2 activation at neutral pH values. Sep Purif Technol 306:122687
119. Yang L et al (2019) Carbon-based metal-free orr electrocatalysts for fuel cells: past, present, and future. Adv Mater 31(13):1804799
120. Low J et al (2017) Heterojunction photocatalysts. Adv Mater 29(20):1601694
121. Habisreutinger SN, Schmidt-Mende L, Stolarczyk JK (2013) Photocatalytic reduction of CO_2 on TiO_2 and other semiconductors. Angew Chem Int Ed 52(29):7372–7408
122. Hou H, Zeng X, Zhang X (2020) Production of hydrogen peroxide by photocatalytic processes. Angew Chem Int Ed 59:17356–17376
123. Chen X, Zhao J, Li G, Zhang D, Li H (2022) Recent advances in photocatalytic renewable energy production. Energy Mater 2(10)
124. Li Q, Guo B, Yu J, Ran J, Zhang B, Yan H, Gong JR (2011) Highly efficient visible-light-driven photocatalytic hydrogen production of cds-cluster-decorated graphene nanosheets. J Am Chem Soc 133:10878–10884, 6
125. Isaka Y, Kato S, Hong D, Suenobu T, Yamada Y, Fukuzumi S (2015) Bottom-up and top-down methods to improve catalytic reactivity for photocatalytic production of hydrogen peroxide using a ru-complex and water oxidation catalysts. J Mater Chem A 3(23):12404–12412
126. Isaka Y, Kawase Y, Kuwahara Y, Mori K, Yamashita H (2019) Two-phase system utilizing hydrophobic metal–organic frameworks (MOFs) for photocatalytic synthesis of hydrogen peroxide. Angewandte Chemie 131(16):5456–5460
127. Kondo Y, Kuwahara Y, Mori K, Yamashita H (2022) Design of metal-organic framework catalysts for photocatalytic hydrogen peroxide production. J Chem 8(11):2924–2938
128. Shiraishi Y, Kanazawa S, Sugano Y, Tsukamoto D, Sakamoto H, Ichikawa S, Hirai T (2014) Highly selective production of hydrogen peroxide on graphitic carbon nitride (g-C_3N_4) photocatalyst activated by visible light. Acs Catal 4(3):774–780
129. Vanduyfhuys L, Vandenbrande S, Wieme J, Waroquier M, Verstraelen T, Van Speybroeck V (2018) Extension of the QuickFF force field protocol for an improved accuracy of structural, vibrational, mechanical and thermal properties of metal–organic frameworks. Wiley Online Library
130. Kim HI, Choi Y, Hu S, Choi W, Kim JH (2018) Photocatalytic hydrogen peroxide production by anthraquinone-augmented polymeric carbon nitride. Appl Catal B: Environ 229:121–129
131. Fu Y, Liu C, Zhang M, Zhu C, Li H, Wang H, Song Y, Huang H, Liu Y, Kang Z (2018) Photocatalytic H_2O_2 and H_2 generation from living chlorella vulgaris and carbon micro particle comodified g-C_3N_4. Adv Energy Mater 8(34):1802525
132. Chen X, Kuwahara Y, Mori K, Louis C, Yamashita H (2020) A hydrophobic titanium doped zirconium-based metal organic framework for photocatalytic hydrogen peroxide production in a two-phase system. J Mater Chem A 8(4):1904–1910
133. Krishnaraj C, Sekhar Jena H, Bourda L, Laemont A, Pachfule P, Roeser J, Chandran CV, Borgmans S, Rogge SM, Leus K, Stevens CV (2020) Strongly reducing (diarylamino) benzene-based covalent organic framework for metal-free visible light photocatalytic H_2O_2 generation. J Am Chem Soc 142(47):20107–20116

Chapter 4
Valorizing Glycerol into Valuable Chemicals Through Photocatalytic Processes Utilizing Innovative Nano-Photocatalysts

Mohamed Tarek Ahmed, Shoeb Azam Farooqui, Sheng-Hsiang Hsu, Lee Daeun, and Siti Khodijah Chaerun

Abstract Significant research efforts focused on the utilization of renewable resources have been increased as a result of the development of environmentally friendly and efficient methods of chemical synthesis. Glycerol, a by-product of biodiesel production, has been identified as a promising renewable feedstock due to its abundant availability. Photocatalysis represents an innovative and sustainable pathway for the conversion of glycerol into a variety of valuable chemicals, including hydrogen, alcohols, aldehydes, ketones, acids, and other high-value compounds. Through the cleavage of either a single C–H or O–H bond, glycerol can be transformed into four distinct intermediates, each leading to different products. This chapter offers an in-depth examination of the photocatalytic valorization of glycerol, emphasizing the role of various nano-photocatalysts in this process. It highlights the critical function of nano-photocatalysts in augmenting the efficiency of glycerol conversion reactions, achieved through increased surface area, optimized charge separation, and enhanced light absorption properties. The complexity of the photocatalytic process is influenced by several catalyst characteristics, such as crystalline structure, surface morphology, hydroxyl group presence, substrate and ion adsorption, all of which directly impact the activity and selectivity of the reaction. Furthermore, the kinetics of the photocatalytic process are intricately affected by substrate concentration, reflecting a complex dynamic interplay between substrate oxidation and carrier recombination (back-reactions), mediated by the substrate itself. This

M. T. Ahmed · S. A. Farooqui · S.-H. Hsu · L. Daeun
Department of Materials Science and Engineering, National Taiwan University of Science and Technology, No.43, Sec. 4, Keelung Road, Taipei 10607, Taiwan

S. K. Chaerun (✉)
Department of Metallurgical Engineering, Institut Teknologi Bandung, Ganesha 10, Bandung 40132, Indonesia
e-mail: skchaerun@itb.ac.id

Geomicrobiology-Biomining and Biocorrosion Laboratory, Microbial Culture Collection Laboratory, Biosciences and Biotechnology Research Center (BBRC) Institut Teknologi Bandung, Ganesha 10, Bandung 40132, Indonesia

H. Abdullah (ed.), *Solar Light-to-Hydrogenated Organic Conversion*,
https://doi.org/10.1007/978-981-99-8114-4_4

comprehensive analysis provides valuable insights into the multifaceted nature of photocatalytic glycerol valorization and the potential of nano-photocatalysts to revolutionize this field. Numerous semiconductor catalyst materials are available for the valorization of glycerol, including TiO_2, MOF, WO_3, Bi_2WO_6, g-C_3N_4, and ZnO, each of which is discussed in detail. Approaches such as elemental doping, co-catalyst hybridization, heterojunction formation, and morphological control can be employed to enhance the photocatalytic activity of these materials. Additionally, synthesis methods such as sol–gel, hydrothermal, and co-precipitation techniques exert influence on the structural, morphological, and optical properties of the photocatalysts. Titanium dioxide (TiO_2) is recognized as an efficient photocatalyst under UV light, but its wide energy bandgap (~3.2 eV), high rate of electron–hole recombination, and limited surface area restrict its photocatalytic performance. Zeolitic imidazolate frameworks, novel photocatalytic materials, are characterized by high specific surface area and abundant structural variability but suffer from poor charge carrier separation efficiency. Tungsten trioxide (WO_3) is notable for its high efficiency, low cost, high catalytic activity, low toxicity, abundant availability, and environmental friendliness. Its narrow bandgap, high electron mobility, and good chemical stability make WO_3 suitable for a wide range of applications, including environmental remediation and solar energy conversion. Bismuth tungsten oxide Bi_2WO_6 (BWO) possesses a relatively low band gap energy (2.70–3.0 eV) and its morphology can be easily controlled, owing to its unique layered crystal structure. Graphitic carbon nitride (g-C_3N_4) features a compatible bandgap (2.7 eV) and can be synthesized from abundant precursors, enhancing its availability and cost efficiency. Its high stability, non-toxicity, and structural versatility make g-C_3N_4 suitable for various applications. Zinc oxide (ZnO) is a widely studied material with a direct bandgap of 3.30–3.37 eV at room temperature. It offers advantages over TiO_2 due to its higher electron mobility, leading to increased electronic transfer efficiency and reduced recombination losses. Additionally, Zinc oxide (ZnO) is attractive due to its cost-effectiveness, non-toxic nature, and the ability to retain high stability and crystallinity. However, its application is limited to ultraviolet (UV) excitation, which constitutes only a minor portion of the solar spectrum. Therefore, this chapter summarizes the current state of knowledge and highlights the potential for nanomaterials to facilitate the sustainable conversion of glycerol into valuable compounds. This chapter also emphasizes recent progress and existing challenges in the field, particularly the creation of innovative nano-photocatalysts with customized characteristics. Serving as a comprehensive guide for researchers and scientists working in the areas of catalysis, renewable energy, and sustainable chemistry, this chapter aims to inspire continued exploration and innovation in the specialized area of glycerol valorization.

4.1 Introduction

As far back as 2800 BC, glycerol, also known as 1,2,3-propanetriol or glycerine, was obtained by heating fats with ash to produce soap. In the late 1940s, synthetic surfactants were developed, leading to the indirect production of glycerol from propylene (derived from fossil oil). This shift occurred as many chemical companies, anticipating a future shortage of glycerol, began manufacturing it synthetically. In recent times, some glycerol plants have closed, while others have opened to utilize glycerol as a feedstock. This surplus of glycerol is a by-product of biodiesel fuel production through the transesterification of seed oils with methanol (CH_3OH). The biodiesel industry generates significant quantities of crude glycerol as a side product, but only a small portion of this is used or converted into other chemicals. As a result, the market price of glycerol has dropped, spurring increased interest in valorizing glycerol into higher-value chemicals.

CO_2 emissions have been continually increasing due to the escalating demand for energy. These emissions are directly linked to human activities, such as the combustion of fossil fuels for transportation, industrial processes, and energy generation. Many efforts have been made to explore lower carbon and renewable fuel alternatives. Upon evaluation, solar energy emerges as the most economical, abundant, and readily accessible energy source, offering a pathway to an environmentally friendly and sustainable industry. In recent years, there has been a significant focus on semiconductor-mediated photocatalytic processes, which are regarded as eco-friendly and cost-effective methods for water and air purification, microbial elimination in water, the creation of self-cleaning and self-sterilizing surfaces, and the storage and conversion of solar energy. Photocatalysis has notably contributed to hydrogen production through water splitting, where photo-generated electrons reduce protons, and photo-generated holes oxidize water. This reaction occurs over irradiated suspensions of semiconductor particles. Considerable efforts have been made in recent years to develop efficient solar light-responsive photocatalysts with suitable optical and electronic properties for glycerol photo-reforming. The use of a hole scavenger helps prevent the recombination of photo-generated species, thereby overcoming the limitation of the short lifetime of conduction band electrons.

4.1.1 Fundamentals and Mechanisms of Photocatalytic Glycerol Oxidation

Oxidation of glycerol

Renewable energy is advancing rapidly in response to global energy shortages and environmental pollution concerns. Biodiesel, a popular form of biomass energy, serves as an eco-friendly and sustainable alternative to non-renewable energy sources. However, the mass production of biodiesel leads to an oversupply of its by-product, glycerol, resulting in a saturated market and declining prices. Many producers now treat crude glycerol as waste, leading to a loss of valuable biomass resources and potential environmental damage. As a result, researchers are investigating methods to convert glycerol into high-value components. Selective oxidation of glycerol is a promising approach to synthesize various fine chemicals, such as glyceraldehyde, glycolic acid, dihydroxyacetone, and mesoxalic acid. Most research in this area has focused on traditional heterogeneous catalysis, often utilizing active centers composed of precious metals[1–6].

The primary objective of our study is to gain a comprehensive understanding of the process by which glycerol is converted into four unique intermediates. These intermediates are formed as a consequence of the cleavage of single C–H or O–H bonds. The dehydrogenation reaction of glycerol involves multiple elementary steps and can lead to a variety of potential products. Our objective is to utilize this information to determine the selectivity towards two desired C_3 products: glyceraldehyde (GALD) and dihydroxyacetone (DHA) (Fig. 4.1). The four glycerol intermediates are categorized based on the group dehydrogenated, such as the primary C–H bond intermediate and the secondary O–H bond intermediate. Our focus is on how the structure of the catalyst can be altered to modify the selectivity between these products. For the purpose of this study, we will disregard the numerous other species that may emerge within the reaction network [7–10].

Esterification of glycerol with acetic acid

A prevalent method for producing additives from glycerin involves the esterification of glycerin with carboxylic acids, resulting in the formation of esters. Triglycerides, found naturally in vegetable oils and animal fats, are compounds formed from glycerol and long-chain fatty acids. In the context of fuel additives, glycerol is often transformed into glyceryl acetate (also known as glyceryl acetate) through an acid-catalyzed esterification process, utilizing acetic acid or sulfuric acid, as depicted in Fig. 4.2. This reaction can produce three distinct compounds: monoacetin, diacetin, and triacetin, alternatively referred to as monoacetylglycerol (MAG), diacetylglycerol (DAG), and triacetylglycerol (TAG), depending on the extent of reaction. By blending mono, di, and triacetins, commercially valuable products with broad industrial applications, including fuel additives, are obtained. The incorporation of these compounds into fuel contributes to environmental sustainability and offers economic advantages [11, 12].

Photocatalytic glycerol oxidation and hydrogen generation

Hydrogen production through the photocatalytic conversion of oxygenated organic compounds derived from biomass has emerged as a compelling alternative, as it mitigates CO_2 emissions linked to anthropogenic activities and the greenhouse effect. Hydrogen, regarded as an energy carrier, offers an environmentally friendly option due to the nature of the by-products formed during its hydrolysis. Titanium dioxide (TiO_2), favored for its stability and affordability, is commonly used as a photocatalyst. By combining or doping TiO_2 with various noble metals, non-noble metals, or non-metal elements, its fundamental properties can be improved, resulting in delayed electron–hole pair recombination or an expanded range of visible light absorption.

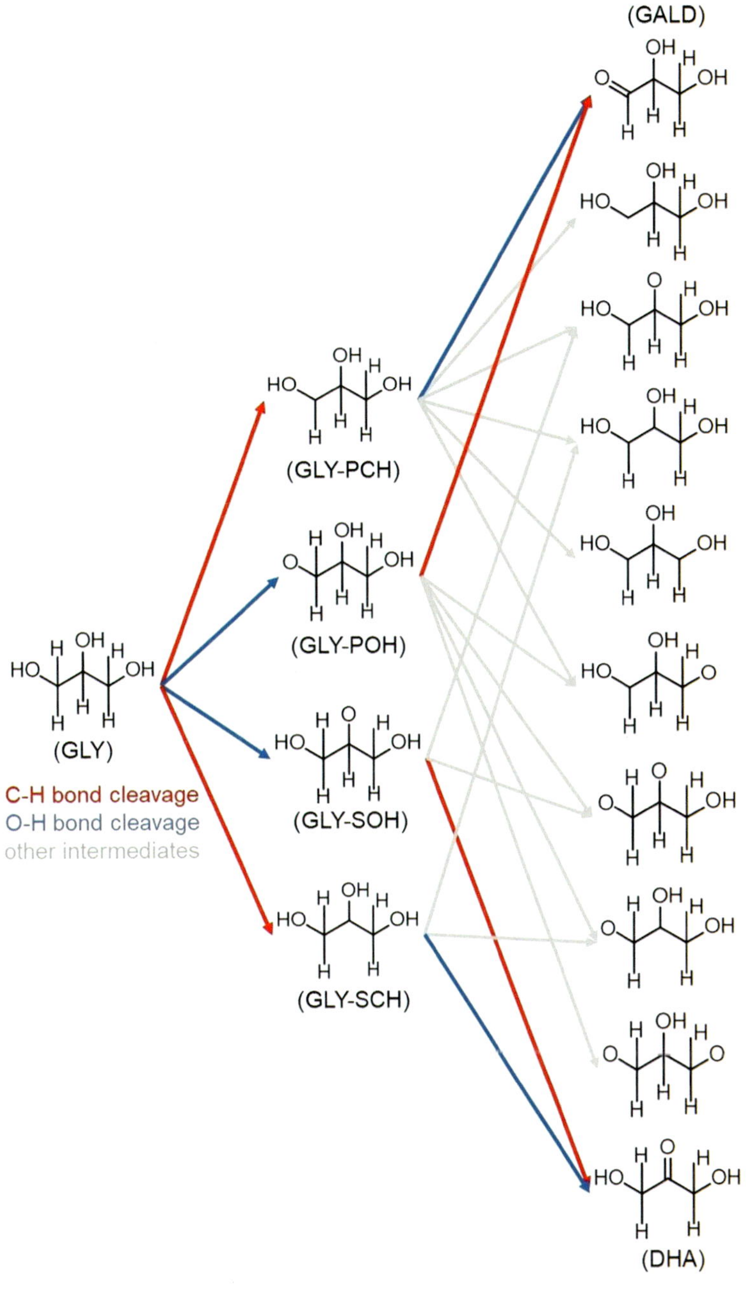

Fig. 4.1 A schematic illustration of the synthesis of dihydroxyacetone (DHA) and glyceraldehyde (GALD) from glycerol through a series of simple chemical reactions. The steps outlined in gray were not explicitly considered in this study. The presence of "H" in the diagram indicates that hydrogen was removed along the reaction pathway, and all instances of hydrogen removal are clearly depicted [10]

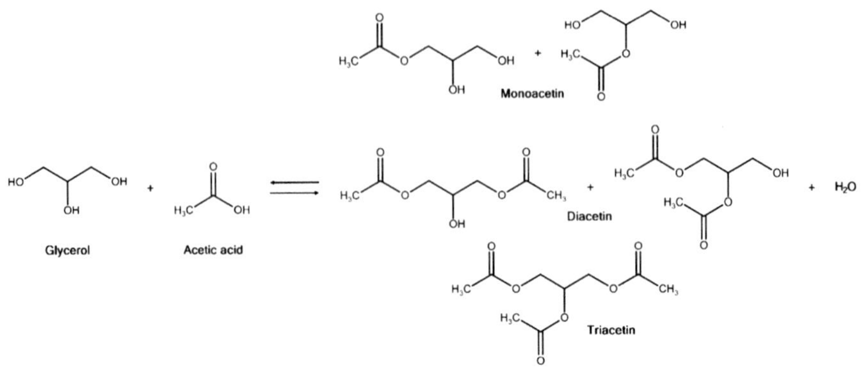

Fig. 4.2 The acetylation of glycerol with acetic acid [12]

Titania-carbon composites may further improve the active surface of the semiconductor, photosensitize the material to substantially reduce the band gap, and minimize charge recombination. Technological innovations, such as the photo-reforming of glycerol aqueous solution, present opportunities to augment biodiesel sustainability and facilitate hydrogen production (Fig. 4.3). Theoretically, the photocatalytic oxidation of glycerol yields an H_2 to CO_2 ratio of 2.33. However, the underlying reaction mechanism involving inorganic semiconductor photocatalysts is complex and remains an active area of research. Some studies have identified multiple concurrent reaction pathways, including oxidative C–C bond cleavage, carbonyl formation through primary or secondary carbon oxidation, or light-induced dehydration. The chosen reaction pathway is vital, as it dictates the H_2 to CO_2 molar ratio, correlating with the extent of carbonyl formation or C–C bond breakage; a higher number of carbonyl groups leads to a larger H_2 to CO_2 ratio. Additional research has documented the formation of various by-products, such as methane, formaldehyde, or carbon monoxide, each capable of yielding different potential products during glycerol photo-reforming. Concurrently, photocatalytic water splitting may also transpire, adding to the complexity of the process[13, 14].

4.1.2 *Kinetics of Glycerol Valorization*

TiO_2 and Pt/TiO_2 were employed to examine the kinetics of glycerol photo-reforming (as shown in Eq. 4.1), as well as to test the photocatalysts for glycerol photo-oxidation reactions (as described in Eq. 4.2). An analysis of the reaction products and intermediates in both gaseous and liquid states confirms that the processes of oxidation and reforming proceed through the same primary reaction pathways. The initial reactions involve glycerol hydrogenolysis, leading to propylene glycol and the formation of glyceraldehyde through glycerol dehydration. Following this initial dehydration step, subsequent dehydrogenation and decarbonylation reactions give rise to multiple

(a) $C_3H_8O_3 + 3\,H_2O + h\upsilon \geq Eg_{photocatalyst} \rightarrow 3\,CO_2 + 7\,H_2$

(b)

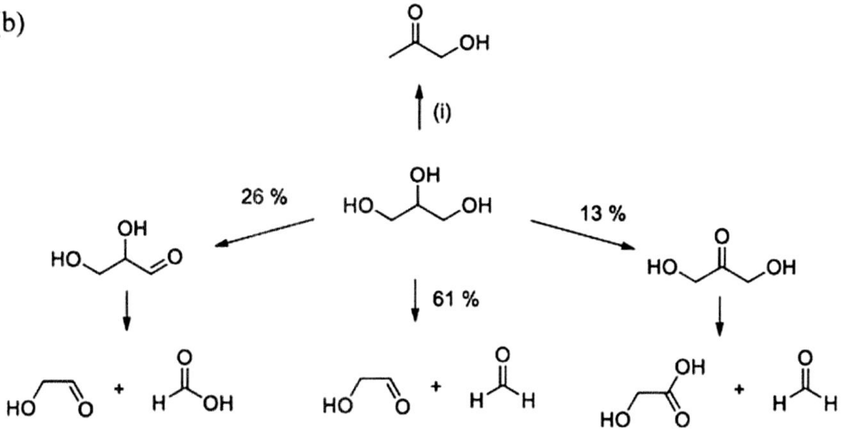

Fig. 4.3 **a** Stoichiometric photo-reforming of glycerol. **b** Estimated initial reaction pathways for the photo-reformation of glycerol over Rh/TiO_2. *Note* A secondary reaction involves light-driven dehydration to hydroxyacetone [14]

intermediates. Under appropriate photo-reforming conditions, these intermediates ultimately yield CO_2 and H_2 gases. The oxidation of glycerol to CO_2 is driven by oxidizing species (H_2O or O_2) in conjunction with the irradiation of the TiO_2 photocatalyst. The overall reaction rate is dependent on the nature of the oxidizing agent and can be enhanced by the inclusion of Pt as a co-catalyst.

$$C_3H_8O_3 + 3H_2O \xrightarrow[\text{Photocatalyst}]{h\nu > E_{bg}} 3CO_2 + 7H_2 \tag{4.1}$$

$$C_3H_8O_3 + \frac{7}{2}O_2 \xrightarrow[\text{Photocatalyst}]{h\nu > E_{bg}} 3CO_2 + 4H_2O \tag{4.2}$$

$$H_2O \xrightarrow[\text{Photocatalyst}]{h\nu > E_{bg}} H_2 + \frac{1}{2}O_2 \tag{4.3}$$

Under the flow of O_2, glycerol is oxidized to CO_2 (as described in Eq. 4.2) through the action of positively generated holes in the valence band of TiO_2, activated by light. Concurrently, the photo-generated electrons reduce oxygen molecules, creating additional oxidizing species (O_2^-) that participate in oxidation reactions. In contrast, the photo-reforming reaction (as shown in Eq. 4.1) takes place in the absence of aeration, where the photo-induced holes oxidize the glycerol, and electrons reduce water molecules to H_2. Under photo-reforming conditions, there is a significantly higher rate of H_2 generation compared to water splitting through a photocatalytic process (as indicated in Eq. 4.3). This enhanced H_2 evolution is attributed to the simultaneous scavenging of photo-generated holes by glycerol, thereby suppressing the recombination rate of the photo-generated species, which otherwise would limit the overall efficiency [15–18].

The rate of photo-oxidation experiences a significant increase when a Pt co-catalyst is dispersed on the surface of the photocatalyst. When aerated, the rate of CO_2 conversion from glycerol is found to be higher, as O_2 acts as a more effective oxidant. This process is further enhanced with the presence of a Pt co-catalyst. The positive influence of Pt coatings on the reaction rate can be attributed to the improved separation of electron–hole pairs. Additionally, the Pt coating promotes the half-reactions occurring at the cathode, which are rate-dependent, including H_2 generation from the photo-reforming of glycerol and O_2 reduction through the photo-oxidation of glycerol [19].

Numerous characteristics of the catalyst, such as its crystalline nature, surface area, morphology, hydroxyl groups, substrate, and ion adsorption, contribute to the complexity of the photocatalytic process in terms of both activity and selectivity. The kinetics of the process are further influenced by substrate concentration, owing to the complex dynamic interplay between substrate oxidation and carrier recombination (back-reactions) mediated by the substrate [20, 21]. Existing kinetic models, whether based on Langmuir–Hinshelwood assumptions or those considering only a single active site, fail to elucidate how glycerol concentration affects the reaction rate [22, 23]. This limitation is particularly pronounced in models that neglect reverse reactions, such as charge carrier recombination mediated by the substrate, which is considered essential for accurate rate interpretation. The works of Lewandowski and Ollis (2003) highlighted the need for a dual-site model to explain the incomplete deactivation observed during the degradation of gas-phase aromatics and alcohols by TiO_2-P25 photocatalyst [24, 25]. However, these works did not provide specific details about the nature of these sites, merely suggesting the existence of a Type I site suitable for adsorbing substrates, water, and reaction intermediates, and a Type II site characterized by its hydrophilic nature due to bridging oxygen or hydroxyl groups and its inability to adsorb substrates. From the evidence provided, it can be deduced that the two catalysts possess unique surface sites that substantially influence their kinetic behavior. Glycerol, with its distinct molecular structure containing two identical lateral hydroxyl groups, undergoes oxidation to form glyceraldehyde. Oxidation of the central hydroxyl group can produce either dihydroxyacetone (a C_3 compound) or break down into two separate components, formaldehyde and glycolaldehyde (C_1 + C_2 compounds).

The experimental time profiles, which measured the disappearance of glycerol and the production of dihydroxyacetone and glyceraldehyde when starting with these pure compounds, exhibited pseudo-first-order kinetics for at least two half-lives. However, to analyze the kinetics through initial conversion rates, the fitting was confined to less than one half-life. The exponentially fitted decay function or first-order kinetic fitting was used exclusively in the raw data analysis to obtain a more accurate assessment of initial rates, without making any implications about the overall kinetics. It is essential to recognize that product degradation and changes in the kinetic system, which alter the time profiles at higher conversions, are observed at low substrate concentrations [26].

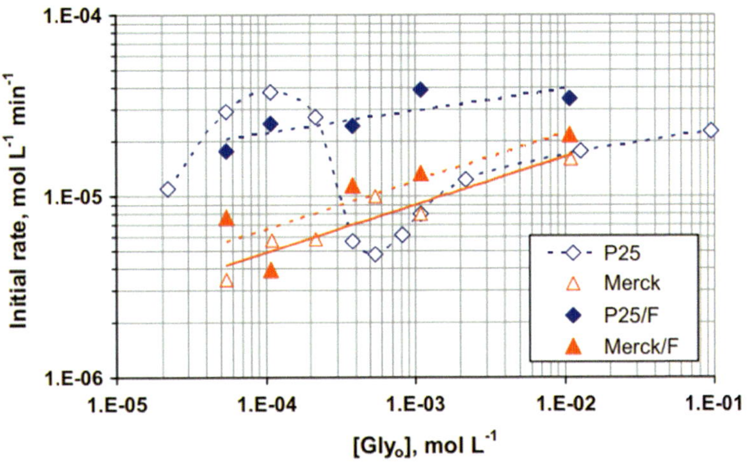

Fig. 4.4 Illustration of the impact of pH values (pH = 5.5 for pristine samples and pH = 3.0 for fluorinated samples at 10 mM fluoride concentration) on the TiO_2 catalyst (at a concentration of 0.5 g L^{-1}) in relation to the photocatalytic transformation rates of glycerol as a function of glycerol concentration [26] (Copyright 2012, Applied Catalysis B: Environmental License Number 5512870456259)

Maurino et al. (2008) partially documented the rate of glycerol's disappearance, and this information was integrated with new data to construct Fig. 4.4, presented on a logarithmic scale for comparison with subsequent kinetic analyses [22]. It is important to note that, despite the pristine and fluorinated samples being at different pH levels, the influence of pH is minimal at acidic levels, with P25 experiencing only a 13% rate decrease when transitioning from pH 3.0 to 5.5 [22]. The primary observation is P25's unusual dependence on substrate concentration for its reaction rate, along with the negligible effect of Merck TiO_2 fluorination. Interestingly, even though P25 has a larger surface area, its rate is comparable to that of Merck TiO_2, following a sharp decrease at glycerol concentrations of 0.4–0.5 mM.

C3 carbohydrates, specifically glyceraldehyde (GAD) and dihydroxyacetone (DHA), are identified as the primary products of glycerol (GLY) phototransformation over TiO_2 Merck, produced in varying ratios ranging from 1.3 to 1.8. Elevated glycerol concentrations lead to additional formation of formaldehyde (FORM) and glycolaldehyde (GLC), with reduced levels of DHA. The decomposition of DHA or a concurrent mechanism may result in FORM production, a pattern that P25 also exhibits even at low concentrations. Conversely, the primary products derived from glycerol on TiO_2 P25 depend on the glycerol concentration. At low glycerol concentrations, the relative ratio of GAD to DHA is approximately 2, with FORM and GLC also detected. As the concentration of glycerol increases, the reaction rate is significantly inhibited, resulting in FORM and GLC becoming the primary products. GAD and DHA are present in substantial quantities before the rate decline, observed at lower glycerol concentrations between 0.02 and 0.1 mM, with only minor amounts of GAD detected at higher concentrations. When fluoride ions are introduced to the slurry, the evolution of intermediates is observed. In the presence of these ions, the primary products derived from glycerol are predominantly GAD and DHA, both of which are C3 carbohydrates. The presence of fluoride leads to minimal changes in the production of GAD and DHA on Merck TiO_2, while the generation of GLC and FORM over P25 is notably reduced. This phenomenon is likely attributable to the fluoride ions acting as effective complexing agents for Ti(IV) cations, inhibiting the adsorption of substrates and consequently altering the oxidative mechanisms and reaction kinetics [27, 28]. Previous studies have demonstrated that redox-inactive fluoride anions on the TiO_2 surface can enhance the transformation rate of organic substrates, especially those reacting through an ·OH-mediated pathway. The bell-shaped pH dependence of this effect suggests that it mirrors the distribution of the ≡Ti-F complex formed via surface complexation [29, 30].

The interactions between fluoride ions and the surface complexation at the ≡Ti–OH sites are directly related to the chemisorbed amount of GLY that forms GLC and FORM. The specific products observed during glycerol oxidation are determined by the particular species responsible for oxidation within the mechanism. These various pathways stem from the ability of the oxidizing agent to facilitate single or multiple electron transfers or to extract hydrogen atoms. Regarding the reactivity of the ·OH radicals, the reported kinetic constants fall within the range typical of diffusion-limited reactions ($k = 2.0 \times 10^9$ L mol^{-1}) [26, 31]. The products resulting from oxidation over Merck TiO_2 are likely produced through a mechanism mediated by ·OH-like species, whereas those observed on P25, following the rate decline depicted in Fig. 4.4, appear to result from a direct two-electron transfer pathway. It is evident that a single mechanism is at work on Merck TiO_2, while two simultaneous mechanisms occur over P25. The possibility of two distinct types of sites being present on P25 cannot be excluded. A model that is based on a specific reaction network for GLY is presented in Fig. 4.5.

Figure 4.6 illustrates the kinetic framework that elucidates the kinetics of glycerol transformation. Sites labeled as R, which neither participate in complex formation nor function as hole traps (meaning they do not oxidize), are considered separately from sites labeled as A. By assuming equilibrium constants similar to K1 and K2, it is possible to both retain specific physical information about the four sites in case A, and to categorize them into two pairs, a and b, that exhibit distinct kinetic behavior. The recombination constants for paths (a) and (b) can be expressed as $k_{ARa} = (k_{ARa1} + K_1 \cdot k_{ARa2})(1 + K_1)$ and $k_{ARb} = (k_{ARb1} + K_2 \cdot k_{ARb2})(1 + K_2)$ [26]. In these equations, k_{ARai} and k_{ARbi} represent the kinetic reactivity constants of the limiting factors, which are assumed to be in "equilibrium" as characterized by K_1 and K_2, respectively. A parallel approach is applied to the other pathways that involve oxidation processes, where $k_{AOa} = (k_{AOa1} + K_1 \cdot k_{AOa2})(1 + K_1)$ and $k_{AOb} = (k_{AOb1} + k_{AOb2} \cdot K_2)(1 + K_2)$. The notation can be succinctly expressed as $[(Ti(\cdot O)(OH))_A] + [(Ti(OH)(O\cdot))_A] = A_{ox}$, and $[(Ti(^{+}\cdot OR)(OH))_A] + [(Ti(OR)(\cdot O))_A] = C_{ox}$ (where C_{ox} represents a chemically adsorbed complex) [26]. The originating sites are denoted as $[(Ti(OH)(OH))_A] = A_o$ and $[(Ti(OR)(OH))_A] = C_o$ [32]. The equilibrium state of C_o with the A_o sites is represented by the equation $C_o = A_o \cdot K_{ads}$ [33]. When integrated with the site number balance, this equation leads to an implied Langmuir isotherm,

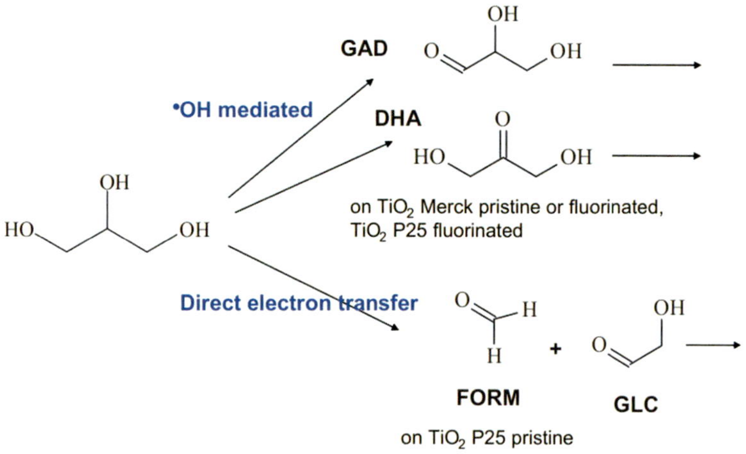

Fig. 4.5 Kinetic pathways for transformation of glycerol [26] (Copyright 2012, Applied Catalysis B: Environmental License Number 5512870456259)

where $C_A = A_o + C_o$ within the system. This notation can also be applied to the sites B, which do not participate in adsorption, represented as $C_B = [(Ti(‡)(OH))_B]$, $B_{ox} = [(Ti(‡)(O))_B]$. Similarly, the notation is relevant to the reductive sites R, denoted as $C_R = [(Ti(O)(O))_R]$, $R_R = [(Ti^{-}(\cdot O)(O))_R]$ [32].

$$C_{sa} = C_A + A_{ox} + C_{ox} \quad (C_{sa} \text{ represents the balance number of all A sites}) \tag{4.4}$$

$$C_{sb} = C_B + B_{ox} \quad (C_{sb} \text{ represents the balance number of all B sites}) \tag{4.5}$$

$$C_{sr} = C_R + R_R \quad (C_{sr} \text{ denotes the balance number of all reducing sites}) \tag{4.6}$$

The rate of disappearance for the compound can be formulated as[33]:

$$\text{Rate}_{HOR} = k_{AOa} \cdot A_{ox} + k_{BO} \cdot B_{ox} + (w_2[h] + w_3)C_{ox} \tag{4.7}$$

where w_2 is defined as $k_{AOb2} \cdot K_2/(1 + K_2)$, w_3 is given by $k_{AOb1}/(1 + K_2)$, and [h] signifies the concentration of free holes [32].

Under the assumption of a stationary state, the following simplifications can be made:

a. The nonexistence of pathway (b) for site A, indicating the absence of chemisorption on the surface, corresponds to the state of P25 prior to the rate drop. Additionally, it is assumed that $k_{eR} = 0$ and $R_R \ll C_{sr}$, suggesting that the adsorption of oxygen primarily governs the rate of electron scavenging.

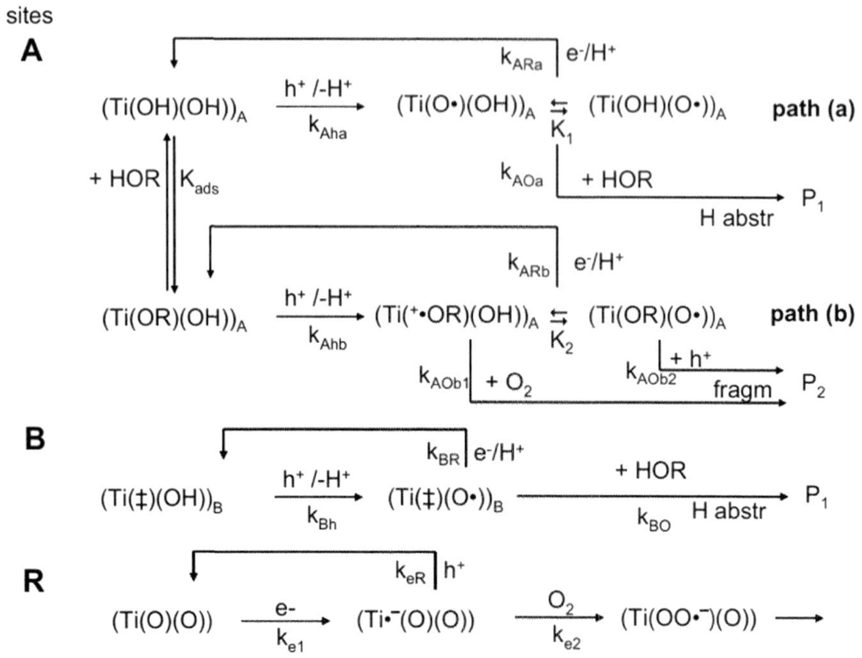

Fig. 4.6 The scheme illustrates potential kinetic pathways for glycerol conversion over active sites A and B, along with spontaneous electron scavenging at the reducing sites R. On site A, two distinct pathways, labeled (a) and (b), may transpire, reflecting the substrate's ability to chemisorb on the surface. In this context, P_1 denotes dihydroxyacetone (DHA) and glyceraldehyde (GAD), while P_2 signifies glycolaldehyde (GLC) and formaldehyde (HCHO) [26] (Copyright 2012, Applied Catalysis B: Environmental License Number 5512870456259)

b. The nonexistence of pathway (a) for site A, indicating the presence of chemisorption on the surface, is clearly related to the P25 rate immediately after the decline. Additionally, it is assumed that $k_{eR} = 0$ and $R_R \ll C_R$.
c. The analysis focuses solely on the pathways occurring on site B, consistent with the rate behavior observed in both fluorinated and pristine Merck TiO_2. Similarly, when the surface of P25 is completely complexed, only B sites are active. This situation is analogous to the one at higher concentrations of GLY, where the rate becomes similar to that of case (b) mentioned earlier. In this context, it is assumed that $k_{eR} = 0$ and that $R_R \ll C_R$.

In conclusion, various TiO_2 specimens demonstrate distinct selectivity in the transformation of glycerol. Glycerol, a symmetric molecule, can undergo hydrogen abstraction to produce oxidized C_3 compounds or directly transfer electrons to proceed through β-fragmentation into C_1 and C_2 compounds. The initial pathway of the reaction occurs when glycerol is adsorbed on surface traps, interacting with these shallow sites. A second pathway emerges from the chemisorption of molecules on the outer layer. Since the reaction rate for chemisorption is slower than that for hydrogen abstraction, the surface's complexity acts as highly effective traps for photo-generated holes, leading to an increase in the back reaction with free electrons. By employing these fundamental principles, the experimental production of various products can be systematically explained. Therefore, due to its unique properties, glycerol can serve as an insightful probe for investigating the surface characteristics of the TiO_2 photocatalyst semiconductor through chemical analysis, as it can operate either by direct hole transfer or through an ·OH radical mechanism. Although the initial electron transfer process within the inner sphere is rapid, the case of glycerol illustrates that this may not necessarily enhance photocatalytic efficiency [26].

4.1.3 Glycerol Photo-Conversion Parameters

The production of biodiesel as an alternative to fossil fuels results in a substantial amount of glycerol as a by-product. The ongoing transformation of glycerol is a subject of development both in academia and industry. Historically, research on glycerol was conducted within the framework of traditional principles, including protection and deprotection strategies, functionalization, and fundamental chemistry [36]. Through chemical conversion, glycerol can be photocatalytically transformed into other valuable products, as illustrated in Fig. 4.7.

Fig. 4.7 Parametric analysis of photocatalytic glycerol conversion [34] (Copyright 2016, Reaction Chemistry & Engineering License Number 1353215)

4.1.3.1 Preparation of Photocatalysts

A variety of semiconductor catalyst materials exist for the valorization of glycerol. Notable examples include TiO_2, MOF, WO_3, and Bi_2WO_6. These catalysts are comprehensively discussed in the section entitled "Semiconducting photocatalysts for glycerol valorization."

4.1.3.2 Source of Light

The existing literature indicates that titanium dioxide (TiO_2) serves as a highly efficient photocatalyst under ultraviolet (UV) light conditions. However, its efficacy is constrained by factors such as a wide energy bandgap (~3.2 eV), limited surface area, and accelerated electron–hole recombination [35]. During photonic excitation exceeding the material's bandgap energy (Eg), electrons transition from the valence band to the conduction band [33]. Generally, glycerol conversion rates are elevated under UV light compared to visible light, given that photocatalysts typically manifest greater activity under UV illumination [34]. Titanium dioxide photocatalysts encounter limitations concerning visible light applications due to two primary factors: (1) swift electron–hole recombination, and (2) preferential absorption under UV light conditions. A plethora of annual publications elucidate these constraints of TiO_2 when exposed to visible light [33]. To mitigate these limitations, ongoing research is focused on modifying semiconductor materials to broaden their light absorption spectrum. Furthermore, the concentration of irradiated photons, which significantly impacts photocatalytic reactions, can be increased by tailored light-focusing techniques [36].

4.1.3.3 Continuous Flow Conditions

The thermal dehydration of glycerol, particularly in the presence of acid catalysts at elevated temperatures, is widely acknowledged to produce a diverse range of products, including acrolein and syngas. Recent studies have demonstrated that zirconium phosphate can serve as an effective catalyst for acrolein generation. The experimental setup entails the concurrent injection of a 10 wt.% glycerol aqueous solution and hydrogen gas at 2 MPa, with respective flow rates of 0.04 and 30 mL/min. The catalytic reaction occurs at a temperature of 315 °C, leading to complete glycerol conversion into acrolein with a high selectivity of up to 82%. Only minor quantities of acetol (11%) were observed, alongside trace amounts of by-products such as propan-1-ol and propan-1,2-diol. Novel methodologies have been explored for the selective ketalization of glycerol to produce solketal and its derivatives. These developments aim to create protected forms of glycerol and oxygen-containing fuel additives. A comprehensive review focused on environmentally sustainable solketal production using both continuous and batch flow reactors. Organic resin-based heterogeneous catalysts, namely Amberlyst and Purolite, were employed in the ketalization process.

A glycerol and acetone mixture (1:5, v/v) was introduced into an acid-catalyst-filled cartridge (0.77 g) at a flow rate of 0.1 mL/min, 20 °C, and 120 bar pressure, leading to solketal production with a yield of 95% and selectivity of 100%. Enhancements in yield and selectivity may be achieved by increasing both pressure and acetone quantity. Glycerol esterification under acidic conditions presents challenges due to the efficacy of triacetin as an anti-knocking agent in bio-additives [37].

4.2 Semiconducting Photocatalysts for the Valorization of Glycerol

4.2.1 TiO_2

Titanium dioxide (TiO_2) is a semiconductor photocatalyst that has been extensively studied for the valorization of glycerol, due to its advantageous characteristics such as non-toxicity, high stability, low cost, and broad commercial availability [38, 39]. Figure 4.8 illustrates the photon-activated oxidation/reduction processes in TiO_2 particles. Upon photon absorption, TiO_2 undergoes electronic excitation, wherein electrons in the valence band acquire sufficient energy to transition to the conduction band. These electron–hole pairs subsequently migrate to the semiconductor's surface, where they engage in photocatalytic reactions through either the reduction or oxidation of adsorbed organic species [40]. Nonetheless, TiO_2's large energy band gap, high rate of electron–hole pair recombination, and limited surface area constrain its photocatalytic efficiency [41]. Both the rutile and anatase forms of TiO_2 possess wide band gaps, with energy levels of 3.0 eV and 3.2 eV, respectively [38]. This wide band gap restricts TiO_2's light absorption to the ultraviolet region, which constitutes only 3–5% of the solar spectrum [38, 42]. In contrast, visible light makes up approximately 45% of the solar spectrum [43].

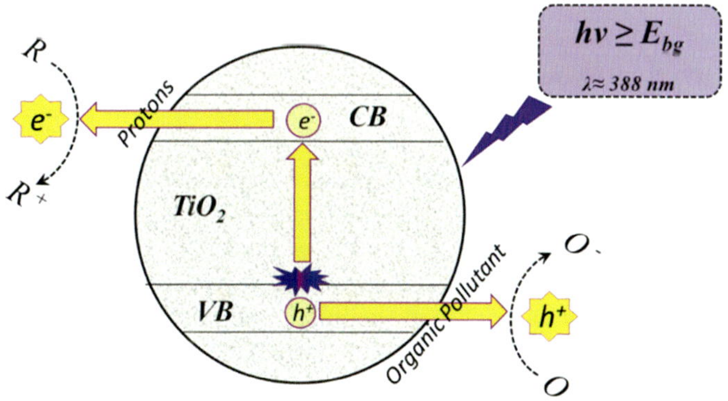

Fig. 4.8 Photon-activated oxidation/reduction processes in TiO_2 particles [40]

Various forms of TiO_2, including rutile, anatase, and brookite, exhibit distinct characteristics that garner significant attention [39]. Compared to the rutile phase, the anatase phase possesses a larger bandgap energy, elevating both the Fermi level and the conduction band edge. Moreover, the lower density of anatase leads to greater Ti–Ti distances, which in turn causes the 3d states to be more localized. Generally, anatase features a larger surface area than rutile, facilitating molecular adsorption due to an abundance of active sites. With its smaller crystal size, anatase enhances the efficiency of electron–hole pair generation, while rutile's higher crystallinity improves carrier mobility [39].

Brookite, another polymorph of TiO_2, also displays photocatalytic properties. Specifically, brookite-based materials used in hydrogen production in aqueous solutions containing ethanol or methanol offer unique advantages, attributed to its larger band gap (3.3–3.4 eV) relative to anatase [44]. When assessing hydrogen production rates in relation to the catalyst surface area, a higher proportion of brookite corresponded to an elevated rate of hydrogen production from glycerol. Conversely, when normalized to the weight of the catalyst, increased proportions of anatase led to higher hydrogen production rates [39]. Mixed-phase TiO_2, containing both anatase and rutile, presents distinct characteristics, such as enhanced transfer of photo-generated species to the surface, thereby improving photocatalytic efficacy.

Preparation of TiO_2

Sol–gel and hydrothermal techniques are predominantly employed for synthesizing precursor materials to achieve specific photocatalytic characteristics. The sol–gel method is renowned for its capacity to produce porous materials through solution-based solidification. In contrast, hydrothermal synthesis is more frequently employed in the fabrication of photocatalysts, owing to its advantages in enhanced crystallization and crystal composition control. Beyond the realm of glycerol conversion, several studies have investigated the formation of TiO_2 in various morphologies, such as nanotubes, nanosheets, hollow spheres, and nanorods. TiO_2 nanotubes, for instance, can be fabricated through the hydrothermal treatment of anatase-form TiO_2 or commercial P25 [39].

4.2.1.1 TiO_2 with Deposited Co-Catalyst

Several approaches exist for augmenting the performance of TiO_2-based materials. Techniques such as elemental doping or composite formation with other semiconductors or metallic nanoparticles aim to diminish the recombination rate of photo-generated electron–hole pairs (e^- and h^+) while broadening the spectrum of visible light responsiveness [44]. Non-metal element doping, specifically with carbon, has proven effective in enhancing TiO_2's absorption capacity in the visible light spectrum. Compared to metal-doped or dye-sensitized TiO_2, carbon-doped TiO_2 demonstrates superior photo-corrosion resistance, thermal stability, and longevity. The incorporation of carbon into the TiO_2 structure enhances its photocatalytic efficiency by

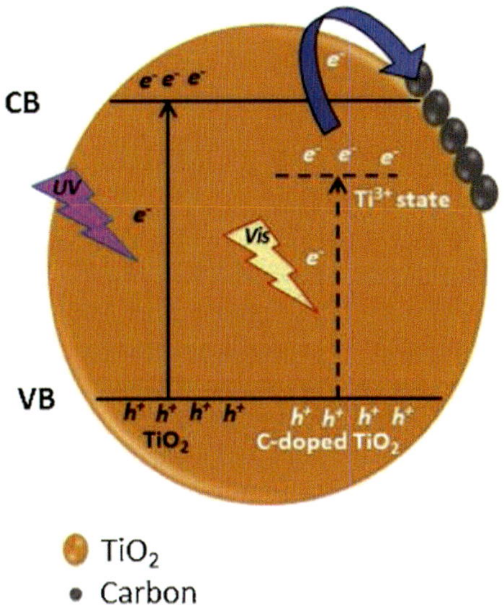

Fig. 4.9 Mechanism of carbon-doped TiO_2 in photocatalytic transformation under visible light [41] (Copyright 2020, Applied Catalysis A: General License Number 5525710530339)

facilitating charge transfer, mitigating the recombination of photo-generated electron–hole pairs, and augmenting electron trapping capabilities [41]. Although noble metals serve as effective co-catalysts in enhancing TiO_2's photocatalytic activity, their application is restricted due to their elevated cost [39]. Figure 4.9 delineates the mechanism of carbon-doped TiO_2 during photocatalytic activation under visible light. Carbon serves as an electron acceptor within the TiO_2 structure, facilitating charge transfer while preventing electron–hole recombination. The presence of carbon atoms and Ti^{3+} ions within the TiO_2 crystalline structure functions as a trapping mechanism, creating intermediate energy states between the valence and conduction bands, thereby narrowing TiO_2's band gap [41].

Metal-doped TiO_2

Doped materials commonly encounter challenges such as thermal instability, photo-corrosion, and elevated rates of electron–hole recombination. Gold (Au), a noble metal, is resistant to corrosion during photo-reactions and exhibits unique localized surface plasmon resonance properties, enabling interaction with light across both visible and infrared spectra. Notably, the Fermi level of Au is relatively lower than that of TiO_2. When exposed to ultraviolet radiation, excited electrons in TiO_2's conduction band can readily migrate to Au's surface, thereby reducing the recombination rate of electron–hole pairs. The integration of Au nanoparticles on TiO_2's surface not only activates photocatalytic behavior under visible light but also enhances its performance under ultraviolet illumination [43]. Similarly, the incorporation of platinum (Pt) nanoparticles as a co-catalyst on TiO_2 significantly elevates the rate of hydrogen production, attributable to the reduced electron–hole recombination. This process enables the production of valuable liquid-phase substances, including glyceraldehyde and glycolaldehyde [45]. Among the co-catalysts studied for glycerol valorization, copper (Cu) has been the focus of extensive investigation [39]. Table 4.1 presents a summary of existing literature on the selective oxidation of glycerol through photocatalysis, utilizing both pure TiO_2 and metal-doped TiO_2 catalysts.

4.2.1.2 Pt/TiO_2

The addition of platinum (Pt) to TiO_2 is widely regarded as enhancing photocatalytic performance. This improvement is attributed to the formation of a Schottky barrier at the metal–semiconductor interface, which traps photo-generated electrons and minimizes electron–hole recombination. This trapping mechanism contributes to the greater efficiency observed in Pt/TiO_2 catalysts during glycerol oxidation compared to using TiO_2 alone. The separation of charge carriers is improved, accelerating the reaction rate [19]. The study focused on photo-reforming glycerol in water using Pt/TiO_2 catalysts and simulated solar light at ambient conditions. Various factors, such as glycerol and catalyst concentration, Pt loading, temperature, and pH, were examined for their impact on the reaction rate. Intermediate products, specifically acetic acid and methanol, were observed during the photo-reformation of glycerol. Ultimate conversion yielded hydrogen (H_2) and carbon dioxide (CO_2). Optimal performance was observed with the addition of 0.1–0.5 wt.% Pt to the TiO_2 catalyst; concentrations of Pt higher than this range had a detrimental effect. The photocatalytic process demonstrated superior efficiency in neutral and basic media compared to acidic conditions. Elevated glycerol concentrations (1 mol L^{-1}) significantly expedited the rate of H_2 evolution relative to the rate in pure aqueous medium. Furthermore, a substantial increase in reaction kinetics was recorded when the temperature was elevated from 40 °C to a range of 60–80 °C, as documented in Fig. 4.10 [18].

4.2.1.3 Cu/TiO_2

On the surface of titanium dioxide (TiO_2), copper oxide nanoparticles are uniformly distributed, forming a Schottky barrier [51]. Copper oxides such as Cu_2O and CuO are gaining increasing attention as photocatalysts. These oxides possess band gaps of 2.1 eV and 1.2 eV, respectively, and act as p-type semiconductors. Their near non-toxicity makes them preferable. However, their instability in electrolytic solutions has been demonstrated due to their susceptibility to photo-oxidation [52]. Cu-doped TiO_2 samples were prepared with varying Cu concentrations (0.2, 0.3, 0.5, 1, 2, and 5) using the impregnation method. The samples underwent calcination at either 350 °C or 450 °C for five hours. Copper ion incorporation allows TiO_2 to extend its light absorption into the visible spectrum. Transmission electron microscopy (TEM) images reveal a fine distribution of copper on the TiO_2 surface (Fig. 4.11). When glycerol and water are mixed, hydrogen production significantly increases. CuO/TiO_2 and Cu_2O/TiO_2 catalysts, with a weight ratio of 0.5%, display the highest hydrogen production rates, attaining 16,500 and 20,060 μmol h^{-1}, respectively, in 5% aqueous glycerol solutions. Cu_2O/TiO_2 maintains its activity and stability under solar

Table 4.1 A summary of existing literature on the selective oxidation of glycerol through photocatalysis, utilizing both pure TiO_2 and metal-doped TiO_2 catalysts [46]

Reaction catalyst		Pt/TiO_2	$Ag\text{-}AgBr/TiO_2$	TiO_2	(Bi, Pd, Pt, Au)/TiO_2	Au/TiO_2
Reaction condition	Catalyst	0.5 g/L	1 g/L	3 g/L	3 g/L	1 g/L
	Glycerol	1 M	4.5 mL	0.3 M	0.3 M	0.05 M
	Temperature	30 ∘C	25 ∘C	25 ∘C	25 ∘C	90 ∘C
	λ	Simulated sunlight	Simulated sunlight	100–600 nm	200–600 nm	> 420 nm
	Reaction time	6 h	4 h	24 h	14 h	5 h
	Reactor type	Batch				
Reaction Products		Glycolaldehyde Glyceraldehyde	Glyceraldehyde Dihydroxyacetone Glycolic acid Glyceric acid	Glycolaldehyde Formic acid Glyceraldehyde Dihydroxyaceton Glycolic acid Hydroxypyruvic acid	Glyceraldehyde Formic acid Hydroxypyruvic acid Dihydroxyacetone Glycolic acid	Glyceraldehyde Dihydroxyacetone Glycolic acid Oxalic acid
Selectivity principal product		Glycolaldehyde	Glyceraldehyde	Glycolaldehyde	Glyceraldehyde	Dihydroxyacetone
Reference		[47]	[3]	[48]	[49]	[50]

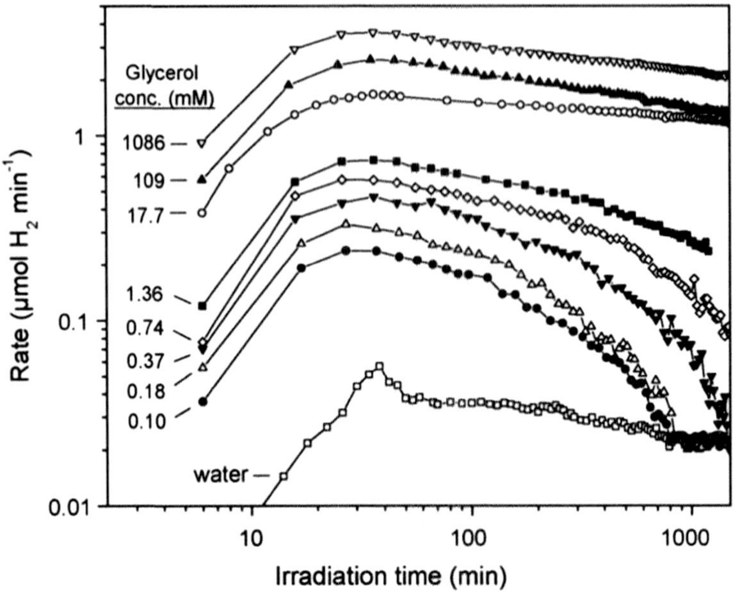

Fig. 4.10 Hydrogen evolution rate as a function of irradiation time for photocatalyst suspensions at indicated glycerol concentrations[18] (Copyright 2009, Catalysis Today License Number 5543741407332)

irradiation. In summary, copper ions in the +1 oxidation state enhance sustainable and stable hydrogen production when interacting with the TiO_2 surface under solar exposure [53].

The performance of Pt/TiO_2 and CuO_x/TiO_2 photocatalysts was assessed for hydrogen production from a glycerol water solution under UV light exposure

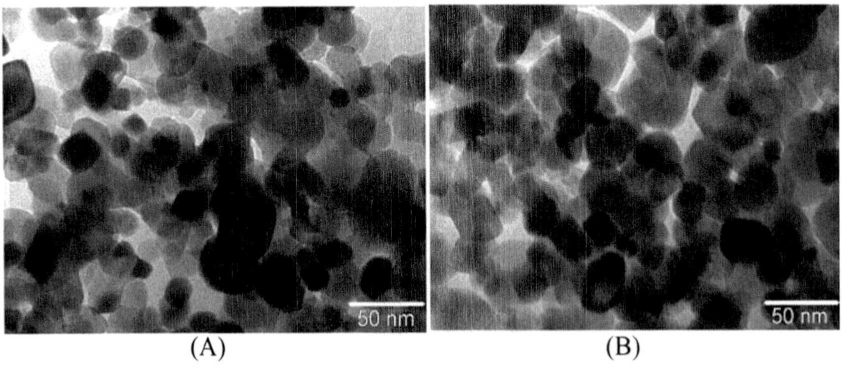

Fig. 4.11 TEM images of CT-2 catalysts with a weight percentage of 0.5%, showing **a** the calcined form and **b** the used form [53]. (Copyright 2010 American Chemical Society)

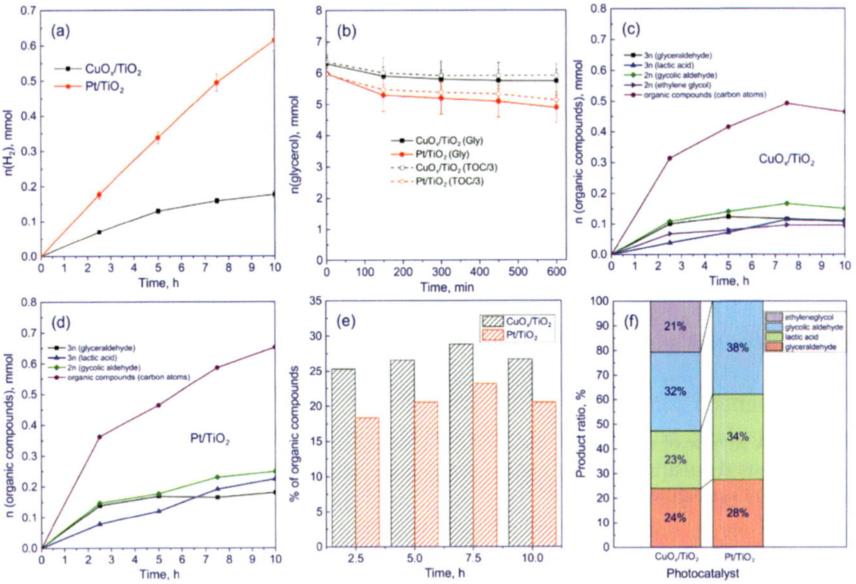

Fig. 4.12 Hydrogen production (**a**) and glycerol breakdown (**b**) in addition to the formation of organic compounds (**c**, **d**) with the proportion of organic compounds (**e**) and the distribution of these organic products (**f**) during the photo-reforming of glycerol using CuO_x/TiO_2 and Pt/TiO_2 catalysts [51] (Copyright 2021, Materials Letters License Number 5543750076339)

(Fig. 4.12). Results suggest that the Pt/TiO_2 catalyst shows superior activity and stability, while the use of CuO_x/TiO_2 broadens the range of products derived from glycerol and promotes a higher yield of organic products [51].

The exploration of bimetallic Cu-Pt coatings on TiO_2 surfaces is attracting substantial attention due to its capability to enhance the separation of photo-generated electron–hole pairs. The effectiveness of TiO_2 embedded with bimetallic Cu-Pt deposits was assessed in the context of H_2 production via glycerol photo-reforming [54]. This assessment was performed under UV–Visible light exposure, using a mixture of glycerol and water. The outcomes showed that the Cu-Pt/TiO_2 catalyst led to the highest H_2 production rate, followed by Pt/TiO_2, Cu-TiO_2, and unmodified TiO_2 in a descending sequence [55].

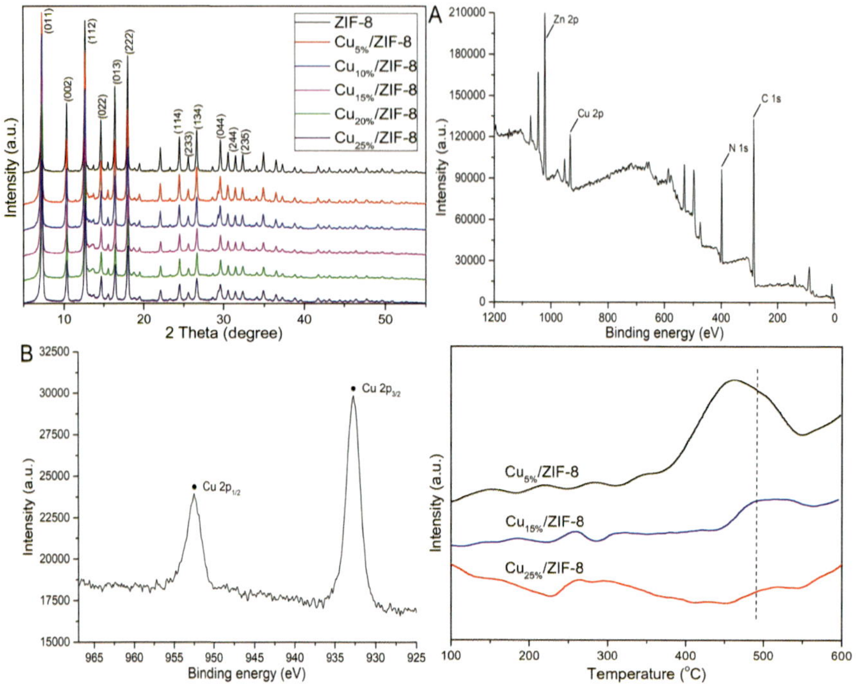

Fig. 4.14 X-ray diffraction (XRD) profiles for ZIF-8 and Cu/ZIF-8 with varying copper load percentages, X-ray photoelectron spectroscopy (XPS) spectra for the Cu/ZIF-8 catalyst (**a**), and the Cu 2p orbital (**b**) specific to Cu/ZIF-8. Additionally, CO_2 temperature-programmed desorption (CO_2-TPD) profiles for catalysts Cu5%/ZIF-8, Cu15%/ZIF-8, and Cu25%/ZIF-8 are presented (**c**) [8] (copyright 2021 Catalysis Letters License Number 5525360017468)

of nanostructured oxides via strategies such as recombination, doping, textural modification, and synergistic coupling with other photocatalysts to enhance performance. In this research, a facile solution-based synthesis technique is employed for the initial fabrication of CO^{2+}-ZIF-8 metal–organic frameworks (MOFs), which subsequently undergo calcination to yield CO_3O_4/ZnO p-n nanoheterojunctions (Fig. 4.16). The resulting architectures demonstrate superior surface characteristics in comparison to those documented in existing literature. Moreover, the integration of CO_3O_4 within the ZnO matrix in proportions ranging from 1.0 to 4.0 wt.% facilitates enhanced absorption across the visible light spectrum by reducing the bandgap energy (E_g). Concomitantly, the CO_3O_4/ZnO p-n nanoheterojunctions manifests superior photoinduced charge separation and mobility, surpassing the performance metrics of pure ZnO derivatives [60, 65].

A facile synthesis method, comprising solution processing and subsequent calcination, was employed to fabricate highly active p-n CO_3O_4/ZnO nanoheterojunctions. These were derived from ZIF-8 metal–organic frameworks by the in-situ

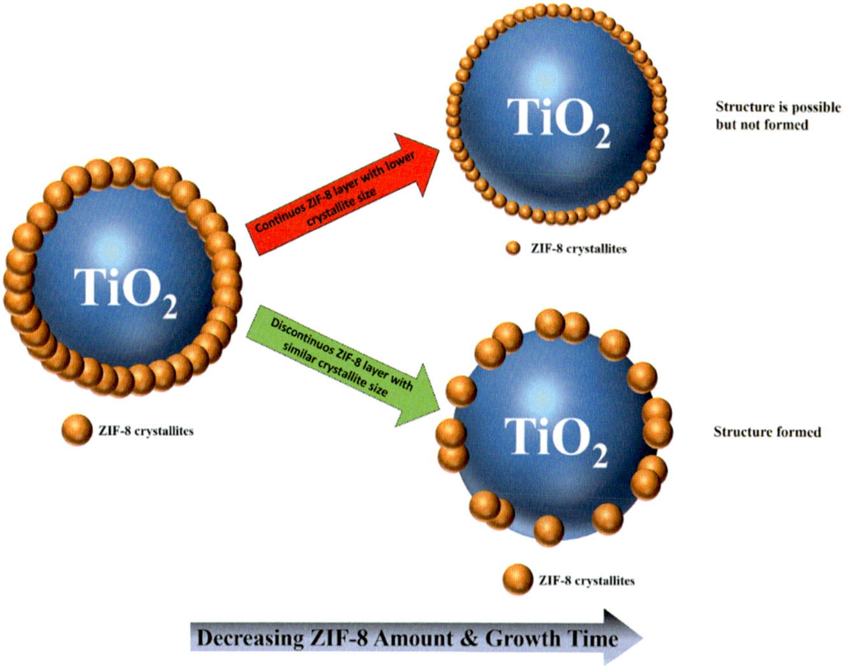

Fig. 4.15 Schematic illustration of the synthesis of ZIF-8/Cu-TiO_2 nanocomposite at low ZIF-8 concentrations [64] (copyright 2021 Colloids and Surfaces A: Physicochemical and Engineering Aspects License Number 5525360205113)

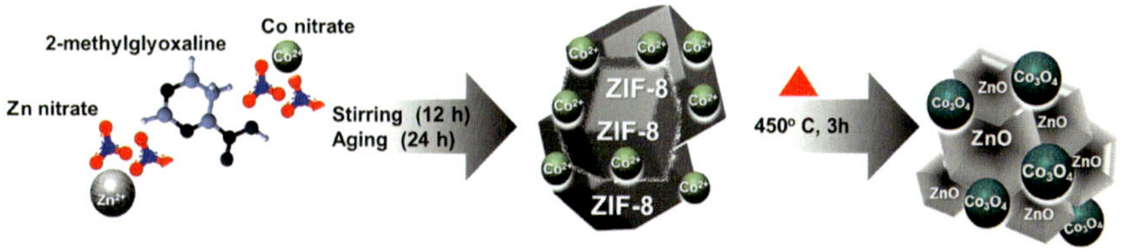

Fig. 4.16 Illustration of the transformation of CO^{2+}-ZIF-8 into CO_3O_4/ZnO nanoheterojunctions [65] (copyright 2023 Optical Materials License Number 5538111260104)

incorporation of CO^{2+} precursors. The resultant nanostructures exhibited an expansive specific surface area, ranging between 1571 and 1691 m^2/g, as well as significant textural attributes. Furthermore, these structures demonstrated a responsivity to visible light, attributable to a reduction in bandgap energy. Upon the incorporation of a 3% weight concentration of CO_3O_4 into ZnO, the rate of photocatalytic hydrogen evolution experienced a 1834-fold increase compared to that of unmodified ZnO, when subjected to conditions featuring a 10% glycerol/water mixture. The optimized 3% CO_3O_4/ZnO composition maintained efficient and sustainable hydrogen production across five consecutive cycles. The enhanced photocatalytic performance of the

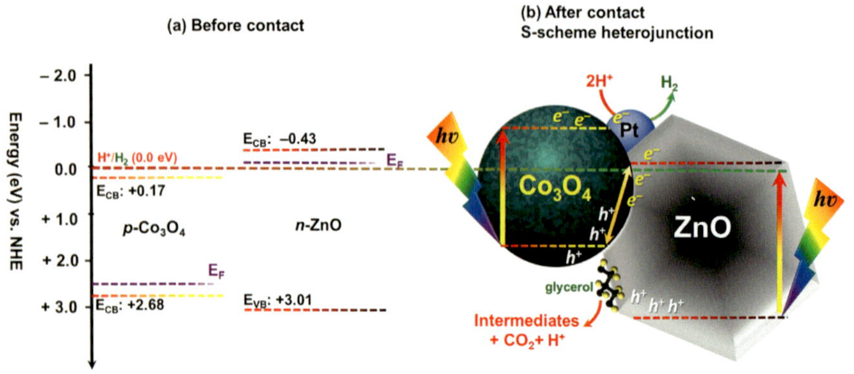

Fig. 4.17 The positions of the bands for p-Co_3O_4 and n-ZnO are shown both prior to (**a**) and following **b** the junction interaction. Figure **b** also includes a schematic that elucidates the generation of hydrogen via photocatalytic processes [65] (copyright 2023 Optical Materials License Number 5538111260104)

CO_3O_4/ZnO nanoheterojunctions was ascribed to the formation of a p-n interface, facilitating effective photo-generated carrier separation and migration, coupled with advantageous surface morphology and a suppressed bandgap energy. This investigation emphasizes the potential of metal–organic frameworks serving as both templates and growth agents for the derivation of oxide p-n heterojunctions, specifically for applications in clean energy production under visible-light conditions (Fig. 4.17) [60, 65].

Fabrication of CO_3O_4/ZnO nanoheterojunctions via synthetic procedure

The synthesis commenced with the amalgamation of 2.0 mmol $Zn(NO_3)_2{\cdot}6H_2O$ and 12.0 mmol 2-methylglyoxaline in 100 mL of methanol, subjected to vigorous stirring for a 12-h period. Following this, the resultant homogenous mixture was left to stand at ambient conditions for an additional 24 h. The precipitate, obtained post-centrifugation, was sequentially washed twice with ethanol and subsequently dried at 70 °C for a duration of 2 h, yielding ZIF-8. During the ZIF-8 formation process, incremental quantities of $Co(NO_3)_2{\cdot}6H_2O$ were introduced into the $Zn(NO_3)_2{\cdot}6H_2O$ mixture. The $Co(NO_3)_2{\cdot}6H_2O$ ratios were varied from 0.01 to 0.04 in the presence of 2-methylglyoxaline while stirring. Subsequently, the dried samples were blended with either 0.01 mmol or 0.04 mmol of $Co(NO_3)_2{\cdot}6H_2O$ individually. The final samples underwent an annealing process in an air atmosphere at 450 °C for 3 h to generate the respective nanoheterojunctions [60, 65].

4.2.3 WO_3

Semiconductor photocatalysts have garnered attention for their potential in environmental remediation and solar energy conversion, owing to their extensive surface area, tunable band gap, and robust photoactivity. In the realm of photocatalysis, semiconductor materials exhibit the propensity for photon absorption, which leads to the formation of electron–hole pairs capable of initiating surface redox reactions. Among the array of semiconductor materials, Tungsten Trioxide (WO_3) has received considerable academic and industrial focus, which is attributed to its distinctive characteristics such as high catalytic efficiency, cost-effectiveness, low toxicity, environmental benignity, and abundant availability. Additionally, due to its narrow band gap, elevated electron mobility, and chemical stability, WO_3 finds extensive applications in environmental purification and solar energy harnessing. In the following discussion, the performance metrics and material characterization of WO_3 will be explored along with its modified photocatalytic variations, with specific emphasis on their potential applications in the process of glycerol valorization.

WO_3 serves as a well-studied photocatalytic material, boasting a bandgap range of approximately 2.6–2.8 eV. This compound manifests in multiple crystallographic phases, including monoclinic, hexagonal, tetragonal, and orthorhombic configurations. A diverse array of synthetic methodologies exists for the fabrication of WO_3 photocatalysts; these encompass hydrothermal synthesis, chemical vapor deposition, electrodeposition, and the sol–gel technique. Hydrothermal synthesis is predominantly employed due to its procedural simplicity, cost-effectiveness, and resultant high-purity WO_3 [66]. The sol–gel process affords granular control over the morphology and particulate dimensions of WO_3 photocatalysts. Chemical vapor deposition operates at elevated temperatures and yields WO_3 thin films of remarkable crystallinity and uniformity. Lastly, electrodeposition represents a straightforward and economically viable methodology for synthesizing WO_3 photocatalysts that exhibit superior crystallinity and enhanced photoactivity [67].

The crystallographic architecture of WO_3 is amenable to modification through elemental doping, employing species such as nitrogen or tin. The photocatalytic efficacy of WO_3 is subject to regulation by multiple variables, including morphological features, crystal structure, doping interventions, and surface modifications. Morphology exerts a direct influence on the material's light absorption and charge transfer characteristics. The morphological attributes of WO_3 photocatalysts are dictated by synthetic conditions, encompassing temperature, pH, and surfactant concentrations. The alteration of particle size can be accomplished by modifying precursor concentrations and reaction time. For instance, WO_3 nanorods exhibit a superior surface area and enhanced charge transfer capabilities in comparison to their nanoparticulate counterparts, thereby enhancing overall photoactivity. Variations in the crystal structure can also impart differential photocatalytic behaviors, as disparate crystal facets manifest different surface energies and reactivities. Elemental doping can induce alterations in the electronic and optical characteristics of WO_3, resulting

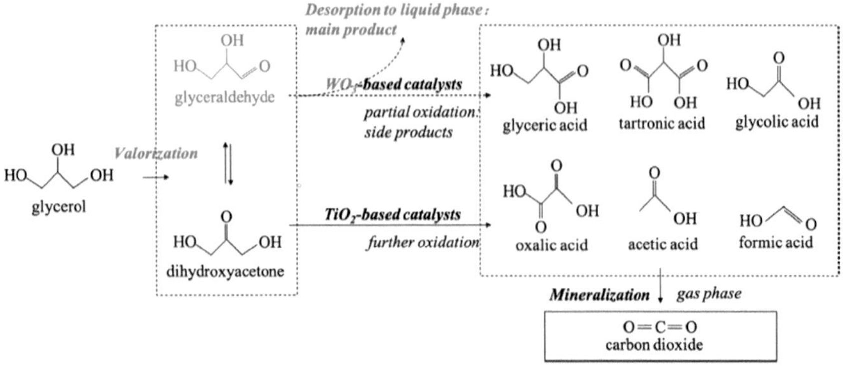

Fig. 4.18 Photocatalytic oxidation of glycerol [72] (Copyright 2021, Applied Catalysis B: Environmental License Number 5550280296779)

in elevated photocatalytic performance. For example, nitrogen-doped WO_3 outperforms its undoped analogue in terms of photocatalytic efficacy. Surface treatments can further ameliorate the photocatalytic attributes of WO_3 by augmenting surface area and reactive sites [68]. Analytical techniques such as X-ray diffraction (XRD), scanning electron microscopy (SEM), transmission electron microscopy (TEM), and UV–Visible spectroscopy serve as indispensable tools for WO_3 photocatalyst characterization. XRD is frequently employed to ascertain crystallographic structures and phase purity of WO_3 nanoparticles [69]. SEM and TEM offer detailed insight into morphological and particulate dimensions [70], while UV–Visible spectroscopy aids in determining the bandgap energy of the material [71].

The incorporation of additional acidic sites into the anatase semiconductor structure has been shown to enhance the selectivity of WO_3-based catalysts [72]. These sites are capable of selectively activating the primary C–O bonds in glycerol, facilitating the synthesis of glyceraldehyde. Moreover, these sites aid in the efficient desorption of aldehyde products in the liquid phase, as illustrated in Fig. 4.18.

Despite the considerable efficacy of WO_3 as a photocatalyst, several impediments persist that limit its practical deployment, including suboptimal quantum efficiency, inadequate selectivity, and rapid charge recombination that adversely affect the photocatalytic performance. Strategies must therefore be formulated to enhance the quantum efficiency, selectivity, and charge separation efficiency of WO_3-based photocatalysts. Additionally, challenges related to cost-effective synthesis and scalability for large-volume production remain to be addressed. These are critical aspects for broadening the application of WO_3 in fields such as water treatment, atmospheric pollution mitigation, and solar energy conversion. Various methodologies aimed at improving the photocatalytic performance of WO_3 are illustrated in Fig. 4.19.

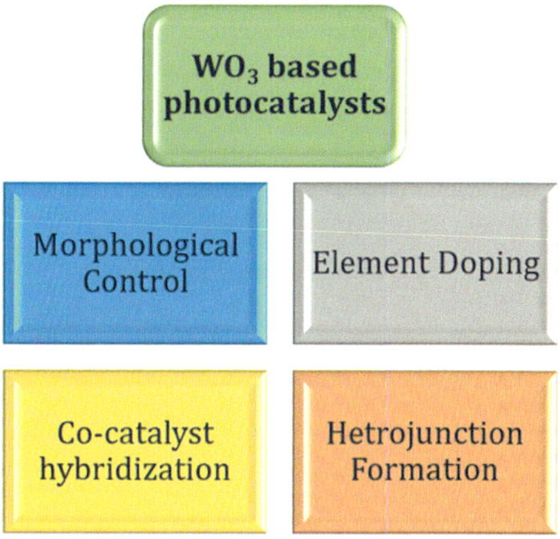

Fig. 4.19 Diverse strategies for augmenting the photocatalytic efficacy of WO3[73]

4.2.3.1 Morphological Regulation of WO_3

Morphological optimization is a highly effective approach for enhancing photocatalytic activity, as catalytic surfaces serve as the principal sites for photocatalytic reactions. Elevating the surface-to-volume ratio can bolster catalytic efficiency. Researchers have thus developed one-dimensional (1-D) nanostructures of tungsten oxide, including nanotubes, nanorods, and nanofibres, which are particularly attractive due to their heightened photocatalytic capabilities and reduced dimensionality. Solvothermal methods have been employed to fabricate WO_3 nanorods and nanowires, and these 1-D configurations have demonstrated notable efficacy in environmental applications [74]. Among these, WO_3 nanofibres are distinguished by their superior mechanical attributes. Electrospinning serves as an effective technique for the synthesis of these nanofibres, permitting large-scale production and tunable mechanical properties [75]. WO_3 nanoparticles have been constructed through a variety of methods, with nanotubes offering expansive surface areas, thereby facilitating accelerated pollutant diffusion and enhanced photocatalytic activity [76]. Nanosheets, characterized by their planar geometry and high aspect ratios, present additional morphological variants. Typically ranging in thickness from 1 to 10 nm and lateral dimensions from sub-micrometer to tens of micrometers, these nanosheets offer exceptional smoothness, minimal turbidity, and excellent substrate adhesion, rendering them promising candidates for diverse environmental applications, particularly in photocatalysis [77].

The annealing temperature serves as a critical variable for enhancing the photocatalytic efficacy of nest-like WO_3 structures through the manipulation of surface area and pore dimensions. Due to increased energy absorption and accelerated carrier generation, the semiconductor photocatalyst exhibits amplified photocatalytic performance [78]. The activity of WO_3 is principally governed by its electronic structure and surface atomic configurations. Tailoring specific crystal facets can result in optimized photocatalytic efficiency for targeted reactions. Facet-engineered WO_3 structures demonstrate markedly superior photocatalytic activity in the degradation of organic pollutants compared to commercially available WO_3 [79].

4.2.3.2 Elemental Doping of WO_3

Elemental doping with metals such as Fe, Mg, Ti, V, Mo, and Zn serves as an efficacious strategy for bandgap reduction. Utilizing a template technique, Fe-doped WO_3 hetero-structures were synthesized to increase the WO_3 performance [80]. The resulting Fe-doped WO_3 specimen demonstrated a minuscule grain size of approximately 10 nm, an ultra-high surface area of 225 m^2/g, and commendable crystalline integrity. Bandgap modulation was achieved through precise control over the concentration of Fe dopants, with Fe^{3+} ions serving to enhance photocatalytic activity on the WO_3 surface by facilitating electron–hole transfer and mitigating photo-generated electron–hole recombination.

Additionally, a WO_3 photocatalyst integrated with nitrogen-doped carbon quantum dots was engineered to bolster photocatalytic activity. Electron–hole pairs were generated within the WO_3 structure, and graphene oxide served as an electron scavenger, thus diminishing the recombination rate. The atomic architecture and lattice configuration of the photocatalyst were pivotal in modulating both the adsorption capacity for reactants and the carrier transfer rates, thereby influencing overall photocatalytic performance. Moreover, faceted WO_3 featuring a <002> crystal facet demonstrated elevated photocatalytic efficacy [81].

4.2.3.3 WO_3 Co-Catalyst Hybridization

The efficacy of semiconductor photocatalysts can be augmented through the incorporation of noble metals, as demonstrated in numerous studies [82]. Employing a nanocasting technique, an Au-loaded WO_3 semiconductor photocatalyst was synthesized [83]. The presence of Au in the WO_3 matrix facilitated enhanced electron–hole pair separation subsequent to WO_3 bandgap excitation. In this configuration, photo-generated holes contributed to the oxidative degradation of methylene blue, while electrons from WO_3's conduction band transitioned to Au's lower Fermi level. As a cost-effective alternative to noble metal co-catalysts such as Pt, Au, and Pd, nanodiamonds (NDs) have been employed. Their addition to WO_3-based visible-light-responsive photocatalysts led to an improvement in the degradation efficiency of volatile organic compounds. The enhanced sequential interfacial electron transmission and charge separation were primarily facilitated by the graphitic carbon shell (sp^2) surrounding the diamond core (sp^3). This graphitic layer effectively amplified the photocatalytic activity of the ND-WO_3 composite by enabling efficient interfacial electron transfer, exhibiting a low reduction potential for O_2, and facilitating rapid charge separation through ND's multi-electron transfer mechanism. Herein, O_2 acted as an electron scavenger in the conduction band, when NDs were employed.

4.2.3.4 WO_3 Heterojunction Formation

In semiconductor systems, Z-scheme based heterojunctions are constructed analogously to type-II heterojunctions; however, the mechanisms governing photo-generated charge transfer differ substantially. Conventional Z-schemes facilitate photo-charge carrier separation, eradicate suboptimal charge carriers, and enhance the redox capacity of the composite heterojunction. For instance, the WO_3/Ag_3PO_4 heterojunction demonstrated superior photocatalytic performance compared to individual WO_3 and Ag_3PO_4 catalysts. The limited efficacy of pristine WO_3 is attributed to its elevated positive conduction band potential, which hinders the generation of potent oxidizing species, such as hydroxyl radicals (·OH) and hydrogen peroxide (H_2O_2).

The enhanced photocatalytic performance of the WO_3/Ag_3PO_4 heterojunction is primarily ascribed to the effective separation of electron–hole pairs and the reduced visible light absorption owing to WO_3's large bandgap. When the WO_3/Ag_3PO_4 heterojunction is formed, electrons in the conduction band of WO_3 migrate to the valence band of Ag_3PO_4, thereby mingling with holes and mitigating charge carrier recombination. As such, the optimization of charge transmission and separation in WO_3 systems provides a pragmatic avenue for augmenting photocatalytic efficiency.

Interestingly, AgI/WO_3 hollow hierarchical nanoflowers exhibited enhanced photocatalytic activity in contrast to pure AgI and WO_3 nanoparticles. This improvement is mainly attributed to three contributory factors: augmented light harvesting, advanced light scattering, and enhanced electron–hole separation facilitated by the Z-scheme heterojunction. The redox potential of OH/H_2O surpasses that of WO_3 (+3.315 eV). Consequently, photo-generated electrons in the conduction band of AgI catalyze the conversion of oxygen to superoxide ions, while photo-induced holes in WO_3 generate hydroxyl radicals (·OH), leveraging their robust oxidizing capacities.

The polymeric semiconductor $g\text{-}C_3N_4$ has garnered interest due to its exceptional chemical stability. Both WO_3 and $g\text{-}C_3N_4$ exhibit low bandgap energies, approximately 2.7 eV, enabling them to absorb the blue component of the solar spectrum when combined in a heterojunction. A Z-scheme $WO_3/g\text{-}C_3N_4$ semiconductor photocatalyst has been synthesized through a facile blending and annealing process, utilizing urea as the precursor [84]. The mechanism for photo-generated electron–hole separation in the $WO_3/g\text{-}C_3N_4$ system is illustrated in Fig. 4.20. The morphological analysis reveals a uniform distribution of WO_3 within the nanocomposite, characterized by a worm-pore-like structure. The photocatalytic efficiency of the heterojunction was found to be significantly influenced by the concentration of WO_3. Enhanced carrier separation rates were observed, contributing to an increase in photocatalytic activity. This improvement is ascribed to favorable alterations in surface properties and electronic band structure. Consequently, the $WO_3/g\text{-}C_3N_4$ heterojunction demonstrated substantially enhanced photocatalytic performance.

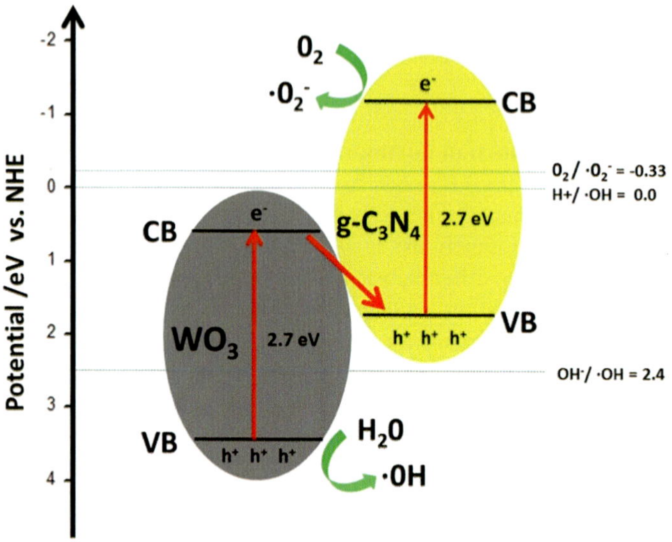

Fig. 4.20 Photo-generated electron–hole segregation technique of Z-scheme $WO_3/g\text{-}C_3N_4$ [84] (Copyright 2017, Applied Surface Science License Number 5550290422399)

Photocatalytic hydrogen production has been observed in catalysts based on $Pd/TiO_2\text{-}WO_3$, with WO_3 serving as the precursor for Pd/WO_3 and tungstic acid (H_2WO_4) formulations [85]. X-ray diffraction (XRD) patterns for Pd/NT-WO_3, Pd/P25-WO_3, and Pd/P25 are presented in Fig. 4.21. All three supports exhibit diffraction lines consistent with anatase phases. The absence of signals corresponding to palladium (Pd) species suggests that Pd is highly dispersed within the support, potentially in nanoparticulate form. Further XRD analyses with elevated Pd concentrations have been performed to investigate any discernible contributions from Pd to the XRD spectra. Peaks corresponding to Pd are only visible in the XRD pattern of the Pd (0.5 wt%)/NT-WO_3 sample, as illustrated in Fig. 4.21a, and are not present in Fig. 4.21b–d. This discrepancy can be attributed to lower Pd loading in the Pd (0.21 wt.%)/NT-WO_3 sample and the presence of a rutile phase with a reflection angle 2θ around 39 degrees in the Pd (0.52 wt.%)/P25-WO_3 and Pd (0.51 wt.%)/P25 samples. Re-analysis of the samples reveals, as shown in Fig. 4.21e–g, that lower Pd loading results in the formation of exceptionally small Pd nanoparticles.

4.2.4 *Bi_2WO_6*

Featuring an Aurivillius structure, Bismuth tungsten oxide (Bi_2WO_6, or BWO) is an n-type semiconductor active in the visible light spectrum, with a moderately low band gap energy ranging between 2.70 and 3.0 eV. BWO exhibits promising

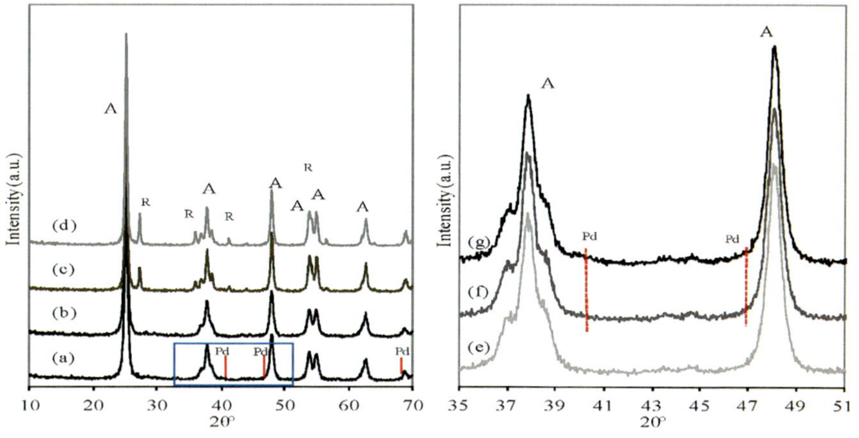

Fig. 4.21 X-ray diffraction patterns: **a** Pd(0.5 wt.%)/NT-WO_3-2m10; **b** Pd(0.21 wt.%)/NT-WO_3-2m10; **c** Pd(0.52 wt.%)/P25-WO_3-2m10; **d** Pd(0.51 wt.%)/P25-2m10; **e** NT-WO_3; **f** Pd(0.21 wt.%)/NT-WO_3-2m10; **g** Pd(0.35 wt.%)/NT-WO_3-2m10 [85] (Copyright 2021, International Journal of Hydrogen Energy License Number 5550250459927)

photoelectrical and photocatalytic properties [86, 87]. Owing to its unique, substantially layered crystalline structure, consisting of interconnected WO_6 octahedra and interspersed $(Bi_2O_2)^{2+}$ layers, the material facilitates the formation of anisotropic structures [88, 89]. Multiple studies have demonstrated its efficacy as a photocatalyst in applications such as dye degradation, wastewater remediation, and the abatement of volatile organic compounds [90–92]. Moreover, BWO has been identified as an effective photocatalyst for the oxidation of various alcohols, including glycerol, propanol, and butanol [93]. The focus of current research is increasingly pivoting towards the selective conversion of glycerol, a by-product of biodiesel production, into value-added compounds. One such approach involves glycerol photo-oxidation. Zhang et al. were the first to document the utilization of Bi_2WO_6 as a highly selective photocatalyst for glycerol oxidation in the presence of oxygen, achieving a selectivity rate of up to 90% in converting glycerol to dihydroxyacetone (DHA) [94].

4.2.4.1 pH-Dependent Spherical Bi_2WO_6

Bi_2WO_6 is regarded as an advantageous semiconductor for glycerol photo-reforming due to its suitable band gap and the ability to enable direct hole transfer to glycerol, thereby circumventing the generation of indiscriminate ·OH radicals, as depicted in Fig. 4.22 [95]. Crystalline samples manifest optimally within a narrow pH range, from neutral to mildly basic (pH 7–9). Conversely, extreme acidic conditions (pH 0–2) lead to an increase in oxygen vacancies within the resultant samples. At a pH of 13, the asymmetric phase $Bi_{3.84}W_{0.16}O_{6.24}$, enriched with Bi^{3+}, is obtained. Comprehensive analyses were executed to assess the morphology and optical properties of all the samples. Photocatalysts fortified with 1 wt.% Pt, under continuous UV–Vis light irradiation in a 5 v/v% aqueous glycerol solution, demonstrated both hydrogen evolution and glycerol oxidation. The sample synthesized under highly acidic conditions (pH = 0) exhibited the highest photonic efficiency (ξ) for hydrogen generation at approximately 1.4%, with a concurrent 99% selectivity for formic acid formation. The non-stoichiometric sample $Bi_{3.84}W_{0.16}O_{6.24}$ displayed comparable activity (ξ ≈ 1.2% and a formic acid selectivity of 88%), whereas the well-crystallized sample synthesized under mildly basic conditions (pH = 9) showed reduced efficiency (ξ ≈ 0.9%) and led to the emergence of various oxidation products. These divergent results were elucidated by the influence of oxygen vacancies, which act as catalytically active sites for redox reactions and adsorption onto the semiconductor surface, thereby establishing a direct correlation between the structural attributes of the semiconductor and its photocatalytic performance.

Preparation of pH-dependent spherical Bi_2WO_6

A hydrothermal synthesis method was employed for sample preparation. Specifically, 2 mmol of $Bi(NO_3)_3 \cdot 5H_2O$ was dissolved in 50 mL of 1.0 mol L^{-1} HNO_3 and stirred continuously. Concurrently, 1 mmol of $Na_2WO_4 \cdot 2H_2O$ was dissolved in 20 mL of deionized water. The tungstate solution was incrementally introduced into the continuously stirred Bi(III) solution, after which the combined solution was mixed for an additional 20 min. Subsequently, to adjust the pH of the mixture, a 3 M NaOH solution was introduced prior to the hydrothermal treatment. This mixture was transferred to a 100 mL Teflon-lined stainless-steel autoclave and maintained at 200 °C for a duration of 3 h. Samples were synthesized at distinct pH values of 0, 2, 7, 9, and 13 and were labelled accordingly. Upon cooling, the resultant powders

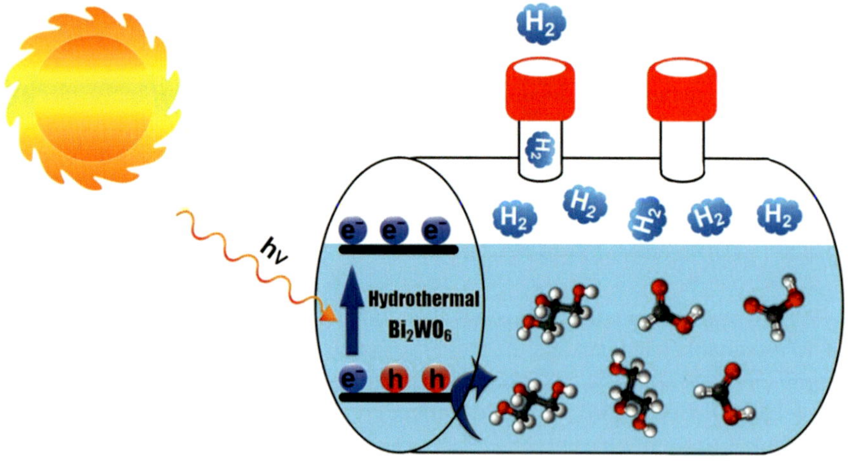

Fig. 4.22 Photo-reforming of glycerol using pH-dependent Bi_2WO_6 photocatalyst prepared via hydrothermal method [95] (Copyright 2022, Photochemical & Photobiological Sciences License Number 5525010560357)

were removed from the autoclave, washed with deionized water and ethanol, and subsequently dried at 80 °C.

Glycerol photo-reforming experiments

The photon flux in the wavelength range of 320–450 nm was quantified utilizing chemical ferrioxalate actinometry [96, 97], yielding a value of 7.4×10^{-8} Einstein·s^{-1}. The photonic efficiency (ξ) was ascertained by the ratio of the hydrogen production rate (r) to the incident photon flux (I), and was computed in accordance with Eq. 4.8. The photo-reactor was sustained at a thermally stable condition of 20 °C via a water bath and subjected to light irradiation for a span of 5 h.

$$\xi = \frac{2r}{I} \tag{4.8}$$

Conversion yields and selectivity of carbon-based products were quantified in accordance with Eqs. 4.9 and 4.10, respectively. In these equations, n_0 signifies the initial number of moles of glycerol, whereas n_f represents the final number of moles. n_{HCOOH} denotes the moles of formic acid generated during the photocatalytic procedure. Control experiments conducted in the absence of a photocatalyst exhibited no degradation of glycerol following ten hours of UV–visible light exposure. The rates of hydrogen evolution for the samples were assessed under UV–visible irradiation in a 20 v/v% aqueous methanol solution, with Pt serving as a co-catalyst, as summarized in Table 4.2.

$$\text{Overall glycerol conversion } (\%) = \frac{(n_0 - n_f)}{n_0} \times 100 \tag{4.9}$$

$$\text{Selectivity } (\%) = \frac{3n_{HCOOH}}{n_0} \times 100 \tag{4.10}$$

Table 4.2 The rates of hydrogen evolution for the samples were assessed under UV–visible irradiation in a 20 v/v% aqueous methanol solution, with Pt serving as a co-catalyst[95] (Copyright 2022, Photochemical & Photobiological Sciences License Number 5525010560357)

Photocatalyst	Co-catalyst (Pt)	Rate (μmol·h^{-1})	ξ of H_2 (%)
$Bi_{3.84}W_{0.16}O_{6.24}$	None	1.2 ± 0.06	0.9 ± 0.1
	1 wt.% Pt	7.8 ± 0.6	5.8 ± 0.1
Bi_2WO_6 (pH 9)	None	1.05 ± 0.1	5.4 ± 0.1
	1 wt.% Pt	7.3 ± 0.4	0.2 ± 0.1
Bi_2WO_6 (pH 7)	None	0.2 ± 0.4	0.2 ± 0.1
Bi_2WO_6 (pH 2)	None	0.1 ± 0.4	< 0.1
Bi_2WO_6 (pH 0)	None	1.7 ± 0.1	1.3 ± 0.1
	1 wt.% Pt	8.2 ± 0.2	6.2 ± 0.1

In summary, using a straightforward hydrothermal method, we obtained bismuth tungstate samples exhibiting sheet-like morphologies. These photocatalysts demonstrated high activity in the UV–visible spectrum for hydrogen generation from aqueous glycerol solutions, while simultaneously achieving a high selectivity for formic acid production. Distinct hydrothermal environments led to photocatalysts with varied morphologies and surface characteristics. Specifically, samples synthesized under neutral to slightly alkaline conditions (pH 7–9) exhibited high crystallinity and an orthorhombic sheet-like structure. Under strongly alkaline conditions (pH 13), a non-stoichiometric $Bi_{3.84}W_{0.16}O_{6.24}$ phase was generated. Conversely, acidic conditions yielded spherical orthorhombic Bi_2WO_6 particles rich in oxygen vacancies. These acidic conditions facilitated increased glycerol adsorption as well as the formation of initial photoreaction intermediates. When platinum was employed as a co-catalyst, these structural and compositional variations contributed to enhanced glycerol conversion rates, resulting in increased photonic efficiencies for hydrogen production. This was accompanied by the selective formation of formic acid.

4.2.4.2 Carbon Quantum Dots-Modified Bi_2WO_6

Utilizing an in-situ method, a marked enhancement in electron coupling and transport was achieved between carbon quantum dots (CDs) derived from biomass waste and Bi_2WO_6 nanosheets. This facilitated the effective integration of CDs onto the Bi_2WO_6 nanosheets. The band gap of the Bi_2WO_6 photocatalyst was effectively reduced from 2.78 to 2.52 eV upon modification with CDs. Subsequently, the CDs-modified Bi_2WO_6 photocatalyst exhibited superior visible light absorbance, augmented electron mobility, and enhanced charge separation efficacy. When assessed for activity and selectivity in the production of glyceraldehyde, the CDs-modified Bi_2WO_6 outperformed its unmodified counterpart. A peak glyceraldehyde concentration of 12.1 mmol/L was achieved with 80% selectivity, whereas glycolic acid was produced at a considerably lower concentration of 2.2 mmol/L.

Synthesis of BWOCD

For the synthesis of Carbon Quantum Dots (CDs), 2 g of de-oiled cake were dissolved in 40 mL of deionized water (DI), followed by 30 min of stirring. The mixture was then subjected to hydrothermal treatment for 12 h at 240 °C. Following the hydrothermal treatment, the mixture was centrifuged at 10,000 rpm for 30 min and filtered through a 0.22 μm syringe to obtain a purified dispersion of CDs. In the synthesis of CDs-modified Bi_2WO_6, a hydrothermal methodology was implemented. Sodium tungstate and bismuth nitrate were amalgamated in an 80 mL aqueous solution at a 2:1 molar ratio. For in-situ modification, a specified volume of CDs was incorporated into the reactant mixture. Cetyl-trimethyl-ammonium bromide (CTAB), utilized as a surfactant, was introduced at a concentration of 7 wt% relative to the amount of bismuth nitrate. An excess of CTAB, exceeding its critical micelle concentration (1.75 CMC), was incorporated to ensure optimal dispersion. Subsequently, the homogeneous mixture was transferred to a hydrothermal reactor and maintained at

180 °C for 24 h. After completing the hydrothermal process, the solution was filtered and cooled. The final product underwent multiple ethanol washes before being dried at 60 °C for 12 h. The Bi_2WO_6 sample modified with CDs is denoted as BWOCD in the subsequent discussion. For comparative purposes, a pure Bi_2WO_6 sample was also synthesized, following the aforementioned procedure but excluding the addition of the CD solution.

Glycerol photo-oxidation tests

Figure 4.23 elucidates the conceivable pathways of glycerol photo-oxidation facilitated by the employed catalysts. As previously noted, hydroxyl radicals (·OH) are generated, interacting subsequently with glycerol's primary alcoholic group to produce the initially oxidized intermediate, GAD. In an alkaline medium saturated with excess · OH radicals, the conversion of glycerol to glyceraldehyde is enhanced [98]. The elevated selectivity of the BWOCD catalyst for glyceraldehyde conversion implies a diminution in glyceraldehyde's secondary reactions, owing to Bi_2WO_6 modification with CDs. A lower yield of glyceric acid and a predilection for glycolic acid when using BWOCD indicate that C–C bond scission is augmented by CD incorporation. Among the products of secondary reactions, this suggests a CD-enhanced C–C bond cleavage. With the unmodified BWO catalyst, trace quantities of glyoxylic acid are present, stemming from additional C–C bond cleavage of glycolic acid[99]. This infers that BWO effectively mitigates secondary oxidation. The absence of glyoxylic acid in the output using the BWOCD catalyst implies a further curtailment of secondary reactions upon CD modification.

Figure 4.24a displays the Mott-Schottky (MS) plot, while Fig. 4.24b delineates the anticipated band alignment. The flat band potentials (V_{FB}) for BWO and CDs

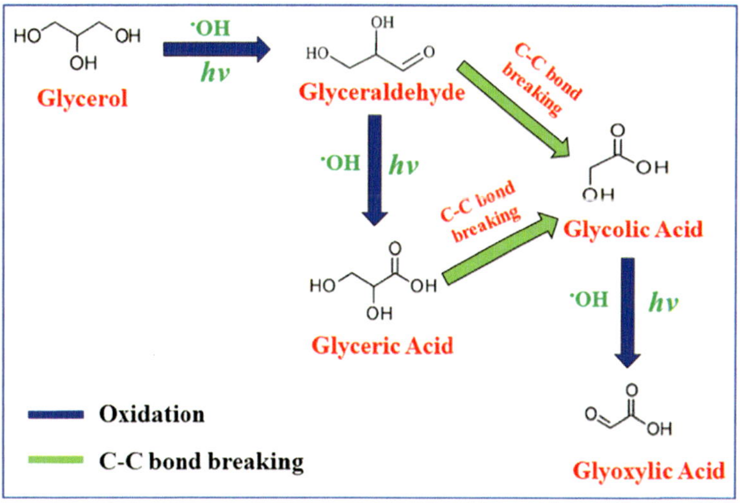

Fig. 4.23 Possible photo-oxidation reaction pathways for employed catalysts [100] (Copyright 2023, Materials Today: Proceedings License Number 5525011148115)

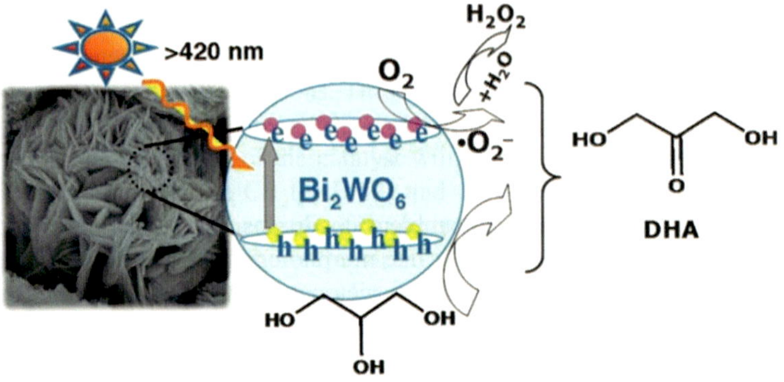

Fig. 4.25 A proposed mechanism for the selective conversion of glycerol to DHA over the Bi_2WO_6 catalyst in aqueous medium under visible-light irradiation [94] (Copyright 2010, Chemical science License Number 1343153-1)

studies of Bi_2WO_6 over six cycles did not display any significant diminution in either the catalytic activity or the selectivity towards DHA during the photo-transformation of glycerol in water under visible-light irradiation.

4.2.4.4 Ag_2O-anchored Bi_2WO_6 Heterostructure

This investigation focused on enhancing the synthesis of nanocrystalline Bi_2WO_6 via an adapted sol–gel technique, while simultaneously introducing trace amounts of silver oxide (Ag_2O) to the material's surface. The resultant heterostructures were employed for the photocatalytic generation of hydrogen (H_2) in an aqueous glycerol solution under visible light irradiation. Comprehensive characterization revealed that the surface deposition of Ag_2O had a synergistic effect, augmenting both the light-harvesting capacity and semiconductor properties of Bi_2WO_6 without significantly altering its crystalline structure or primary surface properties. The specific surface area of Bi_2WO_6 experienced a marginal reduction from 145 to 125 $m^2 \cdot g^{-1}$ upon the incorporation of 3.0 wt.% Ag_2O. Concomitantly, the bandgap energy was reduced from 2.78 to 2.31 eV, thereby enhancing the absorption of visible light. Optimal hydrogen generation was achieved using a concentration of 2.0 g/L of the composite material with 3% Ag_2O/Bi_2WO_6 in the presence of deposited platinum (Pt), yielding a hydrogen generation rate of up to 3.06 mmol/g·h. The process exhibited notable sustainability, retaining 97% of its initial activity after five cycles of use. Such elevated hydrogen evolution rates are attributed to the enhanced light absorption and efficient charge separation facilitated by the Ag_2O-anchored Bi_2WO_6 heterostructure.

Synthesis of Ag_2O-anchored Bi_2WO_6 heterostructures

Initially, nanocrystalline Bi_2WO_6 was synthesized utilizing a modified sol–gel technique, incorporating the F-127 copolymer as a stabilizing agent. In this method, 1.5 g of F-127 was dissolved in 30 mL of ethanol (CH_3CH_2OH) and agitated continuously for 30 min. Subsequently, 12 mmol of hydrochloric acid (HCl) and 40 mmol of acetic acid (CH_3COOH) were introduced into the same flask. Precursors containing Bi^{3+} and W^{6+} ions were then added in respective molar quantities of 20 mmol and 10 mmol, and the mixture was stirred for an additional 120 min. The resultant viscous sol was cast in a Petri dish and subjected to polymerization at 40 °C, followed by overnight drying at 70 °C in a humidity-controlled chamber. Calcination at 600 °C was then performed to obtain the crystalline phase of Bi_2WO_6. In the subsequent step, Ag_2O/Bi_2WO_6 heterostructures were fabricated via a straightforward impregnation method. Specifically, the as-synthesized Bi_2WO_6 was divided into four aliquots, each weighing 0.75 g, and dispersed in 150 mL of ethanol within round-bottom flasks. These solutions were sonicated for 15 min to ensure homogeneous dispersion. Predetermined amounts of Ag^+ ions were then added to each flask to achieve stoichiometric ratios of 1.0, 2.0, 3.0, and 4.0 wt.% relative to the pure Bi_2WO_6. The resulting mixtures were agitated for 60 min to facilitate the impregnation of Ag^+ ions, followed by drying at 80 °C for 12 h. These mixtures underwent calcination at 450 °C for 3 h, yielding four distinct samples: 1.0% Ag_2O/Bi_2WO_6, 2.0% Ag_2O/Bi_2WO_6, 3.0% Ag_2O/Bi_2WO_6, and 4.0% Ag_2O/Bi_2WO_6. For comparative purposes, pristine Bi_2WO_6 and Ag_2O samples were also prepared using identical procedures.

Estimated H_2 production pathway over Ag_2O/BWO

The integration of narrow-bandgap Ag_2O with Bi_2WO_6 (BWO) nanoparticles enhances charge separation efficacy, thereby augmenting photocatalytic H_2 generation and bolstering visible light absorption capabilities. Utilizing the absolute electronegativity of each component, the estimated electronic band diagrams elucidate the anticipated photocatalytic H_2 evolution pathway, as schematically represented in Fig. 4.26. Specifically, Ag_2O and BWO possess valence bands (VB) at energy levels of 1.602 eV and 3.30 eV, and conduction bands (CB) at −0.002 eV and 0.50 eV, respectively. Upon exposure to visible light radiation, electron–hole pairs (e^-/h^+) are generated in both oxides. The intrinsic electric field between p-type Ag_2O and n-type BWO facilitates charge transfer via an S-scheme mechanism [104, 105]. In the presence of a Pt co-catalyst, these photo-generated charge carriers traverse the heterojunction interface. Subsequent to visible light excitation, electrons in the CB of BWO may recombine with holes in the VB of Ag_2O, whilst the VB of BWO and CB of Ag_2O maintain high redox potential carriers. Glycerol undergoes an oxidation reaction facilitated by the holes on the VB of BWO, thereby generating protons (H^+)[106], as articulated in the following chemical equations (Eqs. 4.13 and 4.14):

$$h^+_{VB}(BWO) + glycerol \rightarrow H^+ + CO_2 + intermediates \quad (4.13)$$

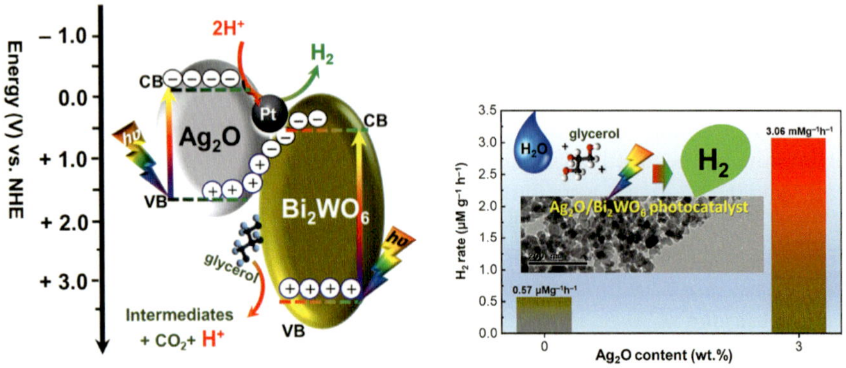

Fig. 4.26 A schematic representation of charge transfer and photocatalytic H_2 generation over the light-responsive Ag_2O/Bi_2WO_6 heterojunction photocatalyst [107] (Copyright 2023, Ceramics International License Number 5525020983787)

Consequently, the interaction between the generated protons (H^+) and molecular hydrogen (H_2) could yield H_2 via the following mechanism, facilitated by the spatially separated electrons (e^-) residing on Ag_2O:

$$2H^+ + 2e^-\left(Ag_2O\right) \rightarrow H_2 \uparrow \tag{4.14}$$

Hence, this innovative Ag_2O/BWO heterostructure demonstrated enhanced efficacy and practical applicability in the generation of hydrogen through visible light-driven photocatalytic processes.

In this study, the efficacy of Ag_2O-anchored Bi_2WO_6 nanocrystals for photogenerated H_2 evolution under visible light irradiation was substantiated. The modification of mesostructured Bi_2WO_6 nanocrystals with varying weight percentages of Ag_2O (ranging from 1.0 to 4.0%) led to an enhanced absorption of visible light, attributed to a reduction in the overall bandgap energy of the resultant heterojunction. Moreover, an optimal Ag_2O loading of 3.0% exhibited a significant augmentation in the rate of photocatalytic H_2 evolution, reaching 3.06 mmol·g^{-1}·h^{-1}, which is 2.65 times greater than that observed for unmodified Bi_2WO_6. The newly developed bimetallic oxide system demonstrated remarkable stability, maintaining 96% of its initial H_2 production capacity over five cycles. This investigation offers promising strategies for addressing challenges associated with the design of metal oxide photocatalytic systems for renewable fuel production under visible light conditions [107].

4.2.5 *g*-C_3N_4

In recent decades, graphitic carbon nitride (*g*-C_3N_4) has garnered significant attention as a promising photocatalyst, a development attributable to its unique electronic and structural characteristics. Possessing an intrinsically compatible and tunable bandgap of 2.7 eV, *g*-C_3N_4 manifests photocatalytic activity in the visible light spectrum [108, 109]. Moreover, it can be readily synthesized from abundant precursors, thus enhancing its availability and cost-effectiveness. Additional attributes include high chemical stability, non-toxicity, and structural versatility, making *g*-C_3N_4 amenable to diverse applications [110]. Structurally, *g*-C_3N_4 is a two-dimensional semiconductor photocatalyst comprising N-bridged tri-s-triazine units that form planar sheets, which are subsequently stacked via van der Waals forces [111, 112]. The valence band (VB) and conduction band (CB) of *g*-C_3N_4 are positioned at approximately + 1.52 eV and −1.13 eV versus the normal hydrogen electrode (NHE), respectively, thus facilitating the photo-reduction of glycerol [113, 114]. Despite its advantageous properties, *g*-C_3N_4 suffers from two primary limitations: a low surface area and suboptimal transfer capabilities for photo-generated charge carriers [115–117]. However, these challenges can be mitigated through strategic doping with various materials to modify the catalyst's intrinsic properties [118, 119].

4.2.5.1 Z-scheme CdS/*g*-C_3N_4

While current catalytic activities remain suboptimal for carbon–neutral objectives, the photocatalytic conversion of organic substrates into carbon monoxide (CO) presents a viable pathway towards sustainability. The process is thermodynamically non-favorable ($\Delta G > 0$) under inert conditions, representing one of its inherent limitations. Although thermodynamic favorability can be achieved in an oxygen-rich atmosphere ($\Delta G < 0$), this leads to substantial degradation into carbon dioxide (CO_2). To address these challenges, we introduce an oxidative photo-reforming methodology conducted under aerated conditions for the conversion of biomass-derived polyols, specifically glycerol, into CO at ambient temperatures. Utilizing a CdS@*g*-C_3N_4 heterojunction in an appropriate oxygen environment, a 48% CO yield from glycerol is achieved, a yield 13-fold higher than that attainable under inert conditions. The heterojunction's Z-scheme and core–shell structure not only facilitate enhanced separation of photo-generated charge carriers but also increase the density of photo-generated electrons on the shell-like *g*-C_3N_4 surface. These attributes contribute to effective dioxygen adsorption and activation. Consequently, this study offers a pioneering oxidative photo-reforming approach for CO generation from biomass under thermodynamically favorable conditions [120].

Synthesis of CdS/*g*-C_3N_4

CdS nanorods were successfully fabricated via a solvothermal process. Specifically, an equimolar solution was prepared by dissolving 4.62 g of $Cd(NO_3)_2$ and thiourea in

60 mL of ethanediamine, followed by stirring for one hour. This mixture was subsequently transferred to a 100 mL Teflon-lined autoclave and maintained at 160 °C for 24 h. Upon cooling to room temperature, the resultant yellow precipitate was isolated via centrifugation, washed thrice with water and ethanol, and dried at 60 °C. Graphitic carbon nitride (g-C_3N_4) was synthesized through a calcination procedure. Precisely, 10 g of urea underwent thermal decomposition in a muffle furnace at 550 °C for a period of 3 h, with a constant calcination rate of 5 °C/min. The outcome was a pale-yellow compound identified as g-C_3N_4. For the exfoliation of g-C_3N_4 nanosheets, the as-synthesized g-C_3N_4 was subjected to acid treatment using a 5 M HNO_3 solution prepared by diluting concentrated HNO_3 in 100 mL of deionized water. Specifically, 0.5 g of g-C_3N_4 was completely dissolved in the acidic medium and refluxed at 125 °C for 12 h. The yielded material was isolated by centrifugation at 10,000 rpm and washed five times to eliminate residual HNO_3, followed by drying at 60 °C for 12 h. The CdS@g-C_3N_4 composite was fabricated through a self-assembly technique utilizing methanol as the solvent. In detail, 8 mg of the exfoliated g-C_3N_4 nanosheets were dispersed in 15 mL of methanol and subjected to ultrasonication for one hour. This colloidal solution was incrementally added to a 10 mL methanol solution containing 92 mg of CdS nanorods under continuous stirring for an additional 24 h. The resultant composite was isolated via centrifugation and subsequently dried under vacuum at 60 °C overnight to yield the final CdS@g-C_3N_4 material.

Photocatalytic CO formation over CdS/g-C_3N_4

In the study, the thermodynamic feasibility of glycerol transformation to carbon monoxide (CO) was initially assessed through Gibbs free energy calculations, as illustrated in Fig. 4.27a. In an inert atmosphere, the process exhibits a positive Gibbs free energy (ΔG) of 4.84 eV, rendering it thermodynamically unfavorable. Conversely, in an oxygen-rich environment, ΔG becomes negative, enhancing the thermodynamic viability of the transformation. However, an excess of oxygen leads predominantly to CO_2 formation through comprehensive photo-oxidation. Manipulation of the O_2-to-glycerol molar ratio can theoretically facilitate selective hydrogen oxidation, thereby enhancing CO selectivity. The study proposes an oxygen-regulated photo-reforming methodology to optimize CO production (Fig. 4.27b, c). While the decarbonylation route is exothermic with a ΔG of −2.48 eV, the dehydrogenation pathway in an oxygen-deficient environment is endothermic, exhibiting a positive ΔG of 3.987 eV. The hydroxyl radical alters this endothermic route into an exothermic process with a ΔG of -2.54 eV (Fig. 4.27b). Literature suggests that thermodynamically favorable photocatalytic reactions occur when ΔG transitions from a positive to a negative slope or remains positive but with a reduced energy barrier [121]. For achieving thermodynamically favorable conversion of glycerol to CO, initial oxidation of the hydroxyl group to an aldehyde is followed by decarbonylation to CO [122]. Hydroxyl radical-mediated cleavage reactions, involving reactive oxygen species, are also explored. Hydroxyl (·OH) radicals possess higher redox potentials compared to the photo-generated holes of CdS, facilitating more efficient oxidation [123]. The simultaneous reduction of O_2 by photo-generated electrons leads to the formation of hydroxyl radicals. The majority of these radicals react with hydrogen to form aldehyde intermediates. Subsequent oxidation to CO_2 is thereby precluded due to the consumption of available oxygen in the dehydrogenation step (Fig. 4.27c). An innovative Z-scheme CdS@*g*-C_3N_4 core–shell photocatalyst structure is thus introduced to enhance the distribution of photo-generated electrons over the *g*-C_3N_4 shell, facilitating efficient interaction with O_2 and the generation of active oxygen radicals.

In an aqueous solution containing a ratio of 0.95:0.05 MeCN to water, the photocatalytic reforming of glycerol was examined under varying light and atmospheric conditions using a range of commonly employed photocatalysts (Fig. 4.28). The reaction was conducted in a 6.5 mL quartz tubular reactor. The most efficacious catalyst was the CdS@*g*-C_3N_4 composite, which achieved a glycerol conversion rate

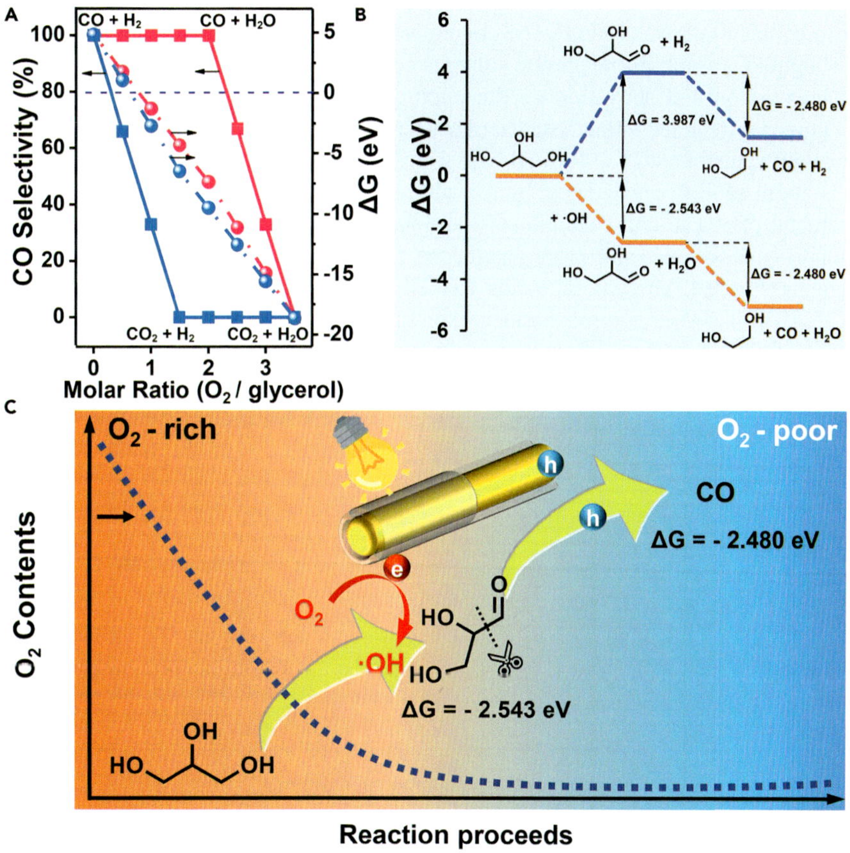

Fig. 4.27 Catalyst design and operating concept [120]; **a** the ratio of O_2 to glycerol significantly influences both the thermodynamic feasibility, represented by ΔG, and the selectivity towards CO formation in the process of glycerol reforming; **b** Various pathways for the oxidative decarboxylation of glycerol are evaluated in relation to the ΔG associated with CO production; **c** A conceptual framework is presented for the photo-reforming of glycerol under controlled oxygen conditions (Copyright 2022, Chem License Number 5541780904583)

of 99% and generated yields of 48% CO, 15% CO_2, 7% formic acid, 2% acetic acid, and 3.8 mmol·g^{-1} of H_2. In comparison to pristine CdS and g-C_3N_4, the CO yield over the CdS@g-C_3N_4 composite was 2.4 and 14.4 times greater, respectively. Performance evaluation indicated that the enhanced activity of the CdS@g-C_3N_4 composite is attributable to the Z-scheme heterojunction rather than mere mechanical contact between CdS and g-C_3N_4 (Fig. 4.29). An oxygenated atmosphere significantly augmented CO production as compared to an argon atmosphere; specifically, the CO yield in an aerated environment was 13 times greater. Other commonly utilized photocatalysts were markedly less effective than CdS@g-C_3N_4 in terms of CO photo-generation under both O_2 and Ar atmospheres when exposed to visible light. These photocatalysts demonstrated some activity under UV light but predominantly produced CO_2 (>50%). An MeCN/H_2O solution with a volume ratio of 0.95:0.05 was found to be conducive for CO generation, potentially due to the higher O_2 solubility in aprotic solvents compared to pure water, thus facilitating O_2 adsorption and activation on the photocatalyst surface [124, 125]. Moreover, CO was not detected in the absence of glycerol, confirming that MeCN degradation does not contribute to CO formation as a by-product. The CdS@$_g$-C_3N_4 composite was successfully recycled four times without noticeable loss of activity. Furthermore, the photocatalytic oxidation of additional biomass-derived compounds, such as monosaccharides and alditols (e.g., glucose, fructose, xylose, erythritol, and xylitol), was investigated. These compounds yielded CO in the range of 11–32% after 24 h of irradiation, while polysaccharides like inulin produced a 9% CO yield under the same conditions.

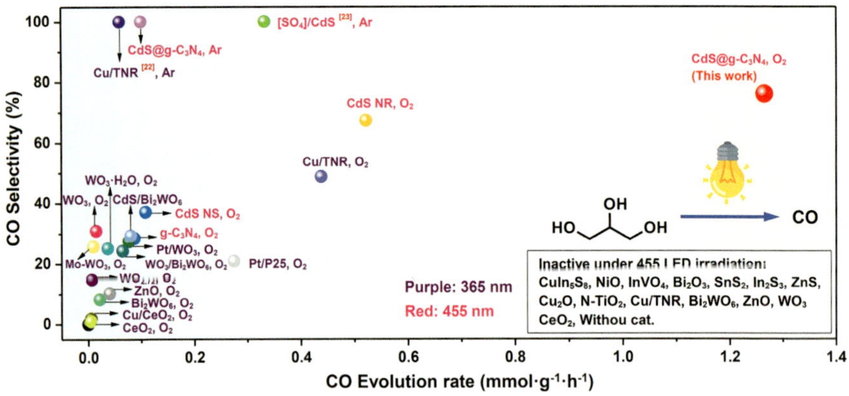

Fig. 4.28 Catalyst screening: Photo-reforming of glycerol across various catalysts. The data represented in purple or red indicate measurements taken under 365 nm LED (UV) and 455 nm LED (visible) irradiation, respectively. Reaction conditions: 10 mg of glycerol, 5 mg of photocatalyst, 0.95 mL of MeCN, 0.05 mL of water, and 1 atm pressure[120] (Copyright 2022, Chem License Number 5541780904583)

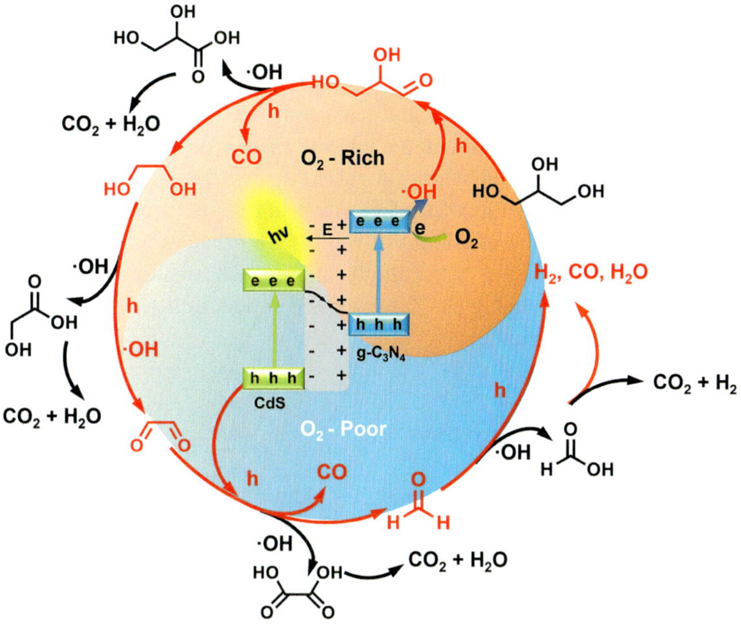

Fig. 4.29 Photo-reforming oxidation of glycerol over CdS@*g*-C_3N_4: proposed reaction pathway [120] (Copyright 2022, Chem License Number 5541780904583)

In light of previously elucidated data, a mechanistic framework is proposed for the oxidative reforming of glycerol over a CdS@*g*-C_3N_4 photocatalyst, as illustrated in Fig. 4.29. Upon photonic excitation, the Z-scheme heterojunction in the CdS@*g*-C_3N_4 system facilitates the migration of photo-generated holes towards the CdS domain and the conduction of photo-generated electrons to the *g*-C_3N_4 domain. These photo-generated holes subsequently engage in the partial oxidation of glycerol, resulting in the formation of aldehydic intermediates. These aldehydic species may then undergo decarbonylation to yield H_2 and CO. Simultaneously, the photo-generated electrons amass on the surface of the catalyst and effectively engage in the reduction of molecular oxygen (O_2) to produce hydroxyl radicals. These hydroxyl radicals facilitate the oxidation of glycerol to aldehydic intermediates, a process that is enhanced by the core–shell architecture of the CdS@*g*-C_3N_4 system. Over-oxidation of these aldehydic intermediates can result in the formation of carboxylic acids and ultimately CO_2. As the reaction progresses, the system's atmosphere transitions to an O_2-depleted state. Consequently, the generation of hydroxyl radicals is curtailed, leading to the conversion of the remaining aldehydic intermediates into carbon monoxide (CO).

A robust method for oxygen-controlled, visible-light-driven photocatalytic reforming has been investigated, with the aim of converting diverse polyols and saccharides to CO under mild conditions. Utilizing a core–shell CdS@*g*-C_3N_4 heterojunction, efficacious photocatalytic activity in glycerol reforming was observed, yielding a 48% CO production. The core–shell structure and the Z-scheme heterojunction enhance the directional migration of photo-generated electrons to the catalyst's surface, thereby accelerating the transformation of molecular oxygen (O_2) into hydroxyl radicals, concomitant with glycerol photo-reforming. The rate and selectivity of CO evolution are significantly influenced by the molar ratio of oxygen to substrate, denoted as the O/S value. A relatively low O/S ratio (2.5) effectively augments the CO evolution rate, with CO emerging as the primary product. This approach presents a resource-efficient utilization of a broad range of biomass feedstocks for the generation of CO. Accordingly, this study introduces a novel oxidative photocatalytic reforming strategy to facilitate the production of CO from biomass under advantageous conditions.

4.2.5.2 Cu_2O-g-C_3N_4 nanocomposites

In the present investigation, mesoporous Cu_2O and *g*-C_3N_4 were synthesized separately via soft and hard-template sol–gel methods, subsequently utilized to fabricate Cu_2O-gC_3N_4 nanocomposites with varying Cu_2O concentrations (0.5–3.0 mol%). The efficiency of these Cu_2O-*g*C_3N_4 nanocomposites is strongly influenced by the molar ratio of Cu_2O to *g*-C_3N_4, as it governs the formation of heterojunctions between these components. Enhanced hydrogen production under simulated light conditions was observed, with an optimum efficiency achieved at a Cu_2O concentration of 2 mol%—a performance 17 and 38 times superior to pristine Cu_2O and *g*-C_3N_4, respectively. The amplified photocatalytic efficiency was ascribed to alterations in the visible-light absorption capacity and the facilitated transport of photogenerated species, directly related to the progressive development of Cu_2O/gC_3N_4 heterojunctions within the nanocomposites.

Synthesis of Cu_2O-*g*-C_3N_4 nanocomposites

Mesostructured *g*-C_3N_4

To synthesize high-surface-area *g*-C_3N_4, hexagonal mesoporous silica (HMS) with a surface area ranging from 500–1000 m^2/g was employed as a hard template [126]. Initially, 50 mL of Milli-Q water and 1 g of HMS were subjected to ultrasonic dispersion for 30 min. Subsequently, 5 g of urea and 3 g of dicyandiamide were incrementally incorporated into the HMS suspension. After thorough mixing, the resultant suspension underwent a drying process for 12 h at 80 °C. This dried blend was then subjected to thermal treatment at a stable temperature of 550 °C for a 4-h duration to facilitate the thermal decomposition and polymerization of dicyandiamide, thereby yielding carbon nitride sheets. These thermally treated precipitates were immersed in a 2 M NH_4HF_2 solution (50 mL) and agitated vigorously for 24 h to remove the HMS template. The filtrate was subsequently subjected to copious rinsing with deionized water to eliminate surface-adhered contaminants, followed by drying at 100 °C overnight, resulting in the fabrication of mesostructured *g*-C_3N_4.

Mesoporous Cu_2O

A soft sol–gel templating approach was utilized to synthesize mesoporous Cu_2O, employing a triblock copolymer, F127, as the structure-directing agent. The molar fractions of the constituents involved, namely $C_{16}H_{36}Cu_4O_4$, F127, C_2H_5OH, HCl, and CH_3COOH, were established at 1:0.02, 50, 2.25, and 3.75, respectively. Initially, F127 (1.6 g) was dissolved in 30 mL of ethanol and subjected to continuous magnetic stirring for 60 min. Subsequent to this, HCl (0.74 mL) and CH_3COOH (2.3 mL) were incrementally introduced into the stirred solution. Copper(I) tert-butoxide was then added to the sol mixture and maintained under intense stirring for an additional 60 min. Following the complete evaporation of ethanol, the resulting sol was stored overnight at 40 °C in a humidity-regulated environment. Ultimately, thermal treatment of the gel was performed at 350 °C in ambient air for a 24-h period to facilitate the removal of the F127 surfactant, yielding the finalized mesoporous Cu_2O structure.

Cu_2O-*g*-C_3N_4 nanocomposites

Cu_2O-*g*-C_3N_4 nanocomposites were synthesized through an ultrasonication-assisted mixing procedure. A quantity of 0.2 g of mesoporous *g*-C_3N_4 was dispersed in 400 mL of deionized water and subjected to ultrasonication at a frequency of 40 kHz for a duration of three hours. Concurrently, varying molar concentrations (0.5–3%) of mesoporous Cu_2O were introduced into the dispersion. The ultrasonic treatment facilitated the exfoliation of *g*-C_3N_4 nanosheets and their subsequent interaction with Cu_2O, thereby yielding thin-layered Cu_2O-*g*C_3N_4 composites. These nanocomposites were isolated via centrifugation and subsequently dried at a temperature of 80 °C. The resultant nanocomposites were designated as "x Cu_2O-*g*C_3N_4," where "x" signifies the weight percentage of incorporated Cu_2O, with values ranging from 0.5 to 3%.

The characteristics and performance of Cu_2O-*g*-C_3N_4

Nitrogen adsorption–desorption isotherms were employed to characterize the textural attributes of *g*-C_3N_4, Cu_2O, and Cu_2O-g-C_3N_4 nanocomposites, with their respective specific surface areas summarized in Table 4.3. Pristine g-C_3N_4 displays a type IV isotherm coupled with an H_2 hysteresis loop, indicative of an intricate mesoporous structure. This material boasts a specific surface area of 170 m^2/g, substantially exceeding conventional graphitic carbon nitrides, which typically exhibit areas ranging from 5 to 15 m^2/g [127]. The surface area of pure Cu_2O, which similarly features a mesoporous structure, is also noteworthy, as evidenced by isotherms and hysteresis patterns presented in Table 4.3. The nitrogen adsorption–desorption isotherm for the $2Cu_2O$-*g*C_3N_4 nanocomposite closely mirrors that of unmodified *g*-C_3N_4, thereby confirming the preservation of the mesoporous structure in the nanocomposite. However, the hysteresis loop for the $2Cu_2O$-*g*C_3N_4 nanocomposite is more pronounced, suggesting the occurrence of particulate stacking, likely instigated during the exfoliation process employed in nanocomposite synthesis. The specific Brunauer–Emmett–Teller (BET) surface area of xCu_2O-gC_3N_4 composites exceeds that of pristine g-C_3N_4, and further increases with greater incorporation of Cu_2O, as delineated in Table 4.3. Such observations emphasize the role of the copper oxide species in facilitating the exfoliation of g-C_3N_4 layers within the composite, thereby contributing to its overall structural integrity. Enhanced mesoporosity and specific surface area in the xCu_2O-gC_3N_4 composites augment reactant transport and light-harvesting capabilities, rendering them conducive for photocatalytic applications.

A proposed mechanism of the operating Cu_2O-g-C_3N_4

In light of their intrinsic characteristics, the formation of Type II p-n heterojunctions can be postulated as a mechanism facilitating the segregation and mobilization of charge carriers between p-type Cu_2O and n-type g-C_3N_4 semiconductors. The relative positions of the valence band (VB) and conduction band (CB) for Cu_2O and g-C_3N_4 can theoretically be determined using the subsequent Eqs. 4.15 and 4.16:

$$E_{CB} = X - 0.5E_g + E_0 \tag{4.15}$$

$$E_{VB} = E_g + E_{CB} \tag{4.16}$$

In this theoretical framework, E_0 represents the hydrogen scale-free electron energy relative to the absolute vacuum scale, set at 4.5 eV, while X signifies the absolute electronegativity of the semiconductor. The relative conduction band (CB) energies for Cu_2O and g-C_3N_4 are documented as 0.69 eV and 1.16 eV, respectively. Additionally, the valence band (VB) positions for these semiconductors are 1.31 eV and 1.47 eV, respectively, as illustrated in Fig. 4.30. It becomes evident that traditional p-n heterojunction formation is precluded by the non-overlapping energy band structures of Cu_2O and g-C_3N_4. Notably, g-C_3N_4 functions predominantly as an n-type semiconductor, its Fermi level (E_f) located proximate to the CB

Table 4.3 Physical and chemical characteristics of g-C_3N_4, Cu_2O, and Cu_2O-$g$$C_3N_4$ nanocomposites, as well as photocatalytic hydrogen generation and photocurrent[128] (Copyright 2020, Journal of Materials Research and Technology License Number 5541790002098)

Photocatalyst	S_{BET} ($m^2 \cdot g^{-1}$)	Absorption edge (nm)	Bandgap energy (eV)	H_2 yield ($\mu mol \cdot g^{-1}$)	Photocurrent ($mA \cdot cm^{-2}$)
g-C_3N_4	170	453	2.63	400	0.158
0.5Cu_2O-g-C_3N_4	180	550	2.18	2400	0.586
1Cu_2O-g-C_3N_4	183	557	2.15	10,500	0.664
2Cu_2O-g-C_3N_4	189	575	2.02	15,150	0.780
3Cu_2O-g-C_3N_4	190	574	2.02	15,075	0.773
Cu_2O	230	581	2.0	900	0.193

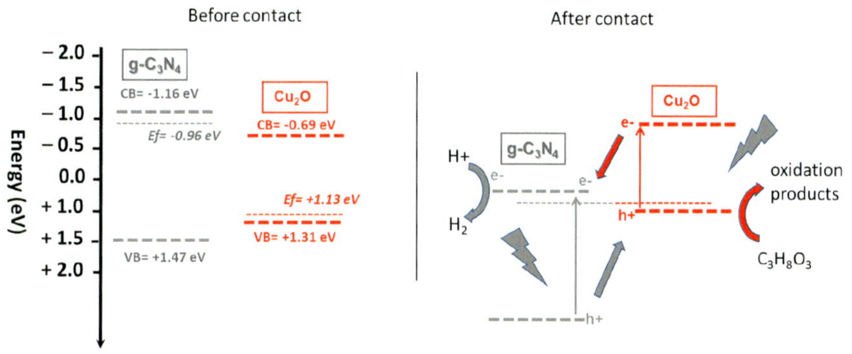

Fig. 4.30 A schematic representation delineates the band alignment of $g\text{-}C_3N_4$ and Cu_2O prior to contact, as well as the formation of the p-n heterojunction within $Cu_2O\text{-}gC_3N_4$ nanocomposites, consequent to the band bending required for Fermi level alignment[128] (Copyright 2020, Journal of Materials Research and Technology License Number 5541790002098)

at 0.96 eV. Conversely, Cu_2O serves as a p-type semiconductor with its E_f adjacent to the VB at +1.13 eV. Upon interfacing, the Fermi levels of the two materials equilibrate, facilitating the emergence of a p-n heterojunction exhibiting a distinct band alignment, as depicted in Fig. 4.30. Subsequent to photoexcitation, the electric field generated at the Cu_2O-g-C_3N_4 heterojunction enables electron migration from Cu_2O's CB to that of g-C_3N_4, while photo-generated holes traverse inversely from g-C_3N_4's VB to Cu_2O's VB. According to this operative mechanism, electrons accumulated in g-C_3N_4's CB react with H^+ ions to produce H_2, while the photo-generated holes in Cu_2O's VB partake in the oxidative degradation of glycerol, as delineated in Fig. 4.30. Consequently, the augmented photocatalytic H_2 evolution observed in xCu_2O-gC_3N_4 nanocomposites, relative to individual Cu_2O and g-C_3N_4 catalysts, can be attributed to the heterojunction formation between g-C_3N_4 and Cu_2O, which amplifies visible-light absorption and mitigates recombination of photo-generated charge carriers.

Mesoporous Cu_2O and g-C_3N_4 were synergistically amalgamated through ultrasonication to engineer mesoporous Cu_2O-gC_3N_4 nanocomposites. Physicochemical characterization affirmed the efficacious dispersion of Cu_2O nanoparticles upon the superficial layers of exfoliated g-C_3N_4 sheets. Owing to the established nanojunctions, the g-C_3N_4 component exhibited enhanced capacity for visible-light absorption, as well as for the generation and effective mobilization of charge carriers. Relative to their individual parent materials, the Cu_2O-gC_3N_4 nanocomposites displayed superior photocatalytic efficiency in the generation of hydrogen from glycerol. The photocatalytic performance of these nanocomposites is contingent upon the molar percentage of Cu_2O, with peak efficacy manifesting at a 2 mol% Cu_2O concentration—17 and 38 times greater than that of pristine Cu_2O and g-C_3N_4, respectively. This augmentation in photoactivity is attributable to the formation of heterojunctions between the g-C_3N_4 and Cu_2O components, which not only amplify the absorption of visible light but also mitigate the recombination of photo-generated charges, thereby enhancing hydrogen production. The efficiency of heterojunction formation within these nanocomposites is modulated by the Cu_2O concentration, with 2 mol% proving the most effective.

4.2.5.3 Carbon@exfoliated *g*-C_3N_4

In the present investigation, Carbon-embedded exfoliated *g*-C_3N_4 was synthesized employing a facile hydrothermal technique. Comprehensive characterization was conducted to assess its crystalline structure, specific surface area, morphological features, photonic attributes, and photo-reforming capabilities. Relative to the specific surface area of bulk *g*-C_3N_4, which is 34 m^2/g, the carbon-coated, exfoliated *g*-C_3N_4 structure demonstrated an enhanced surface area of 43.9 m^2/g. Ultraviolet–visible Diffuse Reflectance Spectroscopy (UV–vis DRS) further revealed augmented light absorption in the visible spectrum. Under optimized conditions, which entailed a reaction time of 8 h, catalyst dosage of 0.75 g/L, formaldehyde concentration of 400 ppm, and light intensity of 150 W/m^2, the catalyst yielded a maximum hydrogen production rate of 997 $mol \cdot g^{-1}$, with an experimental error margin of 1.30%. Given the selected operational parameters, this study could serve as a benchmark for the photo-reforming of glycerol to generate hydrogen at an industrial scale across diverse applications.

Fabrication of carbon@exfoliated *g*-C_3N_4

Exfoliated *g*-C_3N_4

In accordance with the methodology delineated in our previous publications, both bulk *g*-C_3N_4 and exfoliated *g*-C_3N_4 were synthesized [129]. Specifically, the pre-synthesized bulk *g*-C_3N_4 underwent chemical modification to yield exfoliated *g*-C_3N_4, subsequent to thermal treatment of melamine. For this procedure, 0.5 g of the pre-synthesized bulk *g*-C_3N_4 was treated with 8 mL of sulfuric acid (H_2SO_4). An additional 2 mL of deionized water (DI water) was incorporated, and the mixture was agitated for a duration of 60 min at room temperature. Subsequently, the sample underwent multiple rinsing cycles with DI water. Ultrasonication was then performed for 120 min to achieve a uniform dispersion of the material in 10 mL of isopropanol (IPA). After removal of the IPA via decantation, the residual precipitate was air-dried overnight. The resulting dry powder was labelled as exfoliated *g*-C_3N_4.

Carbon@exfoliated g-C_3N_4

Utilizing a hydrothermal approach, Carbon@exfoliated *g*-C_3N_4 was synthesized with citric acid serving as the carbon precursor. In this procedure, 0.5 g of citric acid was amalgamated with 1 g of pre-prepared exfoliated *g*-C_3N_4 in 30 mL of deionized water (DI water). The resultant suspension was agitated for a period of 15 min at ambient conditions. Subsequently, the mixture was transferred to a 50 mL Teflon-lined, stainless-steel autoclave reactor and subjected to a temperature of 180 °C for a duration of 240 min. After this thermal treatment, the reactor was allowed to return to room temperature via natural cooling. The resultant suspension underwent multiple purification cycles with DI water before being air-dried for 12 h. The final solid product was designated as Carbon@exfoliated $g\text{-}C_3N_4$.

Optical properties of Carbon@exfoliated $g\text{-}C_3N_4$

As delineated in Fig. 4.31a, b, the band gap and optoelectronic attributes of the photocatalysts were rigorously evaluated via UV–Vis Diffuse Reflectance Spectroscopy (DRS) and Tauc plots. The Carbon@exfoliated $g\text{-}C_3N_4$ composite manifested a noteworthy augmentation in the intensity across the visible region, specifically between 480 and 800 nm, when contrasted with its bulk and exfoliated $g\text{-}C_3N_4$ counterparts. Concomitantly, a color transition from pale yellow to light beige was observed upon the incorporation of carbon (as indicated in the inset of Fig. 4.31a). Utilizing Kubelka–Munk theory for band gap computation, the intercept of the extrapolated plot of ($\alpha h\upsilon^{1/2}$) versus photon energy was assessed. Subsequently, the estimated band gaps for bulk $g\text{-}C_3N_4$, exfoliated $g\text{-}C_3N_4$, and Carbon@exfoliated $g\text{-}C_3N_4$ were determined to be 2.60 eV, 2.89 eV, and 2.76 eV, respectively. These values are in concordance with those reported for similar materials in existing literature [130]. The elevated band gap in the case of exfoliated $g\text{-}C_3N_4$ is attributable to quantum confinement effects engendered by the ultrathin nanosheet structure. Photoluminescence (PL) spectroscopy, as depicted in Fig. 4.31c, was employed to probe the photo-induced characteristics of the catalyst. A diminished PL peak for the Carbon@exfoliated $g\text{-}C_3N_4$ sample relative to its bulk counterpart signifies efficacious suppression of charge carrier recombination. Additionally, carbon doping serves as an electron buffer, thereby facilitating the activation of excited electrons and augmenting the spatial separation of photo-generated charge carriers.

Photo-reforming results of Carbon@exfoliated $g\text{-}C_3N_4$

Figure 4.32 presents an initial investigation into the photo-reforming mechanisms of formaldehyde to stimulate hydrogen generation, employing bulk $g\text{-}C_3N_4$, exfoliated $g\text{-}C_3N_4$, and Carbon@exfoliated $g\text{-}C_3N_4$ as catalysts. The data substantiates that the Carbon@exfoliated $g\text{-}C_3N_4$ catalyst exhibited superior photocatalytic performance in hydrogen evolution compared to its bulk and exfoliated counterparts, indicative of enhanced physicochemical characteristics. The study elucidates variables such as reaction duration, catalyst concentration, substrate molarity, and irradiation intensity. During the 12-h experimental observation, a marginal increase in hydrogen generation was observed after the 8-h mark, at which point a plateau was reached for all catalysts. This stability in reaction rate beyond 8 h is likely attributable to substrate saturation at the catalyst active sites, thereby inhibiting further photon absorption. The temporal window for the photo-reforming reaction was selected from a 4- to 8-h range, based on previous studies. Additionally, parameter ranges for formaldehyde concentration (200–600 ppm), light intensity (100–200 W/m^2), and catalyst dosage (0.5–1 g/L) were established based on existing literature. Response Surface Methodology (RSM) was employed for an in-depth evaluation of the interrelationship among these variables.

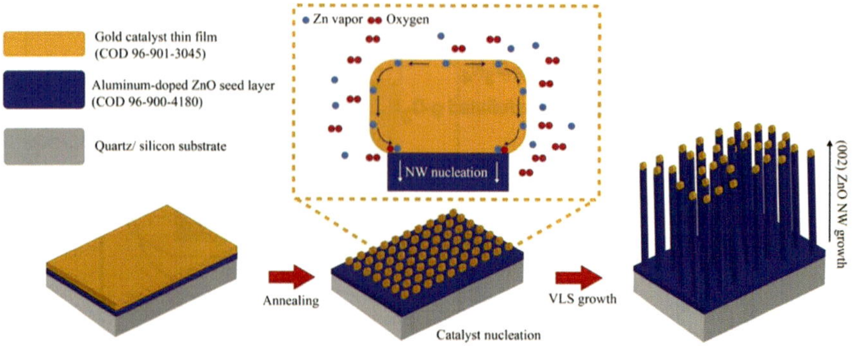

Fig. 4.33 Graphical representation of the vapor–liquid-solid growth process for obtaining ZnO nanowires[136] (Copyright 2020, International Journal of Hydrogen Energy License Number 5540510919684)

Various methodologies for the fabrication of one-dimensional nanostructures have been investigated, with bottom-up techniques gaining prominence due to their capacity for compositional, morphological, and orientational control. While chemical synthesis and various chemical vapor deposition (CVD) methodologies are prevalent for low-temperature NW fabrication, high-temperature physical processes are indispensable for obtaining well-ordered, high-quality crystalline structures. The Vapor–Liquid-Solid (VLS) technique stands out as a versatile approach for the reproducible synthesis of NW arrays. The present study is concentrated on optimizing deposition conditions to yield ZnO nanostructures that are pure, crystalline, and morphologically well-defined [135]. Parameters under consideration encompass substrate temperature, seed layer crystalline orientation, and carrier gas mixtures. Figure 4.33 elucidates the VLS-mediated growth process of ZnO nanowires [136].

The morphological characteristics of the NWs were explored utilizing a JEOL JSM-7600F Scanning Electron Microscope equipped with multiple detectors. For transmission electron microscopy (TEM) investigations aimed at assessing crystalline integrity, spatial orientation, and compositional mapping, a JEOL JEM-2010 equipped with a lanthanum hexaboride (LaB6) filament and scanning transmission electron microscopy (STEM) detector was employed. STEM results were substantiated via energy-dispersive spectroscopy (EDS). Figure 4.34 illustrates electron backscatter diffraction (EBSD) patterns for ZnO NWs synthesized through the vapor–liquid-solid (VLS) growth process. Scherer's equation was applied to determine the grain size of nanocrystals in the NWs. The data corroborated the epitaxial linkage between the seed layer and the resultant nanostructure. ZnO is lauded as a photocatalyst for its plethora of advantageous attributes, including cost-effectiveness, natural abundance, environmental benignity, ease of synthesis, superior electrical conductivity, chemical stability, high quantum efficiency, and robust electron kinetics [137].

The photocatalytic efficacy of ZnO is constrained to a narrow spectral range in the ultraviolet region. Various parameters, such as particulate dispersion, crystallinity,

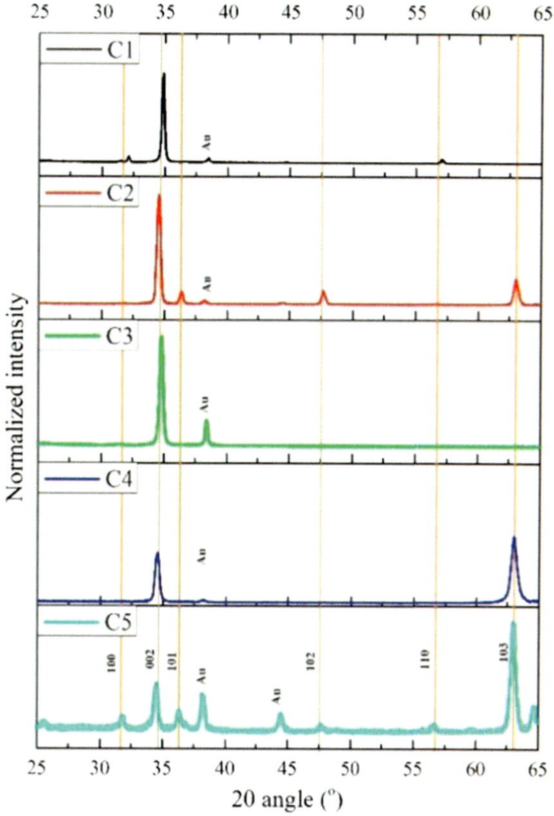

Fig. 4.34 Synthesis scheme of WO_3/GO/NCQDs photocatalyst [136] (Copyright 2020, International Journal of Hydrogen Energy License Number 5540510919684)

and surface area, influence its performance. These variables can be controlled by modulating the solution pH, calcination temperature, and the amount of dispersant introduced [138]. The choice of fabrication methodologies significantly impacts the ZnO microstructure, thereby affecting its photoreactivity. Consequently, the development of a facile synthesis route to yield ZnO with homogeneous morphology is essential [139]. ZnO nanorods exhibit superior electron mobility relative to their nanocrystalline counterparts, enhancing absorption in the visible spectrum and prolonging excited-state longevity—features crucial for high photocatalytic activity [140]. Further augmentation of ZnO nanorod performance can be achieved by incorporation into high-surface-area ceramic substrates. A discussion on the hydrothermal synthesis of pure ZnO nanorods by pH modulation is presented.

Figure 4.35 delineates the mechanistic pathway governing the photo-oxidation of glycerol in the presence of ZnO nanorods under solar irradiance [141]. This process entails the photo-excitation of electrons from the valence band (VB) to the conduction band (CB), where the requisite photon energy for charge carrier generation either matches or exceeds ZnO's bandgap energy (E_g). The photocatalytic hydrogen evolution is facilitated through these ZnO nanorods, activated by solar radiation

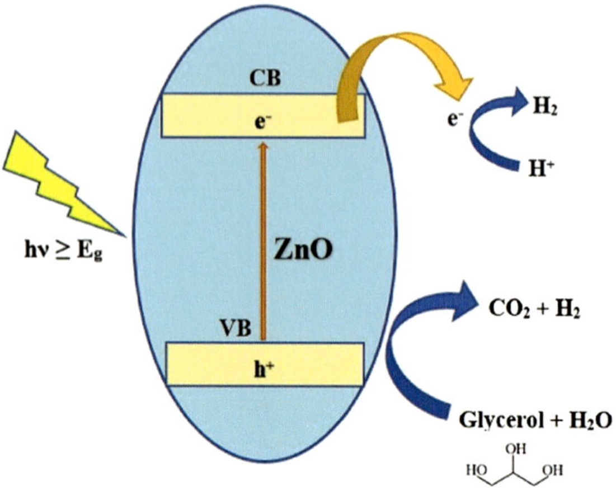

Fig. 4.35 Oxidation of glycerol with ZnO under solar irradiance [141] (Copyright 2022, Journal of Indian Chemical Society License Number 5542871338508)

with a lower energy limit corresponding to E_g. The synergistic interaction between glycerol molecules and hydroxyl radicals serves to mitigate charge carrier recombination, augment the longevity of photo-generated holes, and consequently, amplify the efficiency of photocatalytic hydrogen production.

4.2.6.1 Co-Doped ZnO

The integration of cobalt ions into the ZnO matrix has been shown to augment its photocatalytic efficacy, specifically by enhancing glycerol's surface accessibility and modulating the electronic states within the ZnO structure. The observed enhancements in charge transfer rate and photocurrent density are 2.4-fold and 2.2-fold, respectively, when compared to an undoped ZnO photoanode. Despite its promising attributes, such as a high absorption coefficient, elevated electron mobility, non-toxicity, and cost-effectiveness, ZnO faces challenges including its wide band gap of approximately 3.2 eV, limited glycerol adsorption capability, and sub-optimal glycerol conversion rates in photoelectrochemical biomass oxidation processes. Various mitigation strategies have been suggested, encompassing morphological control and co-catalyst employment to improve photocatalytic properties. A plethora of studies focus on either innovative ZnO morphologies or efficient synthetic methods [142].

To elevate the photo-conversion efficiency of glycerol and to promote effective charge carrier separation, doping strategies involving either noble metals (Pt, Au, Pd) or transition metals (Co, Fe, Cu) have been considered. Among the available strategies, the doping of transition metal ions, particularly cobalt (Co), is identified as an advantageous method. This is attributed to its ionic radius, which closely resembles that of Zn^{2+}, as well as its capacity to tailor the bandgap of ZnO when applied at suitable doping concentrations. Cobalt doping can further induce oxygen vacancies and facilitate charge transfer at reduced band gap energies. The use of a Co-doped ZnO photo-anode in photo-electrochemical glycerol oxidation under AM 1.5 G irradiation

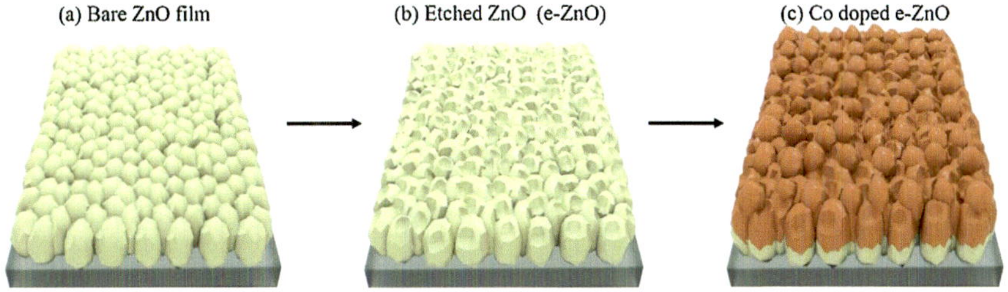

Fig. 4.36 Schematics of **a** bare ZnO film, **b** etched ZnO (e-ZnO), **c** cobalt-doped etched ZnO (Co:e-ZnO) [144] (Copyright 2021, Catalysis Today License Number 5543311276534)

has been documented. To synthesize Co-doped ZnO, an initial ZnO thin-film seed layer is sputter-deposited onto a substrate, as illustrated in Fig. 4.36a. This is followed by surface morphological modification via a chemical etching method employing organic acids, as presented in Fig. 4.36b. Subsequent hydrothermal synthesis allows for the deposition of Co-doped ZnO nanoparticles onto the etched ZnO layer, as indicated in Fig. 4.36c. The resultant Co-doped etched ZnO electrode manifests a markedly higher photocurrent density than its undoped counterpart, attributable to its augmented photocatalytic activity and selectivity towards glycerol. Thus, cobalt doping combined with morphological tailoring offers a novel avenue for enhancing ZnO photoanode's photoreactivity.

In the realm of photoelectrochemical (PEC) oxidation of glycerol, a comprehensive analysis was conducted to scrutinize the morphological and chemical alterations of the ZnO photo-anode for enhancing photo-reactivity. Utilizing citric acid as a surface-modifying agent enabled the facile adsorption of Co-doped ZnO nanoparticles (NPs) onto the ZnO surface, consequently amplifying the density of active sites for glycerol disintegration. Via hydrothermal synthesis, the Co:e-ZnO NP photo-anode was seamlessly integrated, engendering an escalation in both light absorbance and photo-reactivity. Relative to undoped ZnO and etched ZnO (e-ZnO), a marked improvement in glycerol oxidation efficacy was observed for the Co-doped ZnO NPs, specifically Co(3):e-ZnO. This enhanced performance is ascribable to the augmented glycerol adsorption and the efficacious charge carrier transport facilitated by the Co-doped ZnO NPs [144].

4.2.6.2 Cu-Doped ZnO

Zinc oxide (ZnO), characterized by its affordability, environmental stability, and a band gap of 3.2 eV, serves as a notable photocatalyst. To extend its activation into the visible light spectrum, doping strategies involving various metal ions such as cobalt, manganese, and nickel have been explored. Copper-doped ZnO (Cu–ZnO) has emerged as an efficacious material for hydrogen generation from glycerol under visible light conditions [145]. Investigations into its applicability in photoelectrochemical water splitting have revealed a significant photocurrent density of 18 μA/

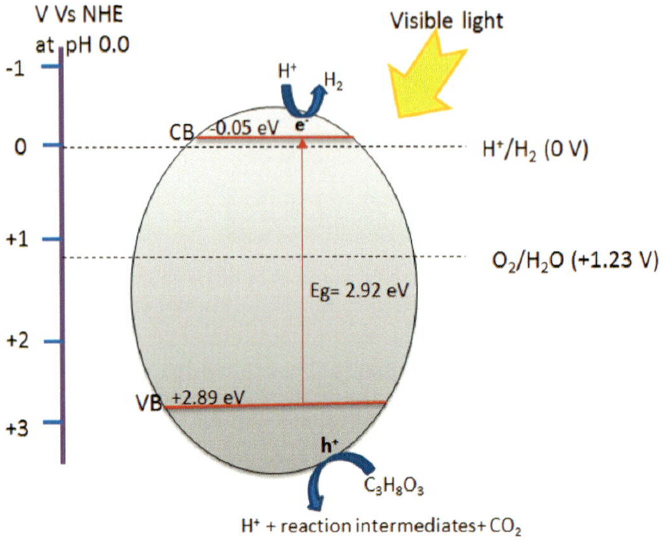

Fig. 4.37 Schematic illustrating hydrogen generation from glycerol via photocatalysis with copper-doped zinc oxide under visible light conditions [146]

cm^2 at a potential of 0.8 V, an enhancement by a factor of 11 compared to undoped ZnO.

Figure 4.37 elucidates that a 1.08 mol% Cu-doped ZnO photocatalyst possesses an electronic structure that is thermodynamically conducive to water splitting. The incorporation of copper ions into the ZnO matrix serves to mitigate the recombination rate, enhancing both thermodynamic and kinetic feasibility for hydrogen production under visible light [146]. Photocatalysts with varying Cu loadings were synthesized via a precipitation method, and 1.08 mol% Cu-doped ZnO was identified as the most efficacious for hydrogen production from a glycerol solution under visible light. Under optimized conditions—namely, an initial glycerol concentration of 5 wt.%, a photocatalyst dosage of 1.5 g/L, and a solution pH of 6—the hydrogen yield reached 2600 μmol/L. When the initial glycerol concentration was increased to 10 wt.%, the hydrogen production rate escalated to approximately 4770 μmol/L. These results highlight the viability of optimized Cu-doped ZnO as a photocatalyst for hydrogen generation in the visible light spectrum, obviating the need for noble metals in the catalyst composition [146].

4.2.6.3 ZnO/ZnS

Zinc oxide (ZnO) offers multiple advantageous properties, making it a material of choice for diverse applications such as photocatalysts, photodetectors, and ultraviolet lasers. Contrarily, zinc sulphide (ZnS) is a well-established photocatalyst possessing a broader band gap ($E_g = 3.6$ eV) compared to ZnO. Photoexcitation in ZnS facilitates the rapid generation of electron–hole pairs, and the associated excited electrons exhibit considerably negative reduction potentials. This characteristic renders ZnS suitable for the photodecomposition of various organic compounds [147]. The hybridisation of these two semiconductors yields a composite material with a reduced photoexcitation threshold energy. Core/shell-structured ZnO/ZnS nanorods exhibit enhanced physical and chemical attributes, thereby finding utility in multiple domains such as electronics, optics, catalysis, and magnetism. Considerable research efforts have focused on the systematic fabrication of these core/shell ZnO/ZnS nanorods using diverse techniques, leading to nanorods with a range of morphological attributes. However, limited investigations have explored the influence of ZnS shell layer thickness and deposition on the photoabsorption and photocatalytic efficacy. In this study, nanorods with varying ZnS/ZnO molar ratios are synthesized using a novel aqueous bath method. Hydrogen production from glycerol solutions under both UV and visible light is examined within a fixed-bed, flow-type reactor system [148].

An n-p heterojunction is formed upon the deposition of ZnS onto ZnO surfaces, altering the inherent charge transfer mechanism within the ZnO nanorods, as illustrated in Fig. 4.38. Upon exposure to UV or solar irradiation, electrons are transferred from the conduction band (CB) of ZnS to that of ZnO, while holes from the valence band (VB) of ZnO migrate to ZnS, driven by potential differentials. Surface-bound hydroxyl or water molecules capture the photo-generated holes, thereby forming highly oxidative hydroxyl radical species capable of oxidizing glycerol ($C_3H_8O_3$) to CO_2. Concurrently, surface-bound H^+ ions on ZnO capture the electrons to generate hydrogen (H_2). Although the specific mechanisms underlying hydrogen evolution in the presence of glycerol remain not fully elucidated, existing literature indicates several potential reaction pathways. The synergistic electron–hole transfer in ZnO/ZnS nanocomposites augments charge carrier yield and longevity while mitigating recombination events. Additionally, the elevated oxidative capacity of surface chemisorbed hydroxyl groups on ZnS, coupled with a narrower energy gap between its VB and the O_2/H_2O energy level compared to ZnO, significantly enhances the generation of photo-generated holes, thereby amplifying the photocatalytic efficiency [148].

4.3 Performance and Mechanisms of Catalytic Materials

The concentration of acidic sites serves as an indicator for material selectivity towards C_3 products; a higher acidity correlates with greater C_3 product selectivity. Acidic characteristics of the sample set were evaluated using pyridine as a probe molecule, followed by Diffuse Reflectance Infrared Fourier Transform Spectroscopy (DRIFTS) analysis. Infrared spectra were acquired at a constant temperature of 100 °C, revealing absorption bands at 1448 cm^{-1} and 1537 cm^{-1}, associated with Lewis and Brønsted acid sites, respectively.

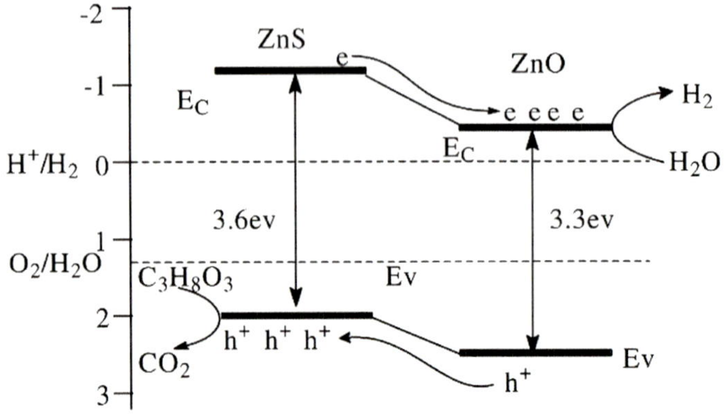

Fig. 4.38 A schematic illustration of charger transfer in ZnO/ZnS core/shell nanorods [148] (Copyright 2022, International Journal of Hydrogen Energy License Number 5542970462600)

Figure 4.39 depicts the DRIFTS spectrum of pyridine adsorption at 100 °C. Observed peaks at 1448 cm^{-1} and 1542 cm^{-1} are attributable to Lewis acid sites and surface-bound protonated pyridine at Brønsted acid sites, respectively. Comparative analysis between P25 (TiO_2) and WO_3 samples reveals a higher prevalence of Brønsted acid sites in WO_3, consistent with expectations given the well-documented acidic nature of tungsten-based catalysts.

In the realm of glycerol dehydrogenation reactions, noble metals such as platinum, palladium, and gold are traditionally used as catalysts due to their remarkable efficacy in cleaving C–H and O–H bonds. However, emerging literature indicates the potential utility of certain non-noble metals and even non-metallic catalytic agents. The majority of these transformations commonly occur at sub-100 °C temperatures in either acidic or alkaline aqueous media. Nevertheless, contemporary research has begun investigating the efficacy of neutral solutions for these reactions.

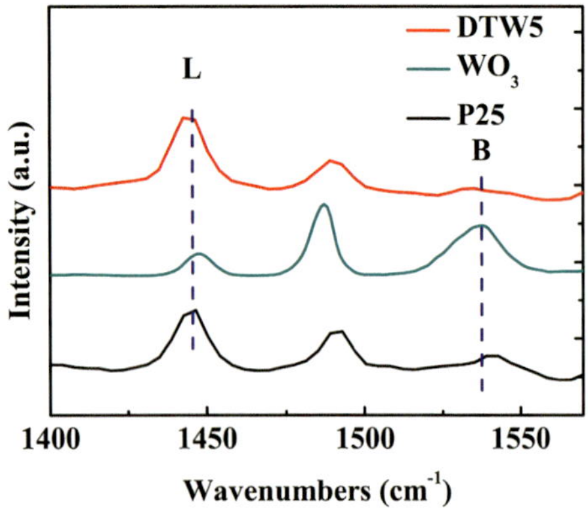

Fig. 4.39 The properties and relative abundances of acid sites in P25, DTW5, and WO_3 were characterized using Diffuse Reflectance Infrared Fourier Transform Spectroscopy (DRIFTS) following pyridine adsorption [5] (Copyright 2021, Applied Catalysis B: Environmental, License Number 5554700134967)

Dehydrogenation of glycerol, which involves the extraction of a hydrogen molecule, predominantly targets either the primary or secondary hydroxyl groups, resulting in the formation of glyceraldehyde and 1,3-dihydroxyethanone, respectively. Utilizing noble metal monometallic catalysts like platinum and palladium primarily yields hydroxy-oxidized species such as glyceraldehyde, along with further oxidized compounds like glyceric acid (step R9), as illustrated in Fig. 4.40. Conversely, the incorporation of catalyst promoters such as bismuth or antimony in conjunction with noble metals shifts the dehydrogenation predominantly towards the secondary hydroxyl group, thereby enhancing the selectivity for 1,3-dihydroxyethanone.

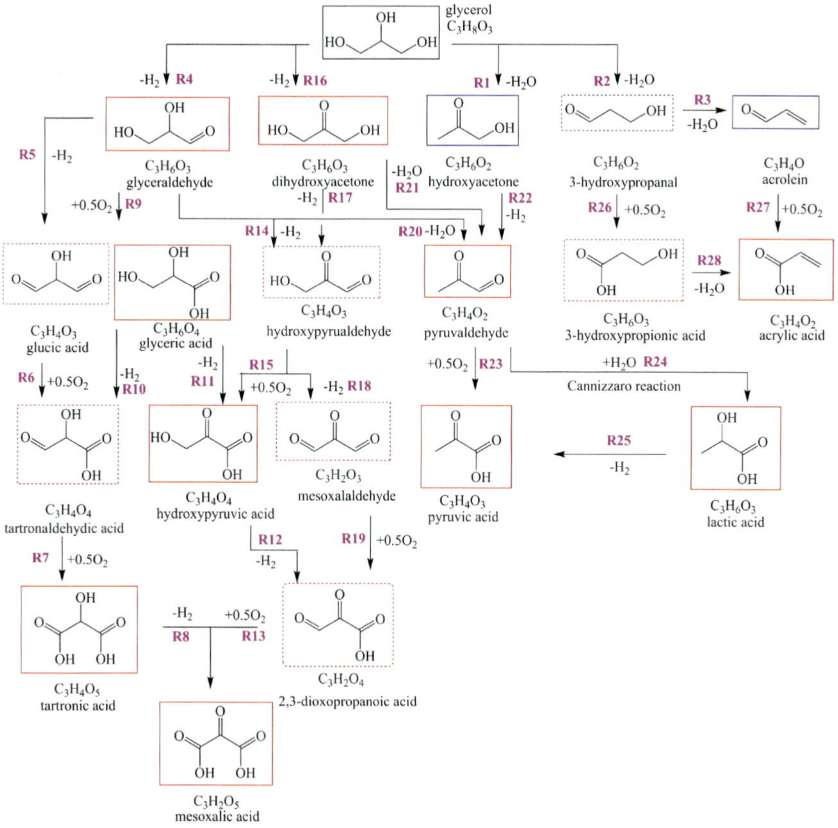

Fig. 4.40 The mechanism of glycerol oxidation [149] (Copyright 2019, Chinese Journal of Chemical Engineering, License Number 5554710061627)

4.4 Product Characterization, Analysis, and Selectivity

Characterization and analysis of the reaction products

It is crucial to emphasize the inherent challenges in analyzing the products resulting from the selective oxidation of glycerol, as findings may be susceptible to error. This has been corroborated in recent investigations focused on the thermal-catalytic oxidation of glycerol over gold-supported catalysts [150]. Exclusive reliance on high-performance liquid chromatography (HPLC) for the assessment of reaction products, specifically by comparing retention times to standard samples, is not recommended, as illustrated in Fig. 4.41 [151]. To corroborate the identity of the produced compounds, supplementary characterization techniques, such as Nuclear Magnetic Resonance (NMR) spectroscopy, should be employed [152]. In this investigation, the ^{13}C NMR spectroscopy was utilized for product characterization, confirming dihydroxyacetone (DHA) as the product of the visible-light-driven photocatalytic aerobic oxidation of glycerol over the Bi_2WO_6 catalyst.

The terms for glycerol conversion (GC), dihydroxyacetone (DHA) yield, and DHA selectivity were delineated based on gas chromatography (GC) analysis:

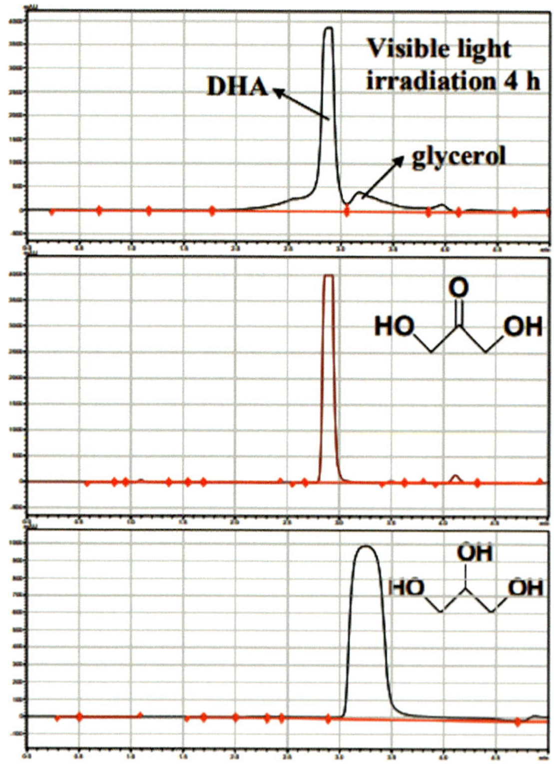

Fig. 4.41 The HPLC spectra of glycerol and dihydroxyacetone (DHA) to discern the selective oxidation of glycerol solution when subjected to continuous visible light irradiation for 4 h in a reaction system featuring a Bi_2WO_6 photocatalyst [94] (Copyright 2010, Chemical science License Number 1357599-1)

$$\text{Conversion } (\%) = [(C_0 - C_{GC})/C_0] \times 100 \tag{4.17}$$

$$\text{Yield } (\%) = C_{DHA}/C_0 \times 100 \tag{4.18}$$

$$\text{Selectivity } (\%) = [C_{DHA}/(C_0 - C_{GC})] \times 100 \tag{4.19}$$

In this context, C_0 represents the initial concentration of glycerol, C_{GC} and C_{DHA} denote the concentrations of glycerol and dihydroxyacetone (DHA), respectively, at a specific temporal point subsequent to the photocatalytic reaction.

Separation of DHA product

The known melting and boiling points for dihydroxyacetone (DHA) and glycerol are 75–80 °C and 213.7 °C, and 17.8 °C and 210 °C, respectively, while water has a boiling point of 100 °C. Consequently, rotary evaporation can effectively separate a glycerol-DHA aqueous mixture post-photocatalytic reaction. Typically, following irradiation from a 300W Xe arc lamp equipped with a UV-cutoff filter (>420 nm) for a 2-h duration, the mixture undergoes centrifugation to eliminate catalyst particulates. The centrifuged solution is then transferred to a round-bottom flask and subjected to rotary evaporation at 328 K under vacuum conditions. This procedure results in a separation of aqueous solvent, leaving the reactants and products in the flask. The remaining liquid is centrifuged once more, yielding a two-phase system: the lower layer predominantly contains DHA, and the upper layer primarily consists of residual glycerol. The upper layer is then removed for ^{13}C NMR analysis, whereas the lower DHA-enriched phase is transferred to filter paper and solidifies at ambient conditions, given DHA's melting point range of 75–80 °C.

DHA purification and analysis by ^{13}C NMR

To achieve further refinement of the dihydroxyacetone (DHA) product, column chromatography was employed utilizing silica gel as the stationary phase and acetone as the eluent. Subsequent to this purification step, the obtained DHA was subjected to ^{13}C nuclear magnetic resonance (^{13}C NMR) spectroscopy for analytical verification. Typically, the sample, quantified at 1 mmol, was solubilized in 0.6 mL of deuterium oxide (D_2O). ^{13}C NMR spectral data were acquired using a Bruker Avance III 400 spectrometer. Figure 4.42, in its right-hand column, displays the pertinent ^{13}C NMR spectra for these processed samples.

Product selectivity

Photo-reforming of glycerol to produce hydrogen

The majority of research in the area of photocatalytic glycerol valorization is oriented towards the generation of gas mixtures, with hydrogen serving as the principal output. Glycerol demonstrates a comparatively rapid rate of photocatalytic hydrogen production in comparison to numerous other alcohols that have been scrutinized. Furthermore, the aqueous-phase conversion of glycerol to hydrogen offers advantages in

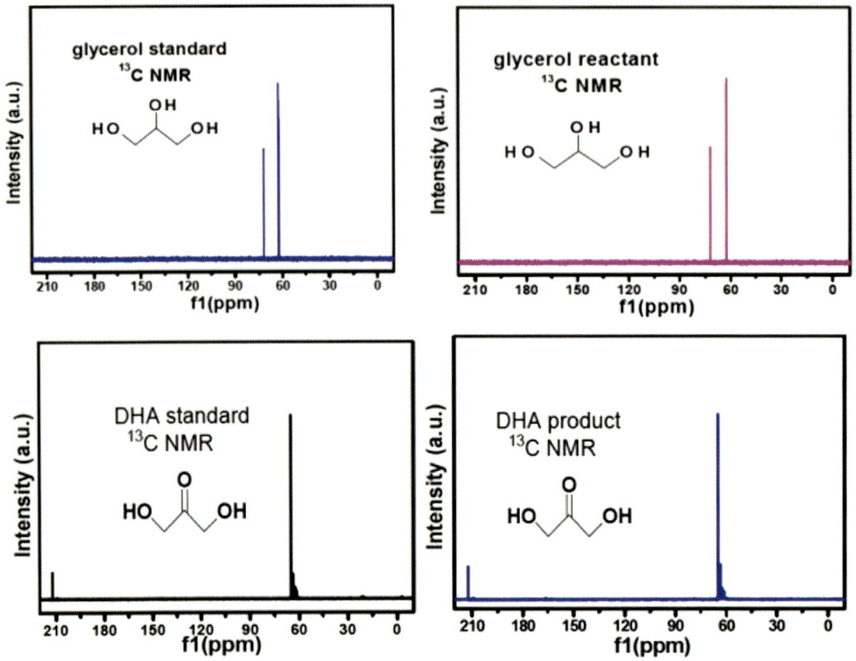

Fig. 4.42 The ^{13}C nuclear magnetic resonance (NMR) spectra for dihydroxyacetone (DHA) and glycerol. The left column displays the ^{13}C NMR spectra of standard samples of glycerol and DHA, while the right column shows the ^{13}C NMR spectra of glycerol and DHA following photocatalytic reaction and subsequent separation of reactants and products[94] (Copyright 2010, Chemical science License Number 1357599-1)

terms of maximizing hydrogen yield while minimizing CO formation. Table 4.4 enumerates the diverse experimental configurations employed in antecedent investigations concerning glycerol-to-hydrogen conversion. Hydrogen gas is isolated and its concentration is quantified using a gas chromatograph furnished with a thermal conductivity detector (TCD).

The efficacy of photocatalytic reactions is often assessed via the Quantum Efficiency (Φ), calculated as the ratio of transported electrons to absorbed photons (Eq. 4.20). However, due to the inherent challenges in quantifying the number of absorbed photons, the Apparent Quantum Efficiency (Φ_{ap}) is frequently employed, which is determined as the ratio of the number of transported electrons to the number of incident photons (Eq. 4.21).

$$\Phi = 100 \times \frac{n(transferred\ electrons)}{n(absorbed\ photons)} \quad (4.20)$$

$$\Phi_{ap} = 100 \times \frac{n(transferred\ electrons)}{n(incident\ photons)} \quad (4.21)$$

Table 4.4 Operating conditions for the valorization of glycerol into hydrogen[153] (Reprinted with permission from [153], Copyright 2020 American Chemical Society)

Photocatalyst	Light source	Photocatalyst load (g/L)	Glycerol concentration (v/v)%	Solution volume (ml)	Quantum efficiency (%)	Ref
Pt/TiO_2	Xe	1.3	0.00073–0.01	60	1.8% (pure water), 70% (glycerol solution) @ λ = 365 nm	[15]
Bi_2O_3-QD/ TiO_2	UV LED	0.625	0.3	80	3.7%@ λ = 365 nm	[154]
Ag_2O/TiO_2	–	2	0–30.24	100	10.9%@ λ = 320–780 nm	[155]
TiO_2, CuO_x/ TiO_2	Hg	2.1	7.3	240	29% @ λ = 365 nm	[156]
M/TiO_2 (M = Ag_2O, Bi_2O_3, ZnO)	UV–vis	0.2	10	100	3.2%@ λ = 320–780 nm	[157]
$Pt/N/TiO_2$ nanotube	Hg	1	50	400	37% @ λ = 365 nm	[158]
Ni/CdS	Hg	0.04	0.15–32.85	10	12.2% @ λ = 410 nm	[159]
ZnO/ZnS	Xe	0.02–4	10	50	13.90%	[160]
ZnO/ZnS	Hg, Xe	30	7	10	22% (UV), 13% (solar) @ λ = 365 nm	[148]
$Pt/Cd_{1-x}Zn_xS$/ ZnO /$Zn(OH)_2$	Hg	0.8	0.365–7.3	–	9.6% @ λ > 420 nm	[161]
Pt/ $Cu_xTi_{1-x}O_{2-\delta}$	Solar, Hg	6.7	33	15	7.5% @ λ = 300–600 nm	[162]
Pt/Bi_4MO_8X (M = Nb, Ta; X = Cl, Br)	Xe	3	20	100	0.4% (visible)@ λ > 420 nm	[163]

The quantum efficiency of a photocatalytic reaction is dictated by the individual efficiencies of each operational phase: photon absorption (η_{PA}), charge separation (η_{CS}), charge transfer (η_{CT}), and charge utilization (η_{CU}). Consequently, variables including the nature of the photocatalyst, the type of light source, the intensity of incident light, the loading of the photocatalyst, the concentration of glycerol in the solution, the volume of the solution, and the reaction temperature can all significantly influence the overall quantum efficiency in the photocatalytic production of hydrogen.

Valorization of glycerol into value-added liquid products

Table 4.5 summarizes diverse experimental configurations employed in previous studies focusing on the valorization of glycerol into value-added liquid products. Generally, the parameters, such as type of photocatalyst, light source, light filter, power range of the light source, quantity of photocatalyst, glycerol concentration, volume of the solution, and reaction temperature, align closely with those pertinent to hydrogen production. Consequently, this segment solely details specific experimental conditions, including the array of liquid products identified, peak glycerol conversion rates, and maximal product selectivity. The liquid-phase photo-conversion of glycerol can yield a broad spectrum of compounds, encompassing dihydroxyacetone, glyceraldehyde, glycolaldehyde, glycolic acid, formic acid, tartaric acid, hydroxypyruvic acid, propylene glycol, methylglyoxal, acetaldehyde, formaldehyde, ethylene glycol, and dimethyl-1,4-dioxane. Within the reaction milieu, only thermally stable compounds are typically discernible. Additionally, glycerol may undergo spontaneous conversion into gaseous species such as CO_2 and H_2.

GAD (glyceraldehyde), DHA (dihydroxyacetone), FORM (formaldehyde), GA (glyceric acid), GCOL (glycolaldehyde), GCOLA (glycolic acid), FA (formic acid), HA (hydroxyacetone), TART (tartronic acid).

4.5 Opportunities and Challenges

Semiconductive photocatalysts present significant potential for the photo-reforming of glycerol into high-value chemicals. Upon light activation, these catalysts facilitate photochemical reactions that transform glycerol into industrially relevant products. Nevertheless, numerous obstacles impede their efficient application. Table 4.6 elucidates the opportunities and challenges tied to the aforementioned semiconductive photocatalysts in the context of glycerol valorization.

4.6 Conclusion and Prospective Applications

The exigency to mitigate the global energy crisis and associated environmental repercussions has led to heightened research for alternative fuels. Biodiesel has garnered attention as a green, renewable energy source; however, its production yields substantial amounts of crude glycerol. Given the limited market utility of this by-product, photocatalysis has been employed for its valorization into higher-value compounds. This review discusses multiple photocatalyst fabrication methods and evaluates their efficacy in glycerol restructuring to generate hydrogen and other high-value liquid products. Despite current advances, challenges persist, impeding photocatalytic efficiency, especially in high-concentration glycerol conversion. Various

Table 4.5 Operating conditions for the valorization of glycerol into value-added liquid products [153] (Reprinted with permission from [153], Copyright 2020 American Chemical Society)

Photocatalyst	Light source	Photocatalyst load (g/L)	Glycerol concentration (v/v)%	Solution volume (ml)	Detected products	Max Glycerol conversion (%)	Max selectivity (%)	Ref
TiO_2	Hg	1–3	2.2	100	DHA, GAD, GA, GCOLA, FA	71	GAD (68), GCOLA (74)	[164]
TiO_2	Hg, fluorescent	0.1–0.4	0.073–1.3	500, 220	DHA, GAD, FA	36	GAD (13)	[15]
Carbon sphere/ TiO_2	Metal halide	1.5	0.0039	200	GAD, FORM, FA	67.5	FA (49)	[33]
M/TiO_2 (M = Pd, Pt, Ru, Rh), Pd/ C_3N_4 Pd/SiO_2	Xe	15.6	7.3–22	1.6	GAD, DHA, GCOLA, FA	20	HA (30)	[165]
M/TiO_2 (M = Bi, Pd, Pt, Au)	Hg	3	6.72	100	DHA, FORM, GAD, GCOL, HPA	92.8	Bi/TiO2: GAD (32.5), Pd/ TiO2: GAD (32.2), Pt/TiO2: FORM (36.8), Au/TiO2: FORM (35.2)	[166]
$Na_4W_{10}O_{32}$	Hg	8	0.073	15	GAD, DHA, GCOLA	37	GAD (59.4)	[167]
Bi_2WO_6	Xe	0.8	0.5	1.5	DHA, GAD	58	DHA (97)	[168]
Bi_2WO_6	Xe	5.3	0.5	1.5	DHA	83	DHA (91)	[94]

(continued)

Table 4.6 (continued)

Photocatalysts	Opportunities	Challenges
Pt/TiO_2	• Pt nanoparticles can separate photo-generated electron–hole pairs, thereby increasing both glycerol conversion rates and catalytic efficiency • Pt nanoparticles can alter reaction pathways, favoring certain products. Optimizing reaction conditions and catalyst properties enables the direct conversion of glycerol to desired chemicals • Pt nanoparticles enhance intermediate electrochemical conversion and hydrogen production, thereby reducing the overpotential required for these reactions and conserving energy • Pt nanoparticles enhance the stability and durability of TiO_2 photocatalysts, thereby preventing the photo-corrosion or degradation of TiO_2 under harsh reaction conditions [18]	• Platinum is a precious metal with a relatively high cost; therefore, the utilization of Pt nanoparticles in large-scale glycerol valorization processes may pose economic challenges • Effective and appropriate synthesis methods should be employed to prepare Pt/TiO_2 photocatalysts, ensuring homogeneous Pt dispersion, appropriate particle size, and controlled loading • Factors such as pH, temperature, catalyst loading, and reaction time must be carefully controlled to achieve the desired outcomes in glycerol valorization [19]

(continued)

Table 4.6 (continued)

Photocatalysts	Opportunities	Challenges
Cu/TiO_2	• Owing to their surface plasmon resonance effect, copper nanoparticles have the capability to broaden the light absorption spectrum of TiO_2 into the visible range, thereby optimizing the harnessing of solar radiation, a significant portion of which comprises visible light • Manipulation of reaction parameters and catalyst characteristics allows for the targeted conversion of glycerol into specific by-products such as hydrogen and glycolic acid • The incorporation of copper nanoparticles enhances charge transfer dynamics and facilitates redox reactions, thereby augmenting the conversion rates and overall catalytic efficiency of glycerol through the effective separation and utilization of photo-generated electron–hole pairs • The inclusion of copper constituents augments the catalyst's resistance to fouling, thus ensuring sustained performance throughout glycerol valorization processes [43]	• While copper offers a cost-effective alternative to noble metals like platinum, the aggregate expenditure for Cu/TiO_2 photocatalysts remains a concern, particularly in large-scale glycerol valorization procedures. Investigating economical synthesis techniques and catalyst recycling strategies is imperative for mitigating this financial constraint • The proficient retrieval and segregation of copper nanoparticles from exhausted Cu/TiO_2 catalysts are critical for sustainable catalyst recycling and waste minimization. The advancement of efficacious recovery methodologies, encompassing solvent extraction, filtration, or magnetic separation, is quintessential for both ecological sustainability and economic feasibility
WO_3	• The WO_3 photocatalyst is characterized by high catalytic efficiency, low cost, elevated catalytic activity, minimal toxicity, abundant availability, and environmental compatibility • Due to its narrow bandgap, elevated electron mobility, and robust chemical stability, WO_3 has found extensive applications in the realms of environmental remediation and solar energy conversion. These applications include the degradation of organic contaminants, the elimination of noxious gases, and hydrogen generation via water splitting	• Limitations in the photocatalytic performance of WO_3 photocatalysts include suboptimal quantum efficiency, lack of selectivity, and a high rate of recombination for photo-generated electron–hole pairs. Thus, targeted strategies are imperative for performance enhancement • Economical synthesis and scalable production of WO_3 photocatalysts are vital for their practical implementation across various domains, such as water purification, air quality management, and solar energy conversion

(continued)

Table 4.6 (continued)

Photocatalysts	Opportunities	Challenges
Bi_2WO_6	• Bi_2WO_6 serves as a semiconductor photocatalyst responsive to visible light, enabling the harnessing of a substantial fraction of the solar spectrum, thereby offering an abundant and sustainable energy source for glycerol valorization [86, 87] • Possessing a suitable band gap, Bi_2WO_6 can directly facilitate the transfer of photo-generated holes to glycerol without the indiscriminate formation of hydroxyl ($\cdot$ OH) radicals. Additionally, it demonstrates elevated photocatalytic activity, thus enabling efficient glycerol conversion [95]	• The modulation of reaction conditions, such as pH, serves as a pivotal factor in steering glycerol conversion towards designated products. Variations in pH during the hydrothermal process engender morphological and surface property distinctions [95] • During the valorization of glycerol, the catalyst may experience phenomena such as photo-corrosion, surface passivation, or aggregation, each contributing to a diminishing performance in its catalytic activities • To alleviate mass transfer restrictions and augment glycerol conversion efficacy, optimization of reactor design, catalyst morphology, and operational parameters is requisite
g-C_3N_4	• Graphitic carbon nitride (g-C_3N_4) possesses a suitable bandgap of 2.7 eV and can be readily synthesized from abundant precursors, thereby enhancing its cost-effectiveness and availability [110, 112] • Owing to its high thermal stability, non-toxic nature, and structural adaptability, g-C_3N_4 finds utility in a diverse range of applications [113]	• Graphitic carbon nitride (g-C_3N_4) is characterized by a limited specific surface area and restricted capabilities in the transfer of photo-generated species [118] • The functional modification of g-C_3N_4 through heteroatom doping can optimize the catalyst's properties, thereby mitigating the aforementioned limitations [120]

(continued)

Table 4.6 (continued)

Photocatalysts	Opportunities	Challenges
ZnO	• ZnO possesses a direct bandgap ranging from 3.30 to 3.37 eV at ambient temperature, coupled with a free exciton binding energy of 60 meV, rendering it compatible with a diverse array of electronic and photonic applications [134] • ZnO is distinguished by elevated electron mobility, thereby enhancing charge transfer efficacy and mitigating recombination-induced energy losses • The material attributes of ZnO include cost-efficiency, non-toxicological nature, and a propensity for sustaining elevated levels of stability and crystallinity	• The applicability of ZnO is constrained predominantly to ultraviolet excitation, which constitutes only a minor fraction of the solar electromagnetic spectrum • The efficacious segregation of photo-generated charge carriers presents a significant hurdle in the implementation of ZnO as a photocatalytic material • The photochemical reactivity of ZnO is substantially modulated by the synthesis techniques and environmental conditions employed, factors which in turn influence the material's morphology, aspect ratio, crystallite dimensions, and crystallographic density [141]
Co-doped ZnO	• The incorporation of cobalt into the ZnO array enhances photocatalytic performance by increasing the surface accessibility of glycerol and modulating the electron states of ZnO [142] • Co-doped ZnO photocatalysts can extend the light absorption range into the visible region, thus enabling them to utilize a larger portion of the solar spectrum • The resulting photocurrent density and charge transfer of the electrode exhibit respective increases of 2.4 and 2.2 times compared to an unmodified ZnO photoanode • Cobalt enhances charge separation efficiency, increases glycerol conversion rates, and improves catalytic performance	• The introduction of cobalt dopants may affect the stability and resistance to photo-corrosion of the catalyst • Achieving a uniform distribution and high dispersion of cobalt within the ZnO matrix is crucial for maximizing catalytic performance • Optimizing the levels of doping, methods of preparation, and conditions for annealing is essential for enhancing the synergistic effects and overall catalytic activity • The separation and purification of products on a large scale, as well as the integration of the photocatalytic process with existing industrial practices, require meticulous engineering and optimization

enhancement strategies have been scrutinized, such as metal co-catalyst deposition, element doping, nanomaterial hybridization, heterojunction system development, pH tuning, and morphological control of the semiconductor photocatalysts. While noble metals like Pt serve as high-performing co-catalysts, their cost has driven research towards more economically viable metals such as Cu. Novel deposition methods for synthesizing nanoparticles of defined sizes and compositions on semiconductor surfaces warrant further exploration. Heterojunction systems, particularly those involving TiO_2, are identified as promising for glycerol valorization.

Despite TiO_2's ubiquity, its performance limitations necessitate metal doping. Metal–organic frameworks (MOFs), specifically Zeolitic Imidazolate Frameworks (ZIFs), have recently drawn attention for their photocatalytic applications, owing to their high surface area and compositional versatility. ZIFs coupled with various semiconductors form heterojunction nanocomposites to address charge separation inefficiency. Cu-loaded ZIF-8, for instance, amplifies the active surface sites, enhancing glycerol valorization performance. Several other semiconductor materials such as WO_3, Bi_2WO_6, g-C_3N_4, and ZnO are also discussed, each with its own set of challenges and solutions. Strategies to improve WO_3 include particle morphological control, elemental doping, co-catalyst hybridization, and heterojunction formation with other semiconductors. Moreover, this review comprehensively illuminates the recent advancements in glycerol valorization for high-value liquid product formation and hydrogen evolution via photo-reforming. The pursuit of solar fuel generation from glycerol valorization, as a sustainable approach, requires additional research. The review also highlights the necessity for synthesizing highly selective photocatalysts. It serves as a comprehensive guide, encapsulating material preparation, characterization, performance metrics, product analysis, and selectivity within the realm of glycerol valorization research.

References

1. Liu C, Hirohara M, Maekawa T, Chang R, Hayashi T, Chiang C-Y (2020) Selective electrooxidation of glycerol to dihydroxyacetone by a non-precious electrocatalyst—CuO. Appl Catal B 265:118543. https://doi.org/10.1016/j.apcatb.2019.118543
2. Vo T-G, Kao C-C, Kuo J-L, Chiu C-C, Chiang C-Y (2020) Unveiling the crystallographic facet dependence of the photoelectrochemical glycerol oxidation on bismuth vanadate. Appl Catal B 278:119303. https://doi.org/10.1016/j.apcatb.2020.119303
3. Rangarajan G, Pal R, Farnood R (2021) Visible light driven selective photocatalytic synthesis of high-value carbonyl compounds from glycerol over Ag-AgBr/TiO_2 nanocomposites in acetonitrile. Mater Lett 292:129535. https://doi.org/10.1016/j.matlet.2021.129535
4. Yan H, Shen Q, Sun Y, Zhao S, Lu R, Gong M, Liu Y, Zhou X, Jin X, Feng X et al (2021) Tailoring facets of α-Mn_2O_3 microcrystalline catalysts for enhanced selective oxidation of glycerol to glycolic acid. ACS Catal 11(11):6371–6383. https://doi.org/10.1021/acscatal.1c01566
5. Yu J, Dappozze F, Martín-Gomez J, Hidalgo-Carrillo J, Marinas A, Vernoux P, Caravaca A, Guillard C (2021) Glyceraldehyde production by photocatalytic oxidation of glycerol on WO_3-based materials. Appl Catal B 299:120616
6. Zhang P, Yue C, Fan M, Haryonob A, Leng Y, Jiang P (2021) The selective oxidation of glycerol over metal-free photocatalysts: insights into the solvent effect on catalytic efficiency and product distribution. Catal Sci Technol 11(10):3385–3392. https://doi.org/10.1039/D1CY00360G
7. Pei YH, Pi ZF, Zhong H, Cheng J, Jin FM (2022) Glycerol oxidation-assisted electrochemical CO_2 reduction for the dual production of formate. J Mater Chem A 10(3):1309–1319. https://doi.org/10.1039/d1ta07119j
8. Qiu KB, Shu YX, Zhang J, Gao LJ, Xiao GM (2022) Effective and stable zeolite imidazole framework-supported copper nanoparticles (Cu/ZIF-8) for glycerol to lactic acid. Catal Lett 152(1):172–186. https://doi.org/10.1007/s10562-021-03610-y

9. Vo T-G, Ho P-Y, Chiang C-Y (2022) Operando mechanistic studies of selective oxidation of glycerol to dihydroxyacetone over amorphous cobalt oxide. Appl Catal B 300:120723. https://doi.org/10.1016/j.apcatb.2021.120723
10. Gupta G, Roling LT (2013) Selective conversion of glycerol to value-added C3 products: effect of catalyst surface structure. Chem Cat Chem 15(2):e202201188. https://doi.org/10.1002/cctc.202201188. Acccessed 23 Feb 2023
11. Keogh J, Tiwari MS, Manyar H (2019) Esterification of glycerol with acetic acid using nitrogen-based brønsted-acidic ionic liquids. Ind Eng Chem Res 58(37):17235–17243. https://doi.org/10.1021/acs.iecr.9b01223
12. Olson AL, Tunér M, Verhelst S (2023) A concise review of glycerol derivatives for use as fuel additives. Heliyon 9(1):e13041. https://doi.org/10.1016/j.heliyon.2023.e13041
13. Escamilla-Mejía JC, Hidalgo-Carrillo J, Martín-Gómez J, López-Tenllado FJ, Estévez-Toledano RC, Marinas A, Urbano FJ (2022) Pt Preferential incorporation onto TiO_2 IN TiO_2-carbon composites for hydrogen production from glycerol photoreforming. Catal Today 113943. https://doi.org/10.1016/j.cattod.2022.11.004
14. Rueda-Navarro CM, Ferrer B, Baldoví HG, Navalón S (2022) Photocatalytic hydrogen production from glycerol aqueous solutions as sustainable feedstocks using Zr-based UiO-66 materials under simulated sunlight irradiation. Nanomaterials (Basel, Switzerland) 12(21). https://doi.org/10.3390/nano12213808. From NLM
15. Kondarides DI, Daskalaki VM, Patsoura A, Verykios XE (2008) Hydrogen production by photo-induced reforming of biomass components and derivatives at ambient conditions. Catal Lett 122(1–2):26–32. https://doi.org/10.1007/s10562-007-9330-3
16. Patsoura A, Kondarides DI, Verykios XE (2006) Enhancement of photoinduced hydrogen production from irradiated Pt/TiO_2 suspensions with simultaneous degradation of azo-dyes. Appl Catal B-Environ 64(3–4):171–179. https://doi.org/10.1016/j.apcatb.2005.11.015
17. Patsoura A, Kondarides DI, Verykios XE (2007) Photocatalytic degradation of organic pollutants with simultaneous production of hydrogen. Catal Today 124(3–4):94–102. https://doi.org/10.1016/j.cattod.2007.03.028
18. Daskalaki VM, Kondarides DI (2009) Efficient production of hydrogen by photo-induced reforming of glycerol at ambient conditions. Catal Today 144(1–2):75–80
19. Panagiotopoulou P, Karamerou EE, Kondarides DI (2013) Kinetics and mechanism of glycerol photo-oxidation and photo-reforming reactions in aqueous TiO_2 and Pt/TiO_2 suspensions. Catal Today 209:91–98
20. Minero C, Vione D (2006) A quantitative evalution of the photocatalytic performance of TiO_2 slurries. Appl Catal B-Environ 67(3–4):257–269. https://doi.org/10.1016/j.apcatb.2006.05.011
21. Minero C (1999) Kinetic analysis of photoinduced reactions at the water semiconductor interface. Catal Today 54(2–3):205–216. https://doi.org/10.1016/S0920-5861(99)00183-2
22. Maurino V, Bedini A, Minella M, Rubertelli F, Pelizzetti E, Minero C (2008) Glycerol transformation through photocatalysis: a possible route to value added chemicals. J Adv Oxid Technol 11(2):184–192
23. Ollis DF (2005) Kinetics of liquid phase photocatalyzed reactions: an illuminating approach. J Phys Chem B 109(6):2439–2444. https://doi.org/10.1021/jp040236f
24. Lewandowski M, Ollis DF (2003) A two-site kinetic model simulating apparent deactivation during photocatalytic oxidation of aromatics on titanium dioxide (TiO_2). Appl Catal B-Environ 43(4):309–327. https://doi.org/10.1016/S0926-3373(02)00310-7
25. Lewandowski M, Ollis DF (2003) Extension of a Two-Site transient kinetic model of TiO_2 deactivation during photocatalytic oxidation of aromatics: concentration variations and catalyst regeneration studies. Appl Catal B-Environ 45(3):223–238. https://doi.org/10.1016/S0926-3373(03)00165-6
26. Minero C, Bedini A, Maurino V (2012) Glycerol as a probe molecule to uncover oxidation mechanism in photocatalysis. Appl Catal B-Environ 128:135–143. https://doi.org/10.1016/j.apcatb.2012.02.014

27. Minero C, Mariella G, Maurino V, Pelizzetti E (2000) Photocatalytic transformation of organic compounds in the presence of inorganic anions. 1. Hydroxyl-mediated and direct electron-transfer reactions of phenol on a titanium dioxide-fluoride system. Langmuir 16(6):2632–2641. https://doi.org/10.1021/la9903301
28. Augugliaro V, Palmisano L, Sclafani A, Minero C, Pelizzetti E (1988) Photocatalytic degradation of phenol in aqueous titanium-dioxide dispersions. Toxicol Environ Chem 16(2):89–109. https://doi.org/10.1080/02772248809357253
29. Maurino V, Minero C, Mariella G, Pelizzetti E (2005) Sustained production of H_2O_2 on irradiated TiO_2—fluoride systems. Chem Commun 20:2627–2629. https://doi.org/10.1039/b418789j
30. Park H, Choi W (2004) Effects of TiO_2 surface fluorination on photocatalytic reactions and photoelectrochemical behaviors. J Phys Chem B 108(13):4086–4093. https://doi.org/10.1021/jp036735i
31. Reuvers AP, Greenstock CL, Borsa J, Chapman JD (1973) Letter: Studies on the mechanism of chemical radioprotection by dimethyl sulphoxide. Int J Radiat Biol Relat Stud Phys Chem Med 24(5):533–536. https://doi.org/10.1080/09553007314551431FromNLMMedline
32. Dodekatos G, Schunemann S, Tuysuz H (2018) Recent advances in thermo-, photo-, and electrocatalytic glycerol oxidation. Acs Catal 8(7):6301–6333. https://doi.org/10.1021/acscatal.8b01317
33. Payormhorm J, Idem R (2020) Synthesis of C-doped TiO_2 by sol-microwave method for photocatalytic conversion of glycerol to value-added chemicals under visible light. Appl Catal a-Gen 590. https://doi.org/10.1016/j.apcata.2019.117362.
34. Estahbanati MRK, Feilizadeh M, Attar F, Iliuta MC (2021) Current developments and future trends in photocatalytic glycerol valorization: process analysis. React Chem Eng 6(2):197–219. https://doi.org/10.1039/d0re00382d
35. Andronic L, Lelis M, Enesca A, Karazhanov S (2022) Photocatalytic activity of defective black-titanium oxide photocatalysts towards pesticide degradation under UV/VIS irradiation. Surf Interfaces 32. https://doi.org/10.1016/j.surfin.2022.102123
36. Gemeda TN, Chang L-H, Liang YT, Phan VHK, Fadhilah G, Prasetyo F, Ahmed MT (2022) A recent review on photocatalytic CO_2 reduction in generating sustainable carbon-based fuels. Photocatal Activities Environ Rem Energy Convers 205–261
37. Varma RS, Len C (2019) Glycerol valorization under continuous flow conditions-recent advances. Curr Opin Green Sust 15:83–90. https://doi.org/10.1016/j.cogsc.2018.11.003
38. Wang M, Nie B, Yee K-K, Bian H, Lee C, Lee HK, Zheng B, Lu J, Luo L, Li YY (2016) Low-temperature fabrication of brown TiO_2 with enhanced photocatalytic activities under visible light. Chem Commun 52(14):2988–2991
39. Karimi Estahbanati M, Feilizadeh M, Attar F, Iliuta MC (2020) Current developments and future trends in photocatalytic glycerol valorization: photocatalyst development. Ind Eng Chem Res 59(52):22330–22352
40. Escobedo S, de Lasa H (2021) Synthesis and performance of photocatalysts for photocatalytic hydrogen production: future perspectives. Catalysts 11(12):1505
41. Payormhorm J, Idem R (2020) Synthesis of C-doped TiO_2 by sol-microwave method for photocatalytic conversion of glycerol to value-added chemicals under visible light. Appl Catal A 590:117362
42. Li ZB, Bian HD, Xiao XF, Shen JD, Zhao CH, Lu J, Li YY (2019) Defective black TiO_2 nanotube arrays for enhanced photocatalytic and photoelectrochemical applications. Acs Appl Nano Mater 2(11):7372–7378. https://doi.org/10.1021/acsanm.9b01878
43. Li X, Chen F, Lian C, Zheng S, Hu Q, Duo S, Li W, Hu C (2016) Au/TiO 2/graphene composite with enhanced photocatalytic activity under both UV and visible light irradiation. J Cluster Sci 27:1877–1892
44. Beltram A, Romero-Ocana I, Jaen JJD, Montini T, Fornasiero P (2016) Photocatalytic valorization of ethanol and glycerol over TiO_2 polymorphs for sustainable hydrogen production. Appl Catal A 518:167–175

45. Maslova V, Fasolini A, Offidani M, Albonetti S, Basile F (2021) Solar-driven valorization of glycerol towards production of chemicals and hydrogen. Catal Today 380:147–155
46. Avilés-García O, Mendoza-Zepeda A, Regalado-Méndez A, Espino-Valencia J, Martínez-Vargas SL, Romero R, Natividad R (2022) Photo-oxidation of glycerol catalyzed by Cu/TiO_2. Catalysts 12(8):835
47. Maslova V, Quadrelli EA, Gaval P, Fasolini A, Albonetti S, Basile F (2021) Highly-dispersed ultrafine Pt nanoparticles on microemulsion-mediated TiO_2 for production of hydrogen and valuable chemicals via oxidative photo-dehydrogenation of glycerol. J Environ Chem Eng 9(2):105070
48. Jedsukontorn T, Ueno T, Saito N, Hunsom M (2018) Narrowing band gap energy of defective black TiO_2 fabricated by solution plasma process and its photocatalytic activity on glycerol transformation. J Alloy Compd 757:188–199
49. Jedsukontorn T, Saito N, Hunsom M (2017) Photocatalytic behavior of metal-decorated TiO_2 and their catalytic activity for transformation of glycerol to value added compounds. Mol Catal 432:160–171
50. Dodekatos G, Tüysüz H (2016) Plasmonic Au/TiO_2 nanostructures for glycerol oxidation. Catal Sci Technol 6(19):7307–7315
51. Kozlova EA, Kurenkova AY, Gerasimov EY, Gromov NV, Medvedeva TB, Saraev AA, Kaichev VV (2021) Comparative study of photoreforming of glycerol on Pt/TiO_2 and CuOx/TiO_2 photocatalysts under UV light. Mater Lett 283:128901
52. Montini T, Gombac V, Sordelli L, Delgado JJ, Chen X, Adami G, Fornasiero P (2011) Nanostructured Cu/TiO_2 photocatalysts for H_2 production from ethanol and glycerol aqueous solutions. Chem Cat Chem 3(3):574–577
53. Lalitha K, Sadanandam G, Kumari VD, Subrahmanyam M, Sreedhar B, Hebalkar NY (2010) Highly stabilized and finely dispersed Cu_2O/TiO_2: a promising visible sensitive photocatalyst for continuous production of hydrogen from glycerol: water mixtures. J Phys Chem C 114(50):22181–22189
54. Jung M, Hart JN, Boensch D, Scott J, Ng YH, Amal R (2016) Hydrogen evolution via glycerol photoreforming over Cu–Pt nanoalloys on TiO_2. Appl Catal A 518:221–230
55. Kumaravel V, Mathew S, Bartlett J, Pillai SC (2019) Photocatalytic hydrogen production using metal doped TiO_2: a review of recent advances. Appl Catal B 244:1021–1064
56. Zhou K, Mousavi B, Luo Z, Phatanasri S, Chaemchuen S, Verpoort F (2017) Characterization and properties of Zn/Co zeolitic imidazolate frameworks vs. ZIF-8 and ZIF-67. J Mater Chem A 5(3):952–957. https://doi.org/10.1039/C6TA07860E
57. Cheng J, Guo H, Yang X, Mao Y, Qian L, Zhu Y, Yang W (2021) Phosphotungstic acid-modified zeolite imidazolate framework (ZIF-67) as an acid-base bifunctional heterogeneous catalyst for biodiesel production from microalgal lipids. Energy Convers Manage 232:113872. https://doi.org/10.1016/j.enconman.2021.113872
58. Hu C, Yoshida M, Chen H-C, Tsunekawa S, Lin Y-F, Huang J-H (2021) Production of glycerol carbonate from carboxylation of glycerol with CO_2 using ZIF-67 as a catalyst. Chem Eng Sci 235:116451. https://doi.org/10.1016/j.ces.2021.116451
59. Varangane S, Vennapoosa CS, Tiwari A, Nataraj SK, Yendrapati TP, Pal U (2022) In situ synthesis of Cu-doped ZIF-8 for efficient photocatalytic water splitting. Appl. Organomet Chem 36(9). https://doi.org/10.1002/aoc.6815
60. Ali RN, Qureshi WA, Yaseen M, Jiang H, Wang L, Yang J, Liu Q (2023) Exploring the recent advancements in metal-organic framework-based photocatalysts for hydrogen production. Mater Today Sustain 22:100337. https://doi.org/10.1016/j.mtsust.2023.100337
61. Zhao L, Gong C, Chen X, He X, Chen H, Du X, Wang D, Fang W, Zhang H, Li W (2023) ZIF-67 derived Mo-CoS2 nanoparticles embedded in hierarchically porous carbon hollow sphere for efficient overall water splitting. Appl Surf Sci 623:157030. https://doi.org/10.1016/j.apsusc.2023.157030
62. El Ouardi M, El aouni A, Ait Ahsaine H, Zbair M, BaQais A, Saadi M (2022) ZIF-8 metal organic framework composites as hydrogen evolution reaction photocatalyst: a review of the current state. Chemosphere 308:136483. https://doi.org/10.1016/j.chemosphere.2022.136483

63. El Ouardi M, El Aouni A, Ahsaine HA, Zbair M, BaQais A, Saadi M (2022) ZIF-8 metal organic framework composites as hydrogen evolution reaction photocatalyst: a review of the current state. Chemosphere 308. https://doi.org/10.1016/j.chemosphere.2022.136483
64. Yurtsever HA, Çetin AE (2021) Fabrication of ZIF-8 decorated copper doped TiO_2 nanocomposite at low ZIF-8 loading for solar energy applications. Colloids Surf, A 625:126980. https://doi.org/10.1016/j.colsurfa.2021.126980
65. Mohamed RM, Shawky A (2022) Visible-light-driven hydrogen production over ZIF-8 derived CO_3O_4/ZnO S-scheme based p-n heterojunctions. Opt Mater 124:112012. https://doi.org/10.1016/j.optmat.2022.112012
66. Wu CM, Naseem S, Chou MH, Wang JH, Jian YQ (2019) Recent advances in tungsten-oxide-based materials and their applications. Front Mater 6. https://doi.org/10.3389/fmats.2019.00049
67. Kondalkar V, Kharade R, Mali S, Mane R, Patil P, Patil P, Choudhury S, Bhosale P (2014) Nanobrick-like WO_3 thin films: hydrothermal synthesis and electrochromic application. Superlattices Microstruct 73:290–295
68. Adhikari S, Sarkar D, Maiti HS (2014) Synthesis and characterization of WO_3 spherical nanoparticles and nanorods. Mater Res Bull 49:325–330. https://doi.org/10.1016/j.materresbull.2013.08.028
69. Thilagavathi T, Venugopal D, Marnadu R, Chandrasekaran J, Alshahrani T, Shkir M (2021) An investigation on microstructural, morphological, optical, photoluminescence and photocatalytic activity of WO_3 for photocatalysis applications: an effect of annealing. J Inorg Organomet Polym Mater 31:1217–1230
70. Zhang Q, Xu MK, Shen ZR, Wei Q (2017) A nanostructured Cr_2O_3/WO_3 p-n junction sensor for highly sensitive detection of butanone. J Mater Sci-Mater El 28(16):12056–12062. https://doi.org/10.1007/s10854-017-7017-0
71. Shahzad K, Tahir MB, Ashraf M, Nawaz T, Khalid NR, Kabli MR (2020) Interfacial growth of activated carbon on WO_3 nanoplates for enhanced photocatalytic activity by surface plasmon resonance. Plasmonics 15(4):1205–1212. https://doi.org/10.1007/s11468-020-01135-5
72. Yu J, Dappozze F, Martin-Gomez J, Hidalgo-Carrillo J, Marinas A, Vernoux P, Caravaca A, Guillard C (2021) Glyceraldehyde production by photocatalytic oxidation of glycerol on WO_3-based materials. Appl Catal B-Environ 299. https://doi.org/10.1016/j.apcatb.2021.120616
73. Dutta V, Sharma S, Raizada P, Thakur VK, Khan AAP, Saini V, Asiri AM, Singh P (2021) An overview on WO_3 based photocatalyst for environmental remediation. J Environ Chem Eng 9(1). https://doi.org/10.1016/j.jece.2020.105018
74. Choi HG, Jung YH, Kim DK (2005) Solvothermal synthesis of tungsten oxide nanorod/nanowire/nanosheet. J Am Ceram Soc 88(6):1684–1686. https://doi.org/10.1111/j.1551-2916.2005.00341.x
75. Atsuhiro T, Keiji H, Hiroshi K (2014) Visible-light-induced hydrogen and oxygen formation over Pt/Au/WO_3 photocatalyst utilizing two types of photoabsorption due to surface plasmon resonance and band-gap excitation
76. Chai F, Tan RX, Cao FH, Zhai FY, Wang XH, Shao CL, Liu YC (2007) Dendritic and tubular tungsten oxide by surface sol-gel mineralisation of cellulosic substance. Mater Lett 61(18):3939–3941. https://doi.org/10.1016/j.matlet.2006.12.087
77. Luo B, Liu G, Wang LZ (2016) Recent advances in 2D materials for photocatalysis. Nanoscale 8(13):6904–6920. https://doi.org/10.1039/c6nr00546b
78. Rong RC, Wang LM (2021) Synthesis of hierarchical hollow nest-like WO_3 micro/nanostructures with enhanced visible light-driven photocatalytic activity. J Alloy Compd 850. https://doi.org/10.1016/j.jallcom.2020.156742
79. Xie YP, Liu G, Yin LC, Cheng HM (2012) Crystal facet-dependent photocatalytic oxidation and reduction reactivity of monoclinic WO_3 for solar energy conversion. J Mater Chem 22(14):6746–6751. https://doi.org/10.1039/c2jm16178h
80. Song H, Li YG, Lou ZR, Xiao M, Hu L, Ye ZZ, Zhu LP (2015) Synthesis of Fe-doped WO_3 nanostructures with high visible-light-driven photocatalytic activities. Appl Catal B-Environ 166:112–120. https://doi.org/10.1016/j.apcatb.2014.11.020

81. Jamila GS, Sajjad S, Leghari SAK, Long M (2020) Nitrogen doped carbon quantum dots and GO modified WO_3 nanosheets combination as an effective visible photo catalyst. J Hazard Mater 382. https://doi.org/10.1016/j.jhazmat.2019.121087
82. Wicaksana Y, Liu S, Scott J, Amal R (2014) Tungsten trioxide as a visible light photocatalyst for volatile organic carbon removal. Molecules 19(11):17747–17762. https://doi.org/10.3390/molecules191117747
83. DePuccio DP, Botella P, O'Rourke B, Landry CC (2015) Degradation of methylene blue using porous WO_3, SiO_2–WO_3, and their Au-loaded analogs: adsorption and photocatalytic studies. ACS Appl Mater Interfaces 7(3):1987–1996
84. Cui L, Ding X, Wang Y, Shi H, Huang L, Zuo Y, Kang S (2017) Facile preparation of Z-scheme WO_3/g-C_3N_4 composite photocatalyst with enhanced photocatalytic performance under visible light. Appl Surf Sci 391:202–210
85. Toledo-Camacho SY, Rey A, Maldonado MI, Llorca J, Contreras S, Medina F (2021) Photocatalytic hydrogen production from water-methanol and -glycerol mixtures using Pd/TiO_2 (–WO_3) catalysts and validation in a solar pilot plant. Int J Hydrogen Energ 46(73):36152–36166. https://doi.org/10.1016/j.ijhydene.2021.08.141
86. Das S, Ohkubo T, Kasai S, Kozuka Y (2021) Deterministic influence of substrate-induced oxygen vacancy diffusion on Bi_2WO_6 thin film growth. Cryst Growth Des 21(1):625–630. https://doi.org/10.1021/acs.cgd.0c01428
87. Noguchi Y, Murata K, Miyayama M (2007) Effects of defect control on the polarization properties in Bi_2WO_6-based single crystals. Ferroelectrics 355:55–60. https://doi.org/10.1080/00150190701515915
88. Withers RL, Thompson JG, Wallenberg LR, Fitz Gerald JD, Anderson JS, Hyde BG (1988) A transmission electron-microscope and group theoretical-study of the new Bi-based high-Tc superconductors and some closely related aurivillius phases. J Phys C Solid State 21(36):6067–6083. https://doi.org/10.1088/0022-3719/21/36/007
89. Shi JW, Liang Y, Li Z, Fang B (2019) Synthesis, microstructure and photodegradation activity of bismuth tungsten oxides. Chem Select 4(17):5010–5018. https://doi.org/10.1002/slct.201901323
90. Chankhanittha T, Somaudon V, Photiwat T, Hemavibool K, Nanan S (2021) Preparation, characterization, and photocatalytic study of solvothermally grown CTAB-capped Bi_2WO_6 photocatalyst toward photodegradation of Rhodamine B dye. Opt Mater 117. https://doi.org/10.1016/j.optmat.2021.111183
91. Lu Q, Dong CC, Wei F, Li JD, Wang Z, Mu W, Han XJ (2022) Rational fabrication of Bi_2WO_6 decorated TiO_2 nanotube arrays for photocatalytic degradation of organic pollutants. Mater Res Bull 145. https://doi.org/10.1016/j.materresbull.2021.111563
92. Qian XF, Yue DT, Tian ZY, Reng M, Zhu Y, Kan M, Zhang TY, Zhao YX (2016) Carbon quantum dots decorated Bi_2WO_6 nanocomposite with enhanced photocatalytic oxidation activity for VOCs. Appl Catal B-Environ 193:16–21. https://doi.org/10.1016/j.apcatb.2016.04.009
93. Lopez-Tenllado FJ, Murcia-Lopez S, Gomez DM, Marinas A, Marinas JM, Urbano FJ, Navio JA, Hidalgo MC, Gatica JM (2015) A comparative study of Bi_2WO_6, CeO_2, and TiO_2 as catalysts for selective photo-oxidation of alcohols to carbonyl compounds. Appl Catal a-Gen 505:375–381. https://doi.org/10.1016/j.apcata.2015.08.013
94. Zhang YH, Zhang N, Tang ZR, Xu YJ (2013) Identification of Bi_2WO_6 as a highly selective visible-light photocatalyst toward oxidation of glycerol to dihydroxyacetone in water. Chem Sci 4(4):1820–1824. https://doi.org/10.1039/c3sc50285f
95. Marinho JZ, Nascimento LL, Santos ALR, Faria AM, Machado AEH, Patrocinio AOT (2022) On the influence of hydrothermal treatment pH on the performance of Bi_2WO_6 as photocatalyst in the glycerol photoreforming. Photoch Photobio Sci 21(9):1659–1675. https://doi.org/10.1007/s43630-022-00249-5
96. Hatchard C, Parker CA (1956) A new sensitive chemical actinometer-II. Potassium ferrioxalate as a standard chemical actinometer. Proc Royal Soc Lond Ser A Math Phys Sci 235(1203):518–536

97. Farnetti E, Crotti C (2016) Selective oxidation of glycerol to formic acid catalyzed by iron salts. Catal Commun 84:1–4. https://doi.org/10.1016/j.catcom.2016.05.014
98. Kwon Y, Schouten KJP, Koper MTM (2011) Mechanism of the catalytic oxidation of glycerol on polycrystalline gold and platinum electrodes. Chem Cat Chem 3(7):1176–1185. https://doi.org/10.1002/cctc.201100023
99. Xu SG, Tian Q, Xiao Y, Zhang WY, Liao SQ, Li JM, Hu CW (2022) Regulating the competitive reaction pathway in glycerol conversion to lactic acid/glycolic acid selectively. J Catal 413:407–416. https://doi.org/10.1016/j.jcat.2022.07.003
100. Kumbhar S, De M (2023) Carbon quantum dots modified Bi_2WO_6 nanosheets for photo-oxidation of glycerol to value-added product. Mater Today: Proc 76:271–278
101. Wan C, Zhou L, Sun L, Xu LX, Cheng DG, Chen FQ, Zhan XL, Yang YR (2020) Boosting visible-light-driven hydrogen evolution from formic acid over AgPd/2D g-C_3N_4 nanosheets Mott-Schottky photocatalyst. Chem Eng J 396. https://doi.org/10.1016/j.cej.2020.125229
102. Gogoi D, Shah AK, Qureshi M, Golder AK, Peela NR (2021) Silver grafted graphitic-carbon nitride ternary hetero-junction Ag/gC_3N_4(urea)-gC_3N_4(thiourea) with efficient charge transfer for enhanced visible-light photocatalytic green H_2 production. Appl Surf Sci 558. https://doi.org/10.1016/j.apsusc.2021.149900
103. Zhang LS, Wang WZ, Chen ZG, Zhou L, Xu HL, Zhu W (2007) Fabrication of flower-like Bi_2WO_6 superstructures as high performance visible-light driven photocatalysts. J Mater Chem 17(24):2526–2532. https://doi.org/10.1039/b616460a
104. Hasija V, Kumar A, Sudhaik A, Raizada P, Singh P, Le QV, Le TT, Nguyen VH (2021) Step-scheme heterojunction photocatalysts for solar energy, water splitting, CO_2 conversion, and bacterial inactivation: a review. Environ Chem Lett 19(4):2941–2966. https://doi.org/10.1007/s10311-021-01231-w
105. Zhang LY, Zhang JJ, Yu HG, Yu JG (2022) Emerging S-scheme photocatalyst. Adv Mater 34(11). https://doi.org/10.1002/adma.202107668
106. Karimi Estahbanati MR, Feilizadeh M, Iliuta MC (2017) Photocatalytic valorization of glycerol to hydrogen: optimization of operating parameters by artificial neural network. Appl Catal B-Environ 209:483–492. https://doi.org/10.1016/j.apcatb.2017.03.016
107. El-Shishtawy RM, Shawky A, Alorfi HS, Hussein MA, Mohamed RM (2023) Efficient visible-light-driven H_2 evolution over sol-gel processed Bi_2WO_6 nanocrystals anchored with Ag_2O support. Ceram Int
108. Maeda K, Wang X, Nishihara Y, Lu D, Antonietti M, Domen K (2009) Photocatalytic activities of graphitic carbon nitride powder for water reduction and oxidation under visible light. J Phys Chem C 113(12):4940–4947
109. Liu X, Ma R, Zhuang L, Hu B, Chen J, Liu X, Wang X (2021) Recent developments of doped g-C_3N_4 photocatalysts for the degradation of organic pollutants. Crit Rev Environ Sci Technol 51(8):751–790
110. Zeng Y, Wang Y, Chen J, Jiang Y, Kiani M, Li B, Wang R (2016) Fabrication of high-activity hybrid $NiTiO_3$/g-C_3N_4 heterostructured photocatalysts for water splitting to enhanced hydrogen production. Ceram Int 42(10):12297–12305
111. Zhang P, Wang T, Zeng H (2017) Design of Cu-Cu_2O/g-C_3N_4 nanocomponent photocatalysts for hydrogen evolution under visible light irradiation using water-soluble Erythrosin B dye sensitization. Appl Surf Sci 391:404–414
112. Qin J, Zeng H (2017) Photocatalysts fabricated by depositing plasmonic Ag nanoparticles on carbon quantum dots/graphitic carbon nitride for broad spectrum photocatalytic hydrogen generation. Appl Catal B 209:161–173
113. Ong W-J, Tan L-L, Ng YH, Yong S-T, Chai S-P (2016) Graphitic carbon nitride (g-C_3N_4)-based photocatalysts for artificial photosynthesis and environmental remediation: are we a step closer to achieving sustainability? Chem Rev 116(12):7159–7329
114. Cao S-W, Liu X-F, Yuan Y-P, Zhang Z-Y, Liao Y-S, Fang J, Loo SCJ, Sum TC, Xue C (2014) Solar-to-fuels conversion over In_2O_3/g-C_3N_4 hybrid photocatalysts. Appl Catal B 147:940–946

115. Tay Q, Kanhere P, Ng CF, Chen S, Chakraborty S, Huan ACH, Sum TC, Ahuja R, Chen Z (2015) Defect engineered g-C_3N_4 for efficient visible light photocatalytic hydrogen production. Chem Mater 27(14):4930–4933
116. Wang Q, Fang Z, Zhang W, Zhang D (2022) High-efficiency g-C_3N_4 based photocatalysts for CO_2 reduction: modification methods. Advanced Fiber Materials 4(3):342–360
117. Wang C-C, Yi X-H, Wang P (2019) Powerful combination of MOFs and C_3N_4 for enhanced photocatalytic performance. Appl Catal B 247:24–48
118. Ma X, Lv Y, Xu J, Liu Y, Zhang R, Zhu Y (2012) A strategy of enhancing the photoactivity of g-C_3N_4 via doping of nonmetal elements: a first-principles study. J Phys Chem C 116(44):23485–23493
119. Patnaik S, Sahoo DP, Parida K (2021) Recent advances in anion doped g-C_3N_4 photocatalysts: a review. Carbon 172:682–711
120. Zhou H, Wang M, Wang F (2022) Oxygen-controlled photo-reforming of biopolyols to CO over Z-scheme CdS@ g-C_3N_4. Chem 8(2):465–479
121. Yasuda M, Matsumoto T, Yamashita T (2018) Sacrificial hydrogen production over TiO_2-based photocatalysts: polyols, carboxylic acids, and saccharides. Renew Sustain Energy Rev 81:1627–1635
122. Bühler W, Dinjus E, Ederer H, Kruse A, Mas C (2002) Ionic reactions and pyrolysis of glycerol as competing reaction pathways in near-and supercritical water. J Supercrit Fluids 22(1):37–53
123. Asghar A, Raman AAA, Daud WMAW (2015) Advanced oxidation processes for in-situ production of hydrogen peroxide/hydroxyl radical for textile wastewater treatment: a review. J Clean Prod 87:826–838
124. Li Q, Batchelor-McAuley C, Lawrence NS, Hartshorne RS, Compton RG (2013) Anomalous solubility of oxygen in acetonitrile/water mixture containing tetra-n-butylammonium perchlorate supporting electrolyte; the solubility and diffusion coefficient of oxygen in anhydrous acetonitrile and aqueous mixtures. J Electroanal Chem 688:328–335. https://doi.org/10.1016/j.jelechem.2012.07.039
125. Millero FJ, Huang F, Graham TB (2003) Solubility of oxygen in some 1–1, 2–1, 1–2, and 2–2 electrolytes as a function of concentration at 25 °C. J Solution Chem 32(6):473–487. https://doi.org/10.1023/A:1025301314462
126. Mohamed R, Aazam ES (2011) Characterization and catalytic properties of nano-sized Au metal catalyst on titanium containing high mesoporous silica (Ti-HMS) synthesized by photo-assisted deposition and impregnation methods. Int J Photoenergy
127. Chen F, Yang H, Wang X, Yu H (2017) Facile synthesis and enhanced photocatalytic H_2-evolution performance of NiS_2-modified g-C_3N_4 photocatalysts. Chin J Catal 38(2):296–304
128. Alhaddad M, Navarro R, Hussein M, Mohamed R (2020) Visible light production of hydrogen from glycerol over Cu_2O-gC_3N_4 nanocomposites with enhanced photocatalytic efficiency. J Market Res 9(6):15335–15345
129. Munusamy TD, Khan MR, Chin SY (2022) Photoelectrochemical performance of P@ MoS2 for hydrogen evolution reaction. Mater Today: Proc 48:766–770
130. Liu GQ, Xue MW, Liu QP, Yang H, Zhou YM (2019) Carbon doped honeycomb-like graphitic carbon nitride for photocatalytic hydrogen production. J Colloid Interf Sci 552:728–734. https://doi.org/10.1016/j.jcis.2019.05.106
131. Munusamy TD, Chin SY, Khan MMR (2022) Optimization of process parameters for photoreforming of hydrogen evolution via response surface methodology (RSM): a study using carbon@exfoliated g-C_3N_4. Chem Eng Res Des 177:513–525. https://doi.org/10.1016/j.cherd.2021.10.028
132. Samadi M, Zirak M, Naseri A, Khorashadizade E, Moshfegh AZ (2016) Recent progress on doped ZnO nanostructures for visible-light photocatalysis. Thin Solid Films 605:2–19. https://doi.org/10.1016/j.tsf.2015.12.064
133. Rokade A, Rondiya S, Sharma V, Prasad M, Pathan H, Jadkar S (2017) Electrochemical synthesis of 1D ZnO nanoarchitectures and their role in efficient photoelectrochemical splitting of water. J Solid State Electrochem 21(9):2639–2648. https://doi.org/10.1007/s10008-016-3427-9

134. Govatsi K, Seferlis A, Neophytides SG, Yannopoulos SN (2018) Influence of the morphology of ZnO nanowires on the photoelectrochemical water splitting efficiency. Int J Hydrogen Energ 43(10):4866–4879. https://doi.org/10.1016/j.ijhydene.2018.01.087
135. Simon H, Krekeler T, Schaan G, Mader W (2013) Metal-seeded growth mechanism of ZnO nanowires. Cryst Growth Des 13(2):572–580. https://doi.org/10.1021/cg301640v
136. Galdamez-Martinez A, Bai Y, Santana G, Sprick RS, Dutt A (2020) Photocatalytic hydrogen production performance of 1-D ZnO nanostructures: Role of structural properties. Int J Hydrogen Energ 45(56):31942–31951. https://doi.org/10.1016/j.ijhydene.2020.08.247
137. Ahmad I, Shukrullah S, Ahmad M, Ahmed E, Naz MY, Akhtar MS, Khalid N, Hussain A, Hussain I (2021) Effect of Al doping on the photocatalytic activity of ZnO nanoparticles decorated on CNTs and graphene: Solvothermal synthesis and study of experimental parameters. Mat Sci Semicon Proc 123:105584
138. Sujaridworakun P, Natrchalayuth K (2014) Influence of pH and HPC concentration on the synthesis of zinc oxide photocatalyst particle from zinc-dust waste by hydrothermal treatment. Adv Powder Technol 25(4):1266–1272
139. Mclaren A, Valdes-Solis T, Li G, Tsang SC (2009) Shape and size effects of ZnO nanocrystals on photocatalytic activity. J Am Chem Soc 131(35):12540–12541
140. Rahimi R, Yaghoubi-Berijani M, Zargari S, Rabbani M, Shariatinia S (2016) SnTCPP-modified ZnO nanorods prepared via a simple co-precipitation method: application as a new photocatalyst for photodegradation and photoreduction processes. Res Chem Intermed 42:4697–4714
141. Wichasilp C, Phuruangrat A, Thongtem S (2022) Influence of pH on the synthesis ZnO nanorods and photocatalytic hydrogen production from glycerol solution. J Indian Chem Soc 99(6):100472
142. Xu L, Guo Y, Liao Q, Zhang J, Xu D (2005) Morphological control of ZnO nanostructures by electrodeposition. J Phys Chem B 109(28):13519–13522
143. Wang B, Xia C, Iqbal J, Tang N, Sun Z, Lv Y, Wu L (2009) Influences of Co doping on the structural, optical and magnetic properties of ZnO nanorods synthesized by hydrothermal route. Solid State Sci 11(8):1419–1422
144. Lee Y, Kim S, Jeong SY, Seo S, Kim C, Yoon H, Jang HW, Lee S (2021) Surface-modified Co-doped ZnO photoanode for photoelectrochemical oxidation of glycerol. Catal Today 359:43–49
145. Wang M, Ren F, Cai G, Liu Y, Shen S, Guo L (2014) Activating ZnO nanorod photoanodes in visible light by Cu ion implantation. Nano Res 7:353–364
146. Vaiano V, Iervolino G (2019) Photocatalytic hydrogen production from glycerol aqueous solution using Cu-doped ZnO under visible light irradiation. Appl Sci 9(13):2741
147. Hu J-S, Ren L-L, Guo Y-G, Liang H-P, Cao A-M, Wan L-J, Bai C-L (2005) Mass production and high photocatalytic activity of ZnS nanoporous nanoparticles. Angew Chem 117(8):1295–1299
148. Sang HX, Wang XT, Fan CC, Wang F (2012) Enhanced photocatalytic H_2 production from glycerol solution over ZnO/ZnS core/shell nanorods prepared by a low temperature route. Int J Hydrogen Energy 37(2):1348–1355
149. Wang Y, Xiao Y, Xiao G (2019) Sustainable value-added C3 chemicals from glycerol transformations: s mini review for heterogeneous catalytic processes. Chin J Chem Eng 27(7):1536–1542. https://doi.org/10.1016/j.cjche.2019.03.001
150. Rodrigues EG, Pereira MFR, Delgado JJ, Chen XW, Orfao JJM (2011) Enhancement of the selectivity to dihydroxyacetone in glycerol oxidation using gold nanoparticles supported on carbon nanotubes. Catal Commun 16(1):64–69. https://doi.org/10.1016/j.catcom.2011.09.009
151. Rodrigues EG, Carabineiro SAC, Delgado JJ, Chen X, Pereira MFR, Orfao JJM (2012) Gold supported on carbon nanotubes for the selective oxidation of glycerol. J Catal 285(1):83–91. https://doi.org/10.1016/j.jcat.2011.09.016
152. Painter RM, Pearson DM, Waymouth RM (2010) Selective catalytic oxidation of glycerol to dihydroxyacetone. Angew Chem Int Edit 49(49):9456–9459. https://doi.org/10.1002/anie.201004063

153. Estahbanati MRK, Feilizadeh M, Attar F, Iliuta MC (2020) Current developments and future trends in photocatalytic glycerol valorization: photocatalyst development. Ind Eng Chem Res 59(52):22330–22352. https://doi.org/10.1021/acs.iecr.0c04765
154. Xu DF, Hai Y, Zhang XC, Zhang SY, He RA (2017) Bi_2O_3 cocatalyst improving photocatalytic hydrogen evolution performance of TiO_2. Appl Surf Sci 400:530–536. https://doi.org/10.1016/j.apsusc.2016.12.171
155. Wang C, Cai X, Chen Y, Cheng Z, Lin P, Yang Z, Sun S (2017) Improved hydrogen production by glycerol photoreforming over Ag_2O-TiO_2 nanocomposite mixed oxide synthesized by a sol-gel method. Energy Procedia 105:1657–1664
156. Gombac V, Sordelli L, Montini T, Delgado JJ, Adamski A, Adami G, Cargnello M, Bernal S, Fornasiero P (2010) CuO_x-TiO_2 photocatalysts for H_2 production from ethanol and glycerol solutions. J Phys Chem A 114(11):3916–3925
157. Wang C, Cai XY, Chen Y, Cheng ZD, Luo XL, Mo SP, Jia LS, Lin PC, Yang Z (2017) Improved hydrogen production from glycerol photoreforming over sol-gel derived TiO_2 coupled with metal oxides. Chem Eng J 317:522–532. https://doi.org/10.1016/j.cej.2017.02.033
158. Slamet, Tristantini D, Valentina, Ibadurrohman M (2013) Photocatalytic hydrogen production from glycerol-water mixture over Pt-N-TiO_2 nanotube photocatalyst. Int J Energ Res 37(11):1372–1381. https://doi.org/10.1002/er.2939
159. Wang JJ, Li ZJ, Li XB, Fan XB, Meng QY, Yu S, Li CB, Li JX, Tung CH, Wu LZ (2014) Photocatalytic hydrogen evolution from glycerol and water over nickel-hybrid cadmium sulfide quantum dots under visible-light irradiation. Chemsuschem 7(5):1468–1475. https://doi.org/10.1002/cssc.201400028
160. Bao D, Gao P, Zhu XY, Sun SC, Wang Y, Li XB, Chen YJ, Zhou H, Wang YB, Yang PP (2015) ZnO/ZnS heterostructured nanorod arrays and their efficient photocatalytic hydrogen evolution. Chem-Eur J 21(36):12728–12734. https://doi.org/10.1002/chem.201501595
161. Lyubina TP, Markovskaya DV, Kozlova EA, Parmon VN (2013) Photocatalytic hydrogen evolution from aqueous solutions of glycerol under visible light irradiation. Int J Hydrogen Energy 38(33):14172–14179
162. Pai MR, Banerjee AM, Rawool SA, Singhal A, Nayak C, Ehrman SH, Tripathi AK, Bharadwaj SR (2016) A comprehensive study on sunlight driven photocatalytic hydrogen generation using low cost nanocrystalline Cu-Ti oxides. Sol Energ Mat Sol C 154:104–120. https://doi.org/10.1016/j.solmat.2016.04.036
163. Wei Z, Liu J, Fang W, Qin Z, Jiang Z, Shangguan W (2018) A visible-light driven novel layered perovskite oxyhalide Bi_4MO_8X (M= Nb, Ta; X= Cl, Br) constructed using BiOX (X= Cl, Br) for enhanced photocatalytic hydrogen evolution. Catal Sci Technol 8(15):3774–3784
164. Jedsukontorn T, Meeyoo V, Saito N, Hunsom M (2015) Route of glycerol conversion and product generation via TiO_2-induced photocatalytic oxidation in the presence of H_2O_2. Chem Eng J 281:252–264. https://doi.org/10.1016/j.cej.2015.06.078
165. Zhou BW, Song JL, Zhou HC, Wu LQ, Wu TB, Liu ZM, Han BX (2015) Light-driven integration of the reduction of nitrobenzene to aniline and the transformation of glycerol into valuable chemicals in water. Rsc Adv 5(46):36347–36352. https://doi.org/10.1039/c5ra06354j
166. Guo L, Sun Q, Marcus K, Hao Y, Deng J, Bi K, Yang Y (2018) Photocatalytic glycerol oxidation on Au $\times$ Cu–CuS@ TiO_2 plasmonic heterostructures. J Mater Chem A 6(44):22005–22012
167. Molinari A, Maldotti A, Bratovcic A, Magnacca G (2013) Photocatalytic properties of sodium decatungstate supported on sol-gel silica in the oxidation of glycerol. Catal Today 206:46–52. https://doi.org/10.1016/j.cattod.2011.11.033
168. Zhang YH, Ciriminna R, Palmisano G, Xu YJ, Pagliaro M (2014) Sol-gel entrapped visible light photocatalysts for selective conversions. Rsc Adv 4(35):18341–18346. https://doi.org/10.1039/c4ra01031k

169. Zhao S, Dai Z, Guo W, Chen F, Liu Y, Chen R (2019) Highly selective oxidation of glycerol over $Bi/Bi_{3.64}Mo_{0.36}O_{6.55}$ heterostructure: dual reaction pathways induced by photogenerated $^{1}O_2$ and holes. Appl Catal B: Environ 244:206–214.
170. Schünemann S, Dodekatos G, Tüysüz H (2015) Mesoporous silica supported Au and AuCu nanoparticles for surface plasmon driven glycerol oxidation. Chem Mater 27(22):7743–7750

Chapter 5
Photocatalysis on Selective Hydroxylation of Benzene to Phenol

Bramantyo Bayu Aji, Ulya Qonita, Fadila Arum Rhamadani, Albertus Jonathan Suciatmaja, Hairus Abdullah, Leonardo Togar Samosir, and Vivi Fauzia

Abstract Phenol is an essential precursor in the industrial world. A promising approach to producing phenol is through the use of photocatalysts. A photocatalyst possesses a significant band gap that can absorb photon energy and excite electrons in valence bands. These excited electrons are then employed to reduce oxidizing agents such as peroxide or water, producing hydroxyl radicals (•OH). These radicals can then oxidize benzene, converting it to phenol. A primary challenge in this process is the over-oxidation of phenol, which arises due to the reactivity of the OH group in its structure. Some strategies to counteract over-oxidation include adjusting the photocatalyst band gap, enhancing photocatalyst conductivity, and diminishing the hydrophilic characteristics of photocatalyst surfaces. This can be achieved by introducing materials like carbon, nitrogen, boron, metal, and organosilanes. This chapter will delve into potential photocatalyst materials suitable for the hydroxylation of benzene reactions to phenol, encompassing topics like material development, underlying mechanisms, methods to procure materials with high stability, activity, and selectivity toward phenol, as well as their applications.

B. B. Aji · F. A. Rhamadani
Department of Materials Science and Engineering, National Taiwan University of Science and Technology, Taipei, Taiwan 106

U. Qonita
Department of Applied Science and Technology, National Taiwan University of Science and Technology, Taipei, Taiwan 106

A. J. Suciatmaja
Department of Chemical Engineering, National Taiwan University of Science and Technology, Taipei, Taiwan 106

H. Abdullah
Department of Industrial Engineering, Universitas Prima Indonesia, Medan, Indonesia

L. T. Samosir · V. Fauzia (✉)
Department of Physics, Universitas Indonesia, Kampus UI Depok, Indonesia
e-mail: vivi@sci.ui.ac.id

H. Abdullah (ed.), *Solar Light-to-Hydrogenated Organic Conversion*,
https://doi.org/10.1007/978-981-99-8114-4_5

Keywords Phenol · Hydroxylation benzene · Photocatalyst · Over-oxidation phenol

5.1 Introduction

Phenol, a vital industrial chemical, is widely used in the production of resin and polymers [1], pharmaceuticals [2], disinfectants [1], and solvents in various applications. Phenols possess unique chemical properties due to the presence of hydroxyl groups coupled with aromatic rings, rendering them more reactive than alcohols. This reactivity positions them as some of the most valuable organic precursors in the industry [3]. In industrial settings, phenol is predominantly produced via the cumene method [4]. The production of phenol often relies on the oxidation of benzene, as this approach is energy-efficient, results in safer intermediate products, and yields fewer acetone by-products. The oxidation reaction of benzene typically involves an oxidizing agents such as H_2O_2, N_2O, or O_2 [5]. Using O_2 directly from the free air is a good alternative for producing phenol from benzene. However, benzene is less reactive with O_2 and must be done in a high-temperature or high-pressure manner, which requires high costs and low safety levels. To reduce production costs, it is critical to investigate alternate phenol synthesis processes that may be conducted in mild conditions. Photocatalyst is an environmentally friendly way that can be used to convert benzene into phenol with the help of O_2 in mild conditions. This process involves materials that can absorb photon energy well to produce holes and radicals such as *OH, $*O^{2-}$, or $*H_2O$ under aerobic conditions [6]. Because of their inherent reactivity, these radical compounds proceed to oxidize benzene, producing the desired phenol result. However, one notable shortcoming of these radical species is their inherent lack of selectivity, which results in uncontrolled reactivity. They can oxidize not only preceding compounds but also the phenol product itself. The O–H binding energy in phenolic compounds is comparably weaker than the C–H bonding in benzene, so the phenolic product will be more easily oxidized than its precursor (benzene) [7]. Consequently, radical compounds will oxidize the phenolic compounds formed to produce by-products such as biphenyls or chain hydrocarbons. Therefore, it is essential to develop materials to control the overoxidation process and minimize unwanted pathways.

In recent years, the photocatalytic method has gained industrial interest due to its low energy consumption and its capacity to utilize high energy via UV light, which is environmentally friendly [8]. The concept of photocatalysis was initially applied primarily to environmental remediation. It is because a series of photocatalysts can quickly decompose various kinds of aromatic organic compounds into the water by solar irradiation absorption [9]. Photocatalysts typically use semiconductor materials to capture photon energy and help break benzene into phenol. Among these materials, TiO_2 is the most favored choice for semiconductors because of its good chemical stability, strong oxidizing ability, cost-effectiveness, and safety. When the TiO_2 photocatalyst gets a higher energy exposure than its energy band gap, a positive charge is generated in the valence band. In contrast, a negative charge is promoted in the conduction band. Positive holes and electrons play an essential role in oxidation and reduction reactions. Positive charges attract water and form hydroxyl radicals ($OH^{\cdot}$), while negative charges can reduce oxygen to produce superoxide ions (O^{2-}). The highly reactive hydroxyl radical compounds can target the aromatic ring of benzene and transfer it into phenol [10].

TiO_2 has two phases, namely, the rutile and anatase phases. The anatase phase exhibits a higher photocatalytic activity because it has a higher surface hydroxyl group. The anatase TiO_2 phase has a wide band gap of 3.2 eV, so it exclusively responds to UV light exposure [11]. However, the most significant component of sunlight is visible light. Therefore, to maximize the absorption of sunlight and maximize its use, the researchers have pursued the development of TiO_2 materials with several additional modifications, such as complex compounds, mesocellular siliceous foam, metals, and transition metals, i.e., Goto & Ogawa developed TiO_2 by modifying the complex compound $[Ru(bpy)_3]^{2+}$ on the surface [12]. The complex compound $[Ru(bpy)_3]^{2+}$ can capture visible light down to a wavelength of ~ 550 nm. The accumulation of the amount of solar energy that TiO_2 can absorb-$[Ru(bpy)_3]^{2+}$ helps oxidize benzene to phenol even better. Unfortunately, this complex compound is difficult to separate from TiO_2 for recycling. On the other hand, the phenol selectivity within this system remains relatively low, thus decreasing the phenol yield. Therefore, it needs to develop a photocatalyst that gains stability and selectivity.

The highly active and selective photocatalysis of benzene to phenol requires the development of photocatalysts that can efficiently activate the C–H bond in benzene and selectively generate the hydroxyl group (–OH) to form phenol. Several critical requirements for achieving highly active and selective photocatalysis of benzene to phenol include visible-light absorption, high surface area, proper band structure, highly selective active site, quiet stability, and scalability. To achieve these objectives, researchers have explored various strategies; one of the most common methods is doping TiO_2 with noble metals such as gold [8, 13], and platinum [14], forming TiO_2-based heterojunctions [15–18], and modifying surface properties and morphology of TiO_2 [18]. These strategies aim to modify the electronic structure of TiO_2 or shift the bandgap to lower energies and allow for the absorption of visible light. Incorporating noble metals, such as platinum, palladium, and gold, into catalysts has indeed been shown to significantly improve their catalytic activity, selectivity, and stability. Ida et al. developed Au-deposited TiO_2 (Au/TiO_2) [19]. Au/TiO_2 has been shown to have the highest phenol yield and selectivity compared to unmodified TiO_2; the yield and selectivity are around 63% and 91%. However, using these metals can be cost-prohibitive and limit the scalability of catalyst production. Another approach involves designing catalysts with a hierarchical structure that combines multiple components, such as metal oxides, carbon materials, and heteroatoms, to create synergistic effects that enhance their catalytic activity.

Metal oxide-based photocatalysts, such as iron-based in a Fenton manner (Fe_2O_4, Fe_3O_4, and $ZnFe_2O_4$) are known to have the ability to facilitate hydroxylation benzene [20–22]. Fe in its structure can undergo a redox reaction to form Fe^{2+}/

Fe^{3+}, which are known to have the ability to oxidize H_2O_2 to •OH radicals. Unfortunately, the formation of Fe^{2+}/Fe^{3+} can trigger metal leaching, which causes low stability and catalytic activity. Furthermore, the presence of Fe^{3+} on the surface is also a drawback in this system because Fe^{3+} tends to be hydrophilic, which prefers phenol over benzene. It causes over-oxidation of phenol and reduces the amount of phenol production. Adding vanadate (V) to Fe ions is proven to help increase the absorbance of photon energy and its stability. However, despite these benefits, the addition of vanadate still shows low selectivity of phenol due to the leaching issue and the hydrophilic surface properties of Fe [23]. To address this issue, it is necessary to develop materials that can withstand the leaching issue by Fe and reduce the hydrophilic properties on the surface. One common method is adding a functional material that can bind to Fe^{3+} and has hydrophobic properties. Some of them are organosilane [24] and carbon shells [25–28]. Wei et al. synthesized $FeVO_4$ with grafted organosilane in tetramethoxylation (TMOS) and dodecyltrimethoxysilane (DTOS) to determine their effect on suppressing metal leaching Fe^{3+} on the surface and its hydrophilicity of $FeVO_4$ by solvothermal method. The result showed that the phenol yield of $FeVO_4$@TMOS and $FeVO_4$@DTOS is determined to be 20% and 13% with a selectivity over 98%. Organosilanes are known to have two side chains that are hydrophilic and hydrophobic. The hydrophilic side can interact with Fe^{3+} on the surface to suppress the leaching issue and the hydrophobic side can facilitate the van der Waals effect if it meets a hydroxyl group in phenol so that it triggers a repulsion effect on phenols and is more likely to attract benzene. Yang et al. synthesized $ZnFe_2O_4$@C via the solvothermal method. Higher phenol yield was obtained by using a $ZnFe_2O_4$@C photocatalyst of 15.5% with a selectivity of 99.4%. Carbon shell has many advantages, including its rigid structure to cover the Fe surface for leaching issues and increase its catalytic stability. In addition, carbon is also hydrophobic due to defects on the surface, which can reduce its interaction with phenol so that its selectivity increases and has a π-conjugated electron system that can facilitate the adsorption and activation of benzene in the hydroxylation process.

Another well-known n-type semiconductor is tungsten oxide (WO_3) [29–32]. It is also widely used as a photocatalyst to facilitate the hydroxylation of benzene into phenol due to its narrow bandgap and enabling photoresponse in the visible region [33]. However, its intrinsic properties do introduce limitations. It is known to undergo structural changes and degradation over time. Moreover, it is classified as a hazardous material, which may pose health and safety risks during its production, handling, and disposal. Usually, a high recombination rate due to a narrow band gap (2.6 eV) is solved by coupling with a p-type semiconductor such as Bi_2O_3 to form a p–n heterojunction structure [30].

$CdWO_4$ is a promising monoclinic wolframite structural material in the tungstate family. Although it has a wide band gap, resulting in limited photo absorption, it can be significantly excited under light energy [34]. Because of its chemical, structural, and optical properties, it has been widely used in the photocatalytic degradation of toxic pollutants. In terms of photocatalytic degradation of organic pollutants, $CdWO_4$ effectively degrades various contaminants, including dyes, phenols, and pesticides. The high photocatalytic activity of $CdWO_4$ can be associated with its ability to absorb a wide range of light wavelengths and generate many electron–hole pairs, hence promoting organic pollutants' degradation. Surface modification and morphology control are well-known effective methods for enhancing the activity of photocatalysts. This is because altering the crystal surface and shape can lead to significant improvements. Additionally, the form of the material plays a role in influencing the photocatalytic activity of WO_3 nanostructures to some degree. Through morphology engineering, Chen et al. successfully prepared novel three-dimensional (3D) $CdWO_4$ micro rods with Bi_2WO_6 nanosheet. It shows that in 4-h irradiation, the benzene is converted about 7.3% to phenol and achieved 99% selectivity [34]. This result is mainly attributed to the hierarchical nanostructured of $Bi_2WO_6/CdWO_4$ (BCW) with suitable band composition and the high specific area, which contributes to the excellent performance by enhancing photon absorption in the visible light region while also suppressing recombination of photo-induced charge carries.

Non-metal elements such as carbon, silicon, phosphate, boron, nitrogen, oxygen, and hydrogen, which are abundant on Earth, can be used as a catalyst due to their unique π-conjugated electron system [35]. The idea of designing carbon-based supported photocatalysts has been proposed to enhance the active area, controlled morphology, and surface engineering manipulation. Unfortunately, carbon could not absorb photon energy due to its more likely insulating properties. However, the introduction of other elements, such as nitrogen and boron, can create binary and ternary systems. These are known to improve the electronegativity and conductivity of carbon.

In a binary system, graphitic carbon nitride (g-C_3N_4), known as *n*-type reduction photocatalysts, was combined by WO_3 to form an S-scheme heterojunction [36, 37]. with this S-scheme heterojunction, once equilibrium is reached, the fermi levels of both semiconductors are grouped, and the photogenerated electrons move towards the oxidized photocatalysts along with the holes migrate toward the reduced photocatalysts. This process creates an internal electric field. Nevertheless, in contrast to alternative heterojunctions like the S-scheme heterojunction, the electrons in the oxidized photocatalytic CB are relatively useless, as are the holes in the reduced photocatalysts VB. The presence of an internal electric field instigates the recombination and subsequent elimination of these electrons and holes, leaving behind only the useful ones. Nonetheless, the efficiency of this mechanism may decline in harsh conditions, and the active area remains relatively low. Addressing this concern through modification control in morphology, structure, and composition could solve the issues [38].

Doping boron atoms into carbon lattice is widely known for improving durability and altering electronic properties [39]. A ternary BCN system provides an adjustable band gap when alloyed with metal or semiconductor photocatalysts [40], in addition to an improved photoinduced charge separation rate [41], abundant active sites, and large surface area [42]. As demonstrated by Liu et al., anchored Cu single atom on boron carbon nitride (BCN) nanosheet exhibited exceptional performance in the oxidation of benzene. This superiority can be attributed to the presence of atomically dispersed active sites and the modulated electronic properties of the photocatalysts [43]. Theoretical studies show that the amplified catalytic efficiency is a result of the optimized adsorption of intermediates, achieved by modulating the electronic structure of isolated copper atoms through the coordination of boron atoms at the $Cu–N_2B_1$ atomic interface.

This study will examine the aspects related to the reaction of benzene hydroxylase to phenol, including material development, underlying mechanisms, ways to obtain materials with high stability, activity, and selectivity to phenol, and their applications. This study is expected to facilitate an understanding of benzene hydroxylation via photocatalysis, provide an overview of designing potential materials, and improve their application in industry.

5.2 Photocatalytic Material Systems in Hydroxylation Reaction

Rapid progress in the field of photocatalysis has shown the potential for using this technique to solve several problems in conventional techniques. This includes the ability to eliminate waste generation, utilize a wide range of materials, minimize energy consumption, and seamlessly integrate with other methods. In photocatalysis, two systems have been reported regarding how the photocatalysts interact with the substrate: homogeneous and heterogeneous systems.

In a homogeneous system, relatively high activity and selectivity have been reported owing to the optimum contact between the photocatalysts and substrate since all the components are in the same phase with the medium. Remarkable results were reported by the Wei group that could achieve more than 99% selectivity of phenol by using supramolecular catalysts consisting of alkoxohexavanadate anions and quinolinium ions [44]. In their work, polyoxoalkoxohevanadates act in a dual role, first as a stabilizer for quinolinium radicals while also being able to reuse H_2O_2 produced by the quinolinium ions. By optimizing the use of this peroxide, high selectivity (> 99%) and 50.1% yield under LED irradiation for 12 h. This result shows how a synergistic effect could significantly enhance the performance compared to a single catalyst system. As reported, although it offers the low-cost option, only a low yield of phenol, about 11.3%, was obtained by metal-free quinoline sulfate (QuH_2SO_4) as the photocatalyst [45]. A low yield of phenol product was attributed to the effect of counter anions that may deactivate the active site complexes and affect the interaction of benzene and the catalyst. In fact, neglected useful by-products such as H_2O_2, known as potential oxidants in photocatalytic systems, make it less efficient for this typical catalyst. Another study highlighted the significance of incorporating metal additives into the quinoline system, such as C_{16}Qu-PW as photocatalysts increase the phenol yield from 14.1% from C_{16}Qu alone up to 20.1% [46]. To summarize, although homogenous photocatalysts show promising results in direct hydroxylation of benzene to phenol, such as high selectivity, low cost, and less energy consumption, the need for co-catalysts and or oxidants is a must to achieve higher activity. It should be noted that the most fundamental difficulty, the separation of the catalyst from the product, remains a formidable challenge for this system.

On the other hand, the solution for these problems has been directed to specific heterogeneous systems. In a heterogeneous system, the phase of photocatalysts is different from the medium phase, making it much easier to collect and separate. Several reports also indicated that heterogeneous photocatalysts could be improved by various strategies, including but not limited to the addition of the dopant [47], supported photocatalysts [25], nanoengineering [48], and composites formation [35].

In a heterogeneous system, the photocatalyst phase differs from the medium phase, making it much easier to collect and separate. Several reports also indicated that various strategies could improve heterogeneous photocatalysts, such as adding the dopant, supported photocatalysts, nanoengineering, and composites formation. All of these strategies aim to systemically change the electronic structures, such as band gap modulation, change the behavior of reaction contact between the substrate and surface of the catalyst, and increase the contact time to improve the conversion yield. For instance, most known metal oxide photocatalysts, such as TiO_2, ZnO, and WO_3, have been proven active when irradiated with UV light and are known as the most well-studied among semiconductor photocatalysts. The fundamental aspect of semiconductor photocatalysts is their ability to generate photoinduced charge pairs (electrons and holes), which are later used in oxidation reactions. In principle, when irradiated with UV light, these semiconductors produce excited electrons in the conduction band while the holes in the valence band react with available oxidants, such as hydroxyl ions from water, to produce hydroxyl radicals. These species then act as the main oxidant in oxidation reactions, such as benzene hydroxylation. However, as reported, hydroxyl radical is a powerful oxidant that causes a common problem with the oxidation of phenol [15]. To prevent this undesired reaction, the common strategy is to dope the semiconductor with the more positive element. The recombination of the charge carrier in WO_3 is unfavorable for hole production. Due to a narrow band gap of WO_3, fast electron–hole recombination may not give enough time to form hydroxyl radicals. As one example of selective reaction, hydroxylation of benzene into phenol requires highly active yet selective photocatalysts since they should be capable of activating stable benzene compounds. In recent years, the development of photocatalysts has focused on improving the visible light absorption ability to react. One strategy that could be done to have visible light-induced photocatalysts is by lowering the band gap while also controlling the recombination of electron–hole pairs. As known, only a limited can be activated by visible light. Among them, the most well-studied semiconductors are TiO_2 [9], ZnO [49], and WO_3 [32]. Although ZnO has a larger band gap than the well-known TiO_2 photocatalysts, it has a wide range of solar light absorption. Thus enabling it to be activated by low photon energy, such as visible light. When solar light photoinduces ZnO with photonic energy (hv) equal to or greater than the excitation energy (Eg), e from the filled valence band (VB) is promoted to an empty conduction band. (CB). This photoinduced mechanism generates electron–hole pairs. The mechanism of photoinduced charge pairs is illustrated in Fig. 5.1.

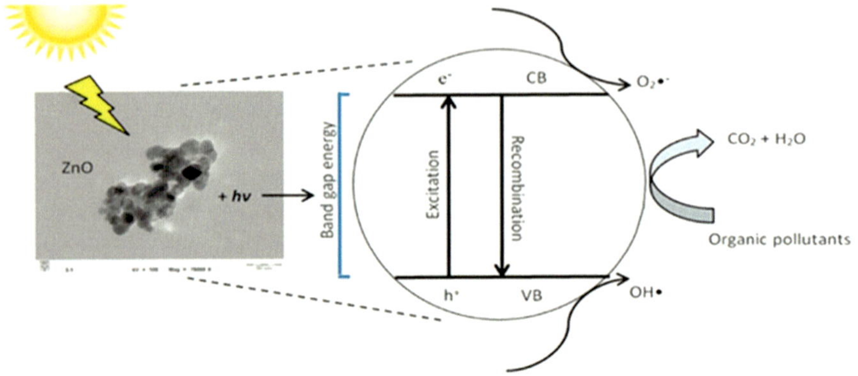

Fig. 5.1 Typical formation of the photoinduced charge pairs in ZnO for organic pollutants degradation. Reprinted with permission from ref. 49. License number 5526430812924. Elsevier

The critical energy-wasting step in the photocatalytic reaction involving ZnO is the electron–hole recombination, which consequently leads to a low quantum yield. Thus, the prevention of electron–hole recombination becomes vital. This objective can be accomplished by introducing a suitable electron donor (or acceptor) into the system. The degradation rate was enhanced by the inclusion of electron acceptors in several ways: (1) The acceptors prevent electron–hole recombination by accepting the conduction band electron. (2) There is an increase in hydroxyl radical concentration. (3) The generated oxidizing species expedite the oxidation rate of intermediate compounds. The degradation process was significantly accelerated when electron acceptors such as H_2O_2, $K_2S_2O_8$, and $KBrO_3$ were present [50].

Meanwhile, in the TiO_2 photocatalytic system, because of the creation of surface disorder, oxygen vacancy, and new band level, there is a need to obtain high visible light absorption. As reported, when titanium hydride and peroxide were employed as Ti sources and oxidants for preparing self-hydroxylated TiO_2 due to defect-rich TiO_2, the primary oxidation reaction for low-valent titanium (Ti^0, Ti^{2+}) and high-valent titanium (Ti^{4+}) reduced by reductant give almost 4.6 times higher photocatalytic performance than commercial P25 [52]. As shown in Fig. 5.2, modification of TiO_2 nanotube (TNT) arrays through electrochemical reaction and calcination under nitrogen atmosphere induce the inclusion of interstitial N, Ti^{3+} states in the lattices can minimize the energy gap from 3.2 eV to roughly 2.0 eV, according to valence-band XPS spectra and the Mott–Schottky plot. Furthermore, TNT cathodic polarization in 0.1 M NaH_2PO_4 exhibits better photoelectrochemical characteristics than TNT cathodic polarization in 0.5 M Na_2SO_4 solution.

Another well-known n-type semiconductor is WO_3. Visible light-driven photocatalysts such as WO_3 and graphitic carbon nitride (C_3N_4) have been intensively studied to maximize the utilization of sunlight in the 400–800 nm range [32]. Under visible light irradiation, new photocatalysts have been developed by adding Pt, Pd, or CuO co-catalysts to WO_3, which can fully mineralize volatile organic compounds (VOCs) such as toluene and acetaldehyde in the gas phase and acetic acid in the liquid

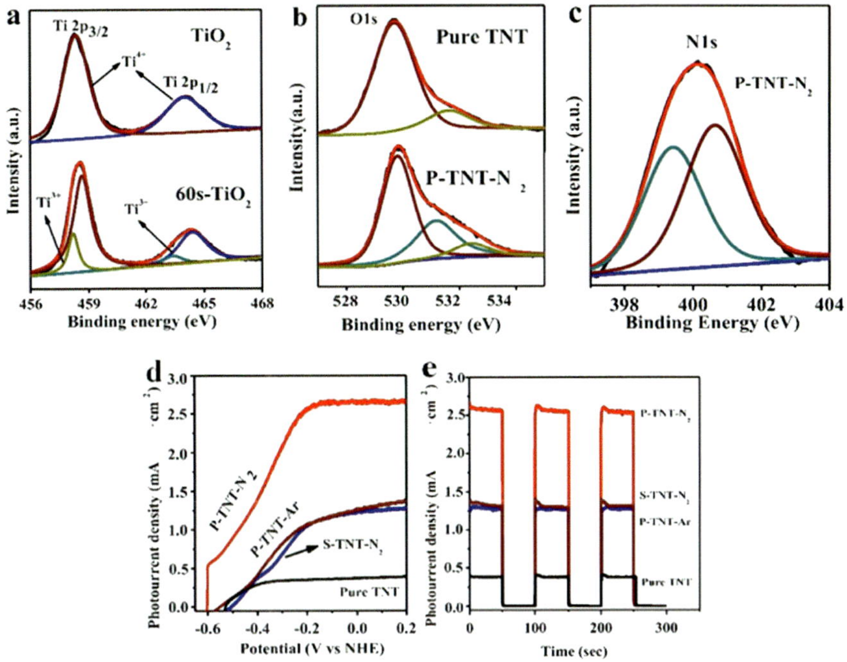

Fig. 5.2 XPS spectra of TNT for **a** Ti 2p, **b** O 1s, **c** N 1s. Linear sweep voltammograms **d** and *I*–*t* curves **e** of TNT. Reprinted with permission from ref. 51. License number 5526431184190. Elsevier

phase. A photocatalytic fuel cell (PFC) that uses a photoanode consisting of an array of $TiO_2/WO_3/W$ nanorods and nanothorns grown in a specific way (epitaxial) has been developed. Compared to a PFC that uses a photoanode made of pure WO_3, this new PFC performed more efficiently in energy output and wastewater treatment. The improved performance is attributed to the addition of a TiO_2 layer, which helps separate and transfer carriers more effectively. Additionally, it provides atomic-level protection against corrosion for the photoanode. The proposed mechanism and route for such an application are illustrated in Fig. 5.3.

In general, it is possible to directly convert benzene to phenol using the strategies mentioned from a photocatalytic material perspective. By using environmentally friendly oxidants like oxygen from the air, it may be possible to improve benzene conversion while maintaining a high level of selectivity for phenol. This chapter provides up-to-date information on the latest developments in well-known semiconductor photocatalysts, along with a brief explanation of the mechanism behind each material. A short discussion will also explore the involvement and mode of interaction between the catalyst and benzene.

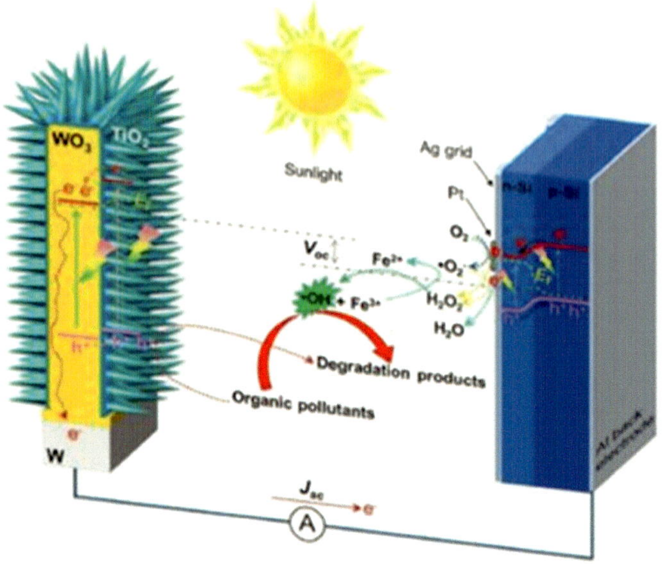

Fig. 5.3 Typical WO_3-based photocatalytic system for both energy and wastewater treatment applications. Reprinted with permission from Ref. 29. License number 5526431399585 (Elsevier)

5.2.1 Metal Alloy Nanocatalysts

Noble metals have high catalytic activity and have been widely applied in various fields. The electron-empty d-layer orbital makes it easy to absorb or capture electrons from the outside. Several noble metal catalysts, such as Pt [32], Ru [53], Ir [54], and Au [13], have been developed. Pd is a popular monometallic photocatalyst for the selective oxidation of benzene to phenol because of the CH activation effect. Metals are known to have the ability to initiate oxidation and reduction reactions, thus enabling them to form several valence species, which are very beneficial in the formation and stabilization of product intermediates. In creating phenol, phenol is an intermediate product easily oxidized into other compounds. To bind and stabilize phenol intermediates, a catalyst effortlessly bridges the interaction with phenol so that the continuous oxidation of phenol can be inhibited. Metal-based catalysts are still being developed because of their extraordinary ability to reduce the activation energy of forming material. The development carried out mostly leads to obtaining catalysts with high performance, catalytic activity, selectivity, and stability. One way to achieve this target is to add a metal alloy to the structure (bimetallic) [27, 55]. This process was adopted from the invention of metal alloys in making steel; metal alloys can convert carbon into a solid and sturdy material. Several Attempts have been made to modify the properties of Pd by combining it with other metals such as Au, Ag, and Cu. These bimetallic nanoparticles are often dispersed on additional materials to enhance catalytic performance. Another development is carried out by depositing it on the surface of carbon-based materials. Carbon is known to have a pi-conjugated system that facilitates electron rotation and has excellent stability. In addition to increasing strength, the metal deposit on the carbon surface aims to increase its conductivity, catalytic activity, and selectivity to phenol [56].

The performance of a photocatalyst is often improved by incorporating two types of metal that could work synergistically with each other. These bimetallic alloy can be synthesized as a composite material or as nanoparticles, which is then dispersed onto a support material. The most commonly used metals for this application are noble metals, such as Au, Pd, and Ag. However, non-noble metals are also used, such as Fe or Cu.

As for alloy nanoparticles, they are most often dispersed onto a support material to augment its catalytic performance. Zhang and Park dispersed CuPd nanoparticles onto holey g-C_3N_4 (HCN) [57]. After deposition of 0.5% weight percentage of CuPd, conversion rate of 98.1%, selectivity of 89.6%, and yield of 87.8% was achieved after 1.5 h in the presence of H_2O_2, which was 32.7 times higher than that of pristine HCN. The increased performance, shown in Fig. 5.4b, was attributed to better charge separation, which was proven by the lower photoluminescence spectrum in Fig. 5.4a, and higher surface area. In addition, as shown in Fig. 5.4c, the band gap of the modified HCN was up-shifted by 0.39 eV, which leads to more efficient charge transfer in CB.

Meanwhile, Hosseini et al. dispersed Au–Pd nanoparticles (NPs) onto carbon fiber felt (CFF) via the wet impregnation method, which is then covered with spring-shaped TiO_2 as a protective layer [58]. In Fig. 5.5, the spring-shaped TiO_2 outer layer is indicated in light blue color, the underlying carbon felt is indicated in yellow and the Au–Pd nanoparticles are uniformly dispersed, which is indicated by the color of purple and orange.

As shown in Table 5.1, the best-performing material was Au–Pd/CFF@TiO_2-450, which achieved 46% phenol yield and 100% phenol selectivity. Meanwhile, under the same conditions, catalysts that were prepared without one of the elements either Au, Pd, or TiO_2, performed far lower than the one where all of them were present. To test the stability of the optimized material, it was run for eight consecutive cycles. The conversion dropped to 41%, the selectivity dropped to 90%, and only tiny amounts of NPs were lost after running the catalyst for eight cycles. The authors propose that the protective TiO_2 layer prevented the Au–Pd NPs from leeching giving the catalyst great stability.

A different strategy involving alloy nanoparticles was used by Zeng et al. In this study, instead of dispersing nanoparticles onto a support material, an Fe and Co alloy nanoparticle core was encapsulated with a carbon shell (FeCo@C) [27]. As shown in Table 5.2, the highest-performing material was FeCo@C-2, which had a phenol yield of 26.4% and selectivity of 96.2%. During testing, the prepared catalysts needed two runs to be fully activated. Similar to the carbon-based support material from the previously mentioned studies, the carbon shell surface favors benzene adsorption rather than phenol, which would decrease the overoxidation of phenol. In addition, the shell protects the metal alloy core from leeching. After six consecutive runs, the performance of the catalyst remains constant and the carbon shell remains intact.

Table 5.2 The catalytic performance of Fe@C, Co@C, and FeCo@C toward liquid-phase benzene hydroxylation reaction. Reprinted with permission from Ref. [27]. License number 5553441223009. 2022. Elsevier

Entry	Catalyst	1st Run		2nd Run	
		Yield (%)	Selectivity (%)	Yield (%)	Selectivity (%)
1	–	t.a	–		
2	Fe@C	6.3 ± 0.5	95.0 ± 1.1	16.0 ± 0.3	94.8 ± 1.7
3	Co@C	t.a.[c]	–	t.a	–
4	FeCo@C-1	14.6 ± 0.4	95.0 ± 0.7	19.6 ± 1	95.2 ± 0.8
5	FeCo@C-2	15.6 ± 0.6	96.8 ± 1.4	26.4 ± 0.7	96.2 ± 1.2
6	FeCo@C-3	10.4 ± 0.3	96.5 ± 0.5	19.0 ± 0.4	96.0 ± 0.7
7	Fe@C–Co@C[b]	5.9 ± 0.2	90.0 ± 1.9	13.9 ± 0.9	86.0 ± 2.3
8	FeCo@C-2[d]	0.8 ± 0.1	98.4 ± 0.5		
9	FeCo@C-2[e]	9.7 ± 0.9	86 ± 2.3		

[a] Reaction conditions: catalyst (30 mg), benzene (250 μl), H_2O (3 mL) + CH_3CN (3 mL), and H_2O_2 (2 mL); reaction temperature, 60 °C; reaction time, 4 h

[b] Fe@C (15 mg) and Co@C (15 mg), which were physically mixed and calcined at 300 °C

[c] This means trace amount

[d] Quenched with 1 mL ethanol

[e] Reaction performed in pure water

Table 5.3 Photocatalytic Benzene to Phenol under Different Conditions. Reprinted with permission from ref. 13. License number 5520580828061 2014 ACS Catalysis

Catalyst and Irradiation Conditions	t (h)	Benzene Conversion (%)	Phenol	
			Yield (%)	Selectivity (%)
TV_2 with UV	6	3.0	3.0	100
TV_2 with UV	18	9.0	8.2	91
TV_2 with UV	24	12.7	10.8	85
$1Au/TV_2$ with UV	18	18.0	15.9	88
$3Au/TV_2$ with UV	18	5.0	4.5	90
TiO_2 with UV	24	0.3	0.3	100
TV_2 with visible	24	0.4	0.4	100
$1Au/TV_2$ without visible	24	2.0	2.0	100
No catalyst with UV	18	5.5	3.3	61
$1Au/TV_2$ with UV[b]	18	3.0	2.9	97
No catalyst with UV ($\lambda >$ 300 nm)	18	1.2	0.6	50

[a] Unless mentioned otherwise, the reaction was conducted with H_2O_2, H_2O, and CH_3CN

[b] Dissolved oxygen was removed by argon bubbling for 30 min

of electrons in the valance band of $FeVO_4$ to the conduction band, facilitating the oxidation reaction of benzene to phenol. Unfortunately, the Fe^{2+}/Fe^{3+} redox reaction on the surface triggers the metal leaching issue. It can destroy the bulk structure of $FeVO_4$ so that the reversible reaction in the system is disrupted and the photocatalytic performance decreases [61]. In addition, Fe^{3+} on the surface is more hydrophilic because it prefers phenol with a hydroxyl group in its structure compared to benzene. It triggers the phenol over-oxidation reaction and reduces its production. To overcome this problem, researchers are developing materials that can suppress metal leaching and reduce the hydrophilic properties of the $FeVO_4$ surface by the silylation strategy. The silylation strategy is adding an organosilane molecule (OS) to the surface of $FeVO_4$ [7]. Organosilanes are known to have two side chains that are hydrophilic and hydrophobic. This hydrophobic side can facilitate the van der Waals effect if it meets a hydroxyl group so that it triggers a repulsion effect on phenols and is more likely to attract benzene.

Wei et al. synthesized $FeVO_4$ with grafted organosilane in tetramethoxylation (TMOS) and dodecyltrimethoxysilane (DTOS) to determine their effect on suppressing metal leaching Fe^{3+} on the surface of $FeVO_4$. $FeVO_4$ was synthesized by solvothermal method and then grafted with TMOS and DTOS with regular stirring using a stirrer machine for 24 h and form a yellow precipitate. FTIR analysis confirmed the formation of $FeVO_4$@TMOS and $FeVO_4$@DTOS, as evidenced by the stretching of the Si–O bond at wave number 1085 cm^{-1} (Fig. 5.6). The XPS data in Fig. 7a also reinforces the success of the grafted OS on the $FeVO_4$ surface. In the pristine sample of Fe spectra, the dominant peaks at 711.8 and 725.4 eV belong to the Fe $2p_{1/2}$ and Fe $2p_{3/2}$ spectra of Fe^{3+}, and the peaks at 710.8 and 723.9 eV belong to Fe^{2+}. The appearance of the Fe^{2+} peak indicates that some species were reduced. After silanization with the OS functional group, there was a slight shift of about 0.1 eV in the positive direction, indicating the Fe^{3+} species belonging. It indicates an electron interaction between $FeVO_4$ and the OS functional group. Whereas in the Si 2p spectra, peaks at 101.8 eV belong to the Si–O bond (Fig. 7b), proving that the OS has been successfully grafted onto the $FeVO_4$ surface [24].

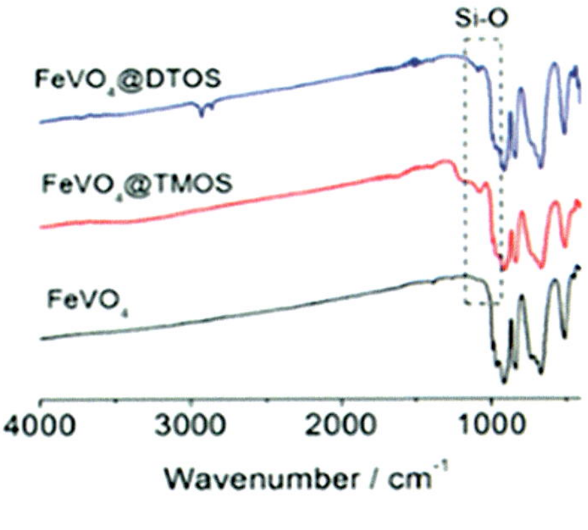

Fig. 5.6 FTIR analysis of $FeVO_4$, $FeVO_4$@TMOS, and $FeVO_4$@DTOS. Reprinted with permission from ref. 62. License number 1345720. 2021 Catalyst Science and Technology

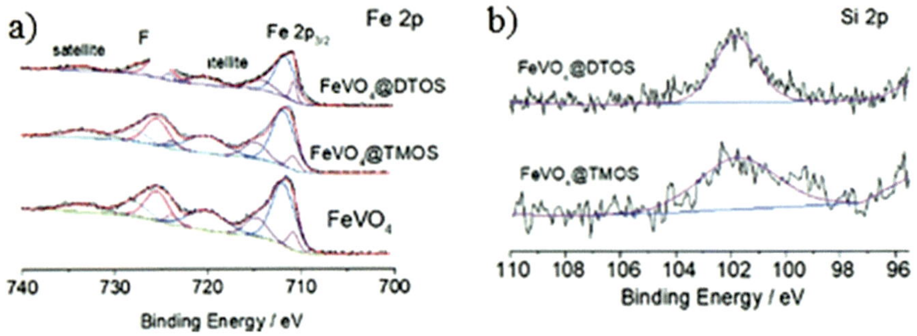

Fig. 5.7 XPS analysis of $FeVO_4$, $FeVO_4$@TMOS and $FeVO_4$@DTOS, Fe spectra (**a**), and Si spectra (**b**) Reprinted with permission from ref. 62. License number 1345720. 2021 Catalyst Science and Technology

The surface affinity of $FeVO_4$ before and after adding organosilane was tested using the contact angle measurement. The hydrophobic nature of organosilanes can be identified from their water contact angles. Figure 5.8 describes the water contact angles of $FeVO_4$, $FeVO_4$@TMOS, and $FeVO_4$@DTOS, respectively 18.7, 28.8, and 60.1°. The hydrophobic properties of $FeVO_4$ with the addition of organosilanes such as TMOS and DTOS were proven to increase. The best results were seen in the acquisition of DTOS, the long alkyl chain in DTOS helped increase the tailoring surface affinity of $FeVO_4$. Furthermore, this test was also carried out on benzene. The results are found at angles 15.9, 11.2, and 4.5 for the $FeVO_4$, $FeVO_4$@TMOS, and $FeVO_4$@DTOS samples. The addition of TMOS and DTOS is proven to increase the concentration of benzene that can be adsorbed to the surface and increase the formation of phenol transformation (Fig. 9a,b). It demonstrates that the large concentration of adsorbed benzene is associated with an increase in the production of phenol. The phenol yield of $FeVO_4$@TMOS and $FeVO_4$@DTOS is determined to be 20% and 13%, with selectivity over 98%.

5.2.3 $ZnFe_2O_4$

Photocatalyst is an efficient method used in several applications such as waste degradation, fuel energy, organic compound transformation, etc. Phenol is an organic compound obtained from the conversion of benzene initiated by the formation of the •OH radical from H_2O_2. OH radicals can attack the aromatic ring of benzene to form phenols. Iron-based photocatalysts in a Fenton manner, such as $ZnFe_2O_4$ (ZFO) can facilitate the hydroxylation of benzene [62]. Fe in its structure can undergo a redox reaction to form Fe^{2+}/Fe^{3+}, which are known to have the ability to oxidize H_2O_2 to •OH radicals. Unfortunately, the formation of Fe^{2+}/Fe^{3+} can trigger metal leaching, which causes low stability and catalytic activity [63]. Moreover, the appearance of Fe^{3+} on the surface is also a drawback in this system due to its hydrophilic nature,

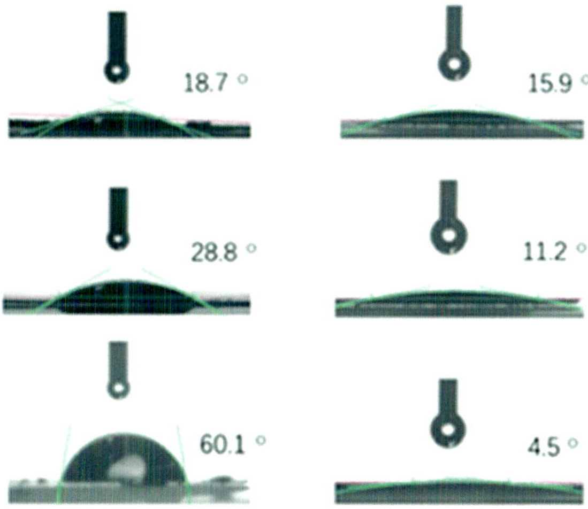

Fig. 5.8 Water contact angle analysis of $FeVO_4$, $FeVO_4$@TMOS, and $FeVO_4$@DTOS. Reprinted with permission from ref. 62. License number 1345726. 2021 Catalyst Science and Technology

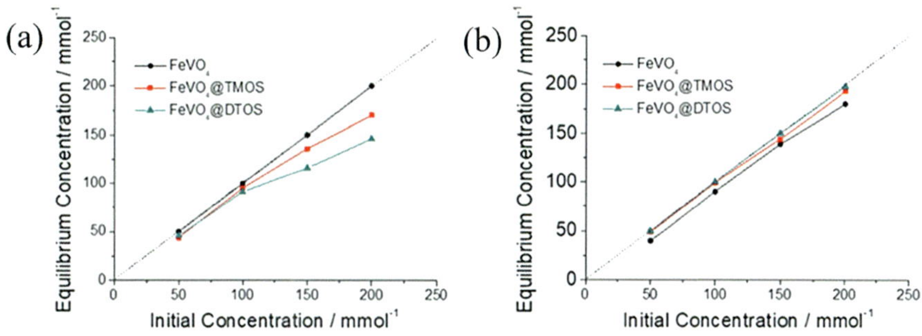

Fig. 5.9 **a** Benzene concentration adsorption and **b** fenol concentration production reprinted with permission from ref. 62. License number 1345726. 2021 Catalyst Science and Technology

which prefers phenol over benzene. This inclination causes over-oxidation of phenol and reduces the amount of phenol production. Several methods have been carried out to overcome this problem, namely by coating the surface with a material with hydrophobic properties, such as organosilane and carbon shell. Carbon shell has many advantages, including its rigid structure to cover the Fe surface for leaching issues and increase its catalytic stability. In addition, carbon is also hydrophobic due to defects on the surface, which can reduce its interaction with phenol so that its selectivity increases and has a π-conjugated electron system that can facilitate the adsorption and activation of benzene in the hydroxylation process.

Yang et al. synthesized ZFO@C by adding glucose to a mixture of $Zn(NO_3)_2$ and $Fe(NO_3)_3$ precursors using solvothermal method [28]. In Fig. 5.10, compared to the pristine ZFO sample, there is a decrease in the intensity of the IR absorption peak at 1640 cm^{-1}, which belongs to O–H bonding, which means the hydrophilic group on the surface was successfully decreased. It also has two new peaks at 1390 and 1610 cm^{-1} belonging to the C–C bond and or C = C bond vibrations due to π-conjugated on the carbon shell, indicating the success of carbon shell grafting on the ZFO surface. This is also supported by XPS data, which shows a peak at the bond energy of 288.7 eV on C1 spectra belonging to C = C or C = O. There is also a decrease in intensity in the Zn 2p and Fe 2p spectra in the ZFO@C sample, proving that the ZFO surface has been covered with carbon shells (Fig. 11a, b).

Its surface properties to water are then tested using a contact angle measurement. Figure 8a shows that the water contact of ZFO and ZFO@C is 12.9° and 38.2°. Adding a carbon shell on the ZFO surface is proven to help improve the surface's hydrophobic properties. Benzene contact ZFO and ZFO@C at 36.8° and 16.1°. The decreasing angle shows that ZFO@C can adsorb more benzene (Figs. 5.12 and 5.13). Because ZFO@C can adsorb more benzene, the benzene conversion to phenol increases.

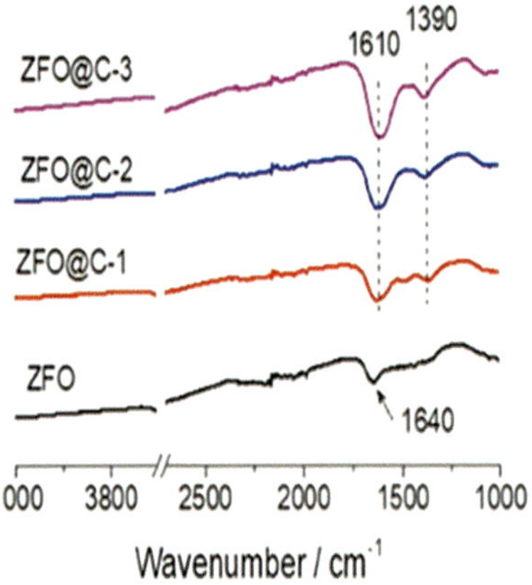

Fig. 5.10 FTIR spectra of ZFO and ZFO@C. Reprinted with permission from ref. 28. License number 5531191501703. 2021 Elsevier

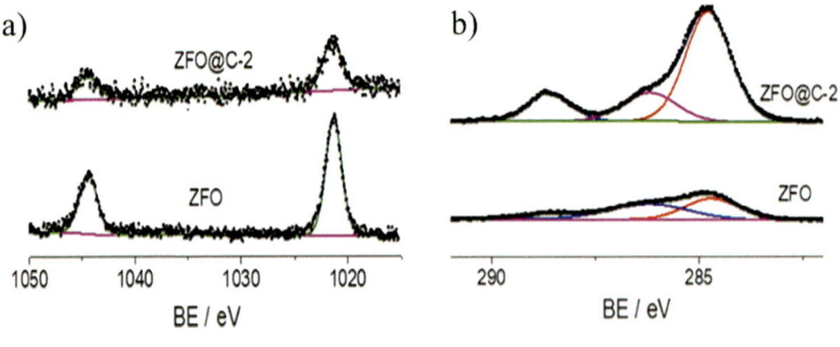

Fig. 5.11 XPS spectra of ZFO and ZFO@C in Z 1s spectra **a**, and C 1s spectra **b**. Reprinted with permission from ref. 28. License number 5531191501703. 2021 Elsevier

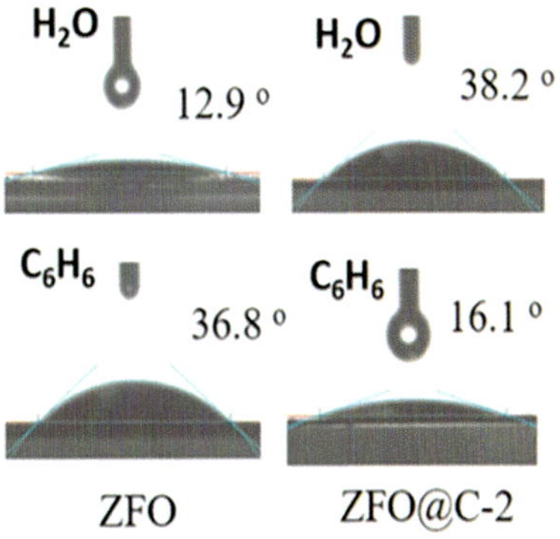

Fig. 5.12 Water and benzene contact angle of ZFO and ZFO@C. Reprinted with permission from ref 5,531,200,798,555 [28]. 2021 Elsevier

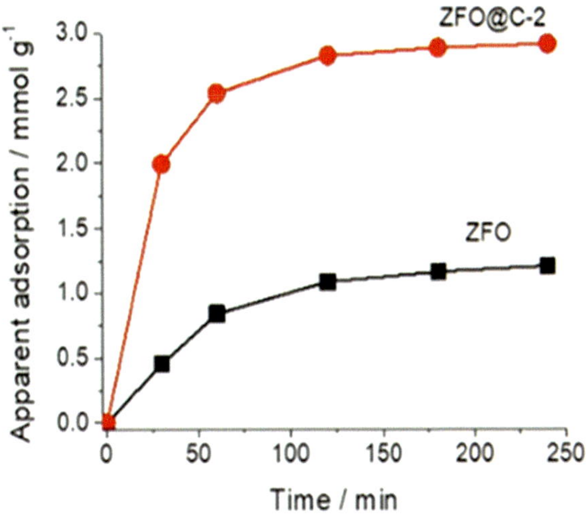

Fig. 5.13 Adsorption of benzene concentration of ZFO and ZFO@C. Reprinted with permission from ref 5531200798555 [28]. 2021 Elsevier

Higher phenol yield was obtained by using a ZFO@C photocatalyst of 15.5% with a selectivity of 99.4%.

5.2.4 *Modified* TiO_2

Semiconductor materials as photocatalysts show promise in facilitating the clean and direct process of synthesizing phenol from benzene. Among these materials, TiO_2 is a promising option due to its low cost, lack of toxicity, remarkable chemical stability against photo corrosion, and wide bandgap of around 3.0 eV [64]. TiO_2 can generate pairs of electrons and holes that can decompose organic material. The energy of photons, rather than their intensity, is a critical factor in the TiO_2 effectiveness of a photocatalyst. Unfortunately, the amount of solar radiation that TiO_2 absorbs is limited, as only a tiny fraction falls within the band gap range of the material [64, 65]. Nonetheless, TiO_2 remains a top contender for photocatalytic phenol synthesis from benzene.

TiO_2 exhibits two distinct photo-induced phenomena: the first is photocatalysis, which breaks down organic materials, while the other involves high wettability. The most crucial aspect of TiO_2 in photocatalytic phenol synthesis is its ability to break down organic materials. To improve TiO_2 effectiveness as a photocatalyst, some studies have tried to extend its photo response into the visible wavelength range. One such method involves using ruthenium (II) complex ($[Ru(bpy)_3]^{2+}$) deposited onto the TiO_2 surface. The complex compound $[Ru(bpy)_3]^{2+}$ can capture visible light down to a wavelength of ~ 550 nm. The accumulation of the amount of solar energy that TiO_2 can absorb-$[Ru(bpy)_3]^{2+}$ helps oxidize benzene to phenol even better. Unfortunately, this complex compound is difficult to separate from TiO_2 for recycling. The use of sodium-type synthetic saponite (SSA) can be solved to separate the $[Ru(bpy)_3]^{2+}$ complex from TiO_2. Another method involves synthesizing meso-porous TiO_2 ($mTiO_2$); compared to non-porous TiO_2 ($nTiO_2$), $mTiO_2$ has a catalytic performance in photocatalytic hydroxylation of phenol from benzene under UV light that produces a ten times greater selectivity phenol (81%) [66]. The strong adsorption ability of benzene on the catalyst surface was found to be much stronger than that of phenol, and phenol's rapid desorption inhibited further oxidation, leading to higher phenol selectivity. To avoid excessive oxidation of phenol that has just been formed (Fig. 5.14), $H_2Si_{14}O_{29}$ (H-mag) was used to increase the adsorption selectivity of the phenol. Although the exact benzene conversion was obtained in the presence of H-mag, the phenol product reaches up to 100% selectivity [67].

The highly active and selective photocatalysis of benzene to phenol requires the development of photocatalysts that can efficiently activate the C–H bond in benzene and selectively generate the hydroxyl group (–OH) to form phenol. To achieve this target, researchers have explored various strategies; one of the most common methods is doping TiO_2 with noble metals such as gold and platina, forming TiO_2-based

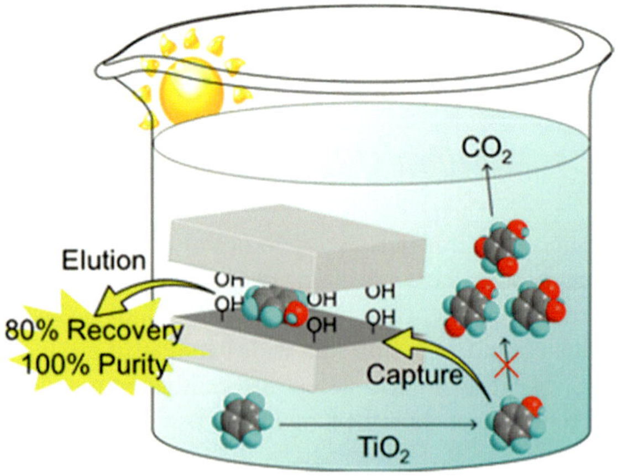

Fig. 5.14 Illustration of effective and selective recovery of phenol by H-mag in photocatalytic hydroxylation of benzene under solar radiation. Reprinted with permission from ref. 68. 2013 American Chemical Society

heterojunctions, and modifying surface properties and morphology of TiO_2. These strategies aim to modify the electronic structure of TiO_2 or shift the bandgap to lower energies and allow for the absorption of visible light.

Ide et al. developed Au-deposited TiO_2 (Au/TiO_2) [19]. This method involves incorporating Au nanoparticles in place of excited electrons to promote benzene oxidation. Additionally, Au nanoparticles have been placed in the interlayer space of a layered titanate to create a clay-like substance with molecular sieving capabilities that can isolate the desired product [68]. Adding an excess of seed phenol and immersing CO_2 into the reaction can increase phenol yield (62%) without producing undesired products [19]. Au/TiO_2 has been shown to have the highest phenol yield and selectivity compared to unmodified TiO_2; the yield and selectivity are around 63% and 91%. By dispersing TiO_2 powder in ethanol and subjecting it to UV light irradiation, Ti^{3+} ions can be produced on the surface. In turn, this allows for the uniform deposition of Au nanoparticles on the surface of the TiO_2 and enhances the yield and selectivity of phenol [69]. TiO_2-supported Au–Pd (shell and core) nanoparticles have also been studied and have demonstrated the highest photocatalytic performance [70]. In addition, a new method has been developed involving doping V into the TiO_2 lattice and depositing Au, resulting in the formation of the $Ti_{0.98}V_{0.02}O_2$ (TV_2) catalyst. This method can significantly increase the conversion of benzene to phenol by promoting electron and hole separation and migration on the catalyst surface under UV light [8].

Apart from gold, other noble metals, such as platinum, were investigated in combination with TiO_2 nanoparticles. The study explored the impact of modifying the TiO_2 surface with platinum and combining it with a polyoxometalate (POM) system on the photocatalytic hydroxylation of benzene. The results showed that Pt deposition generated free radicals on the catalyst surface, significantly increasing the yield and electron acceptor, leading to a phenol yield of 11% [71]. This experiment was conducted using the photo deposition method to create platinum-loaded titanium dioxide from TiO_2 [72]. It was discovered that optimizing the platinum loading amount improved phenol selectivity.

Incorporating noble metals (such as Au, Pt, and Pd) into catalysts has been demonstrated to enhance their catalytic performance, selectivity, and stability significantly. The use of these metals, however, can be costly and limit the scalability of catalyst production. Therefore, exploring and developing transition metal-based photocatalysts is an alternative strategy to achieve green energy, low-cost, and high-performance goals to increase productivity and selectivity for phenol. Transition metals, such as Ni, Cu, Co, Fe, etc., can provide vacant d-orbitals (electrophiles) or lone pairs of electrons (nucleophiles) to form intermediate products, thereby reducing the activation energy of the reaction and improving the reaction kinetics. Singha et al. investigated that Fe^{3+} ions were incorporated into the TiO_2 structure. The optimized catalyst (5%Fe–TiO_2) was able to produce phenol with high selectivity (99%) at a temperature of 70 °C (selectivity around 31.2%), and it could be reused multiple times [73].

Researchers developed a nanostructured catalyst using TiO_2, CeO_2, and Pd nanoparticles to improve the catalytic activity and selectivity of benzene to phenol

Table 5.4 Activities for different catalysts of benzene. Reprinted with permission from ref. 75 2020 Chemical Engineering Science

Catalyst	C_B (%)	Selectivity (%)				Yield (%)
		OH	Catachol Cat	Hydroquinone HQ	others	
Pd	23	38	52	16	4	8.7
CeO_2	12	26	56	18	–	3.1
TiO_2	18	32	60	6	2	5.8
Pd/CeO_2	26	68	20	10	2	17.7
Pd/TiO_2	31	44	36	18	2	13.6
CeO_2/TiO_2	24	34	36	14	16	8.2
$Pd/CeO_2/TiO_2$[com]	40	56	32	6	6	22.4
$Pd/CeO_2/TiO_2$[b]	73	95	3	2	1	69.4
$Pd/CeO_2/TiO_2$[c]	71	93	4	2	1	66.0
$Pd/CeO_2/TiO_2$[d]	97	99	–	–	1	96.03
No catalyst	4	–	–	–	–	–

[a] Typical reaction conditions: solvent (MeCN) = 10 mL, substrate (benzene) = 1 g, catalyst = 0.1 g, benzene: H_2O_2 (molar ratio) = 1:5, reaction temperature = 80 °C, time = 10 h
[b] Fresh Catalyst
[c] Spent Catalyst
[d] Benzene (1 mmol), Ch_3CN (5.0 mL), H_2O_2 (1.1 mmol), 20W domestic cool LED bulb, 30 mg of catalyst, time 30 min

hydroxylation. This $Pd/CeO_2/TiO_2$ catalyst with a size of 5–20 nm showed high efficiency, producing phenol with a selectivity of 95% at 80 °C and a benzene conversion and benzene conversion rate of 73% [74], as shown in Table 5.4. The combination of the active metal and the supports in the catalyst had a synergistic effect, essential for achieving high catalytic activity.

5.2.5 *$CdWO_4$*

Tungsten oxide (WO_3), a well-known n-type semiconductor with a direct band-gap excitation at 2.42.8 eV, is also active for the photocatalytic synthesis of phenols from benzene, allowing photo response in the visible region [33]. In recent years, hierarchical nanostructures with specific morphology have received increased attention through morphology engineering [75]. Three-dimensional (3D) $Bi_2WO_6/CdWO_4$ (BCW) with $CdWO_4$ micro rods decorated with Bi_2WO_6 nanosheets. The unique hierarchical nanostructures can facilitate the absorption of visible light and photogenerated carriers. Under light irradiation with O_2 as an oxidant, BCW has a high phenol selectivity (>99%) and a benzene conversion of up to 7.3% [34]. The elements' surface composition and chemical states were analyzed using X-ray photoelectron spectroscopy (XPS) (Fig. 5.15).

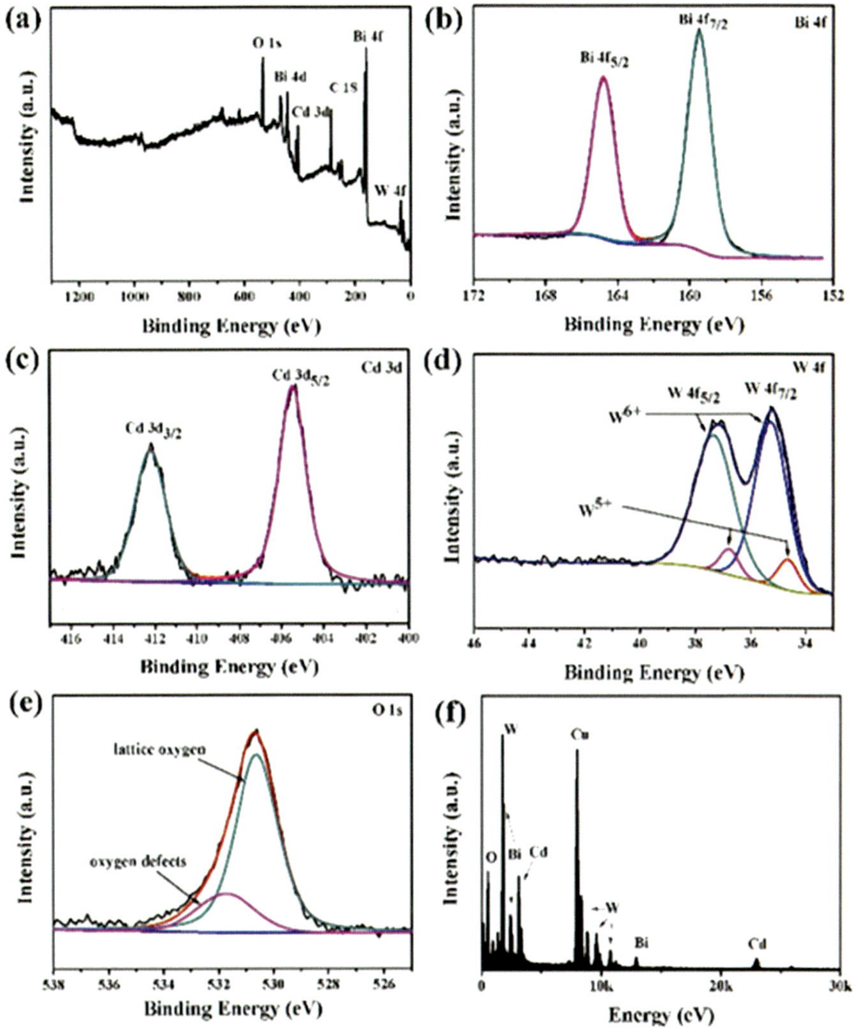

Fig. 5.15 XPS spectra for **a** survey, **b** Bi 4f, **c** Cd 3d, **d** W 4f, **e** O 1s spectra, and **f** EDS spectrum of BCW^{-4}. Reprinted with permission from ref. 34. License number 5521850534956 2018 Elsevier

The composition and chemical states of elements on the surface were investigated using X-ray photoelectron spectroscopy (XPS). The survey spectrum shows the presence of Bi, O, Cd, and W, while the high-resolution Bi 4f peaks indicate that Bi is mainly in the + 3-chemical state. The two peaks observed in Fig. 4c correspond to Cd^{2+} signals in $CdWO_4$. The W 4f XPS spectrum showed the presence of W^{6+}. The XPS spectrum in Fig. 4e is attributed to the sample's O^{2-} ions and oxygen defects. The XPS results confirm the presence of Bi_2WO_6 and $CdWO_4$ in the BCW^{-4} sample, which is consistent with the EDS analysis results. The excellent performance is primarily due to the hierarchical nanostructured $Bi_2WO_6/CdWO_4$ (BCW), which has a high specific area with an appropriate band composition that promotes photo absorption in the visible light region while suppressing photo-induced charge carrier recombination. This strategy outperforms noble metal doping, as previously reported [72].

5.2.6 BCN Nanosheet

Phenol is an important intermediate chemical for many industrial products, such as resins and medicine [76]. The traditional process of making phenol via cumene hydroperoxide has several drawbacks. The process consists of multiple complex steps, requires harsh conditions, involves an explosive intermediate, and creates large amounts of byproduct [77]. Due to these flaws, researchers are investigating a simpler one-step conversion method of converting benzene to phenol.

Carbon-based materials are a popular choice as a catalyst due to their abundance, high stability, and other interesting properties. It can be used as an active material or a support material to catalyze many organic reactions, including benzene hydroxylation. Among various carbon-based materials, reduced graphene oxide (RGO) has been shown to generate hydroxyl radicals (•OH), which can react with benzene to produce phenol [78]. Because •OH is highly reactive, the solvent used in benzene hydroxylation is composed of water and an organic solvent to extract phenol before it is further oxidized. Cai et al. developed a modified RGO (RGO-Cys) by adding L-cysteine during the reduction of GO [79]. As shown in Fig. 5..16, after 4 h of irradiation, benzene conversion of RGO-Cys reached 0.5%, while regular RGO only reached 0.1%. In addition, the selectivity of phenol for both materials is 99%. Prolonged irradiation up to 20 h, increases conversion to 3.1% for RGO-Cys and 1.0% for RGO. However, selectivity decreases to 90% for RGO-Cys and 87% for RGO. The authors attributed the increased performance to the hydrophobicity of the materials and not because of the surface area, since the BET surface area of RGO-Cys is smaller than RGO. The increased hydrophobicity allows RGO-Cys to suspend in the interface between water and benzene during the benzene hydroxylation reaction. Furthermore, when the additive was changed to L-lysine to increase the hydrophilicity of the material, the material (RGO-Lys) did have lower performance compared to RGO.

Catalytic activity can be further improved by doping the material with iron. Chen et al. analyzed the performance of Fe-doped porous graphitic carbon nitride (Fe-g-C_3N_4) with the presence of H_2O_2 [36]. The bare mesoporous g-C_3N_4 (mpg-C_3N_4) itself was able to catalyze the benzene hydroxylation reaction, reaching a benzene conversion of 2.0% and turnover frequency (TOF) of 0.16 h^{-1} after 4 h of illumination. Meanwhile, after modification, Fe-g-C_3N_4 was able to attain a conversion of 4.8% and TOF of 0.43 h^{-1}. To further increase the performance, Fe-g-C_3N_4 was coated onto SBA-15, a silica catalyst support, to increase the surface area of the material. Fe-g-C_3N_4/SBA-15 was able to achieve a benzene conversion of 11.9% and TOF of 14.84 h^{-1}. Meanwhile, Ye et al. pursued a different strategy to increase

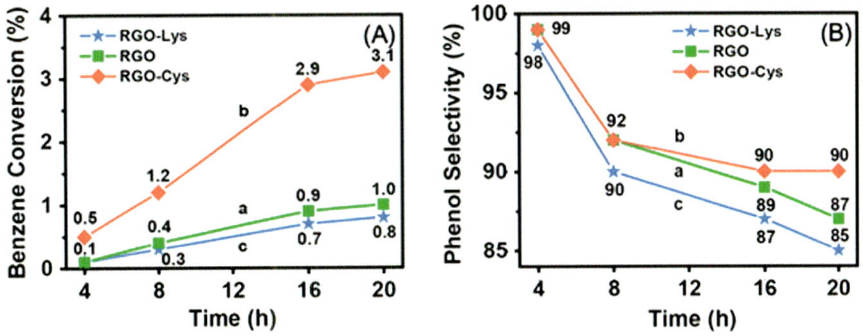

Fig. 5.16 Benzene conversion and phenol selectivity of RGO, RGO-Cys, and RGO-Lys. Reprinted with permission from ref. [79]. Copyright 2018 American Chemical Society

the surface area of the catalyst. They synthesized mesoporous g-C_3N_4 (MCN) via the templating method with SiO_2 nanoparticles as the template, which is also modified with ferrocene carboxyaldehyde (Fc-CHO) to create a hybrid material dubbed Fc-MCN [80]. With the bare MCN, phenol yield only reached 0.7% after 4 h of illumination. Increasing the amount of Fc-FCO incorporated into the material leads to higher performance. The highest performing material, Fc-$MCN_{1.5}$–5, contained the highest amount of Fc-FCO, reaching a phenol yield of up to 16.5%. In both of the g-C_3N_4-based materials, Fe ions played an important role in the benzene hydroxylation process. The Fe^{2+} ion facilitates the Fenton reaction that releases the hydroxyl radical. Meanwhile, the carbon nitride absorbs the photon to generate an electron to activate the inert Fe^{3+} to the reactive Fe^{2+} ion. In addition, the hydrophobic carbon nitride attracts the benzene to be converted into phenol.

Recently, hexagonal boron carbon nitride (h-BCN) nanosheets have been gaining interest because they combine the advantages of hexagonal boron nitride (h-BN) with graphene. Graphene has high tensile strength, high electrical conductivity, and low bandgap (0 eV). Meanwhile, h-BN has high transparency and high thermal conductivity but low electrical conductivity and high bandgap (5.5 eV) [42]. By combining both materials, h-BCN demonstrates exceptional mechanical, electrical, and thermal properties. Moreover, the conductivity and band gap of the material can be tuned by changing the ratio of the carbon, boron, and nitrogen [81]. Therefore, h-BCN is considered to be an excellent photocatalyst and has been used for many reactions, including hydrogen production [82], carbon dioxide reduction [83], and oxidation of benzene to phenol.

Wang et al. prepared h-BCN nanosheets via pyrolysis by heating a precursor made of boric acid, urea, and varying amounts of glucose for 5 h at 1523 K with a flow of ammonia vapor [84]. The resulting products were designated as h-BCN_x, where x is the weight percentage of glucose to boric acid. In the presence of H_2O_2 and $FeCl_3$, h-BCN_{30} was able to reach a conversion rate of 15.9% and selectivity of 88.3%, while hexagonal boron nitride (h-BN) only reached a conversion rate of 5.91% and selectivity of 69.0%. The increased catalytic activity was attributed to a larger surface area, higher light absorption ability, lower band gap, increased charge transfer and separation, and preferential absorption towards benzene. As more carbon content is added to h-BCN_x, the Brunauer–Emmett–Teller (BET) specific surface area and single point total pore volume increase. However, the BJH pore size diameter also decreases as carbon content increases, indicating that more microporous structures are formed that can impede the conversion of benzene. As shown in Fig. 5.17a, increasing carbon content also broadens the absorption spectra of h-BCN_x towards the visible range. Furthermore, the Tauc plots (Fig. 5.17b) show that the bandgap of h-BCN_x is much lower than h-BN, from 3.90 to 2.45 eV. Electrochemical impedance spectroscopy (EIS) was used for h-BN and h-BCN_{30}, which shows that the charge transfer and separation of h-BCN_{30} was better than h-BN, as shown in Fig. 5.17c. Next, in Fig. 5.18, the preferential absorbance of h-BCN_{30} towards benzene instead of phenol also contributes to the increased benzene conversion. Benzene easily attaches to h-BCN_{30} while the converted phenol quickly detaches from the catalyst.

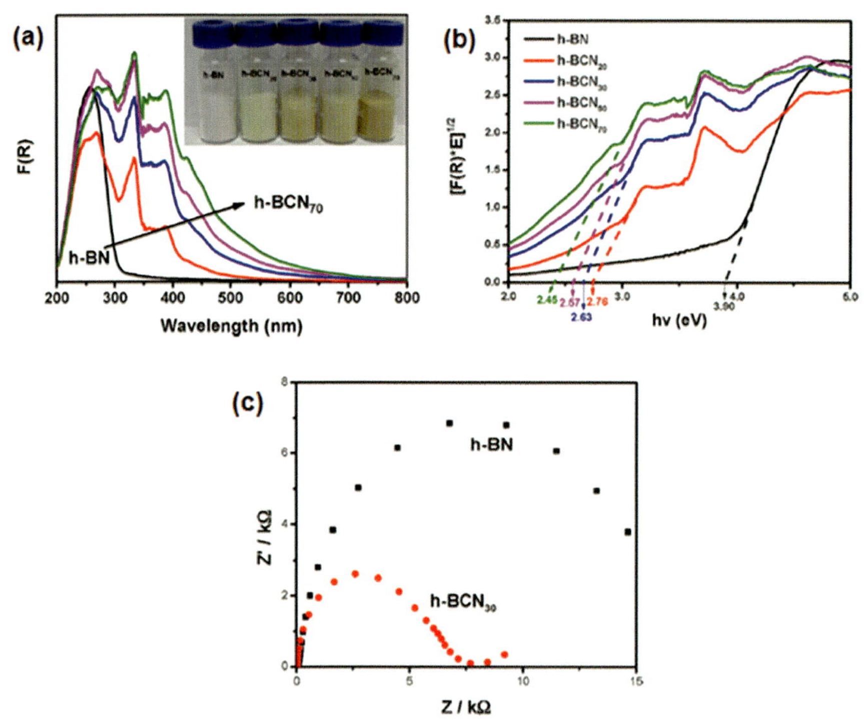

Fig. 5.17 **a** Ultraviolet–visible diffuse reflectance spectra (UV–Vis DRS) of h-BN and h-BCN_x; **b** Tauc plots of h-BN and h-BCN_x; **c** Nyquist plots of h-BN and h-BCN_{30}. Reprinted with permission from ref. [84]. License number 5553431327768. 2019. Elsevier

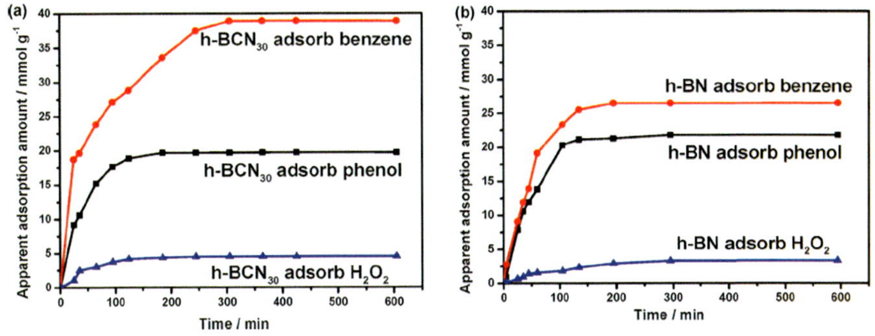

Fig. 5.18 Apparent absorbtion of benzene, phenol, and H_2O_2 by h-BCN_{30} **a** and h-BN **b**. Reprinted with permission from ref. [84]. License number 5553431327768. 2019. Elsevier

5.3 Performance and Mechanism of Catalyst Materials

Phenol is one of the most valuable chemicals widely used as a precursor for dyes, resins, pharmaceuticals, and pesticides. Phenol can be obtained from coal extraction or benzene or benzoic acid oxidation. In recent years, much of the production of phenol has been obtained from the oxidation of benzene because it does not require a large amount of energy, is a safer intermediate product, and has fewer acetone by-products formation. The benzene oxidation reaction involves an oxidizing agent of H_2O_2, N_2O, or O_2 [5]. Using O_2 directly from the free air is a good alternative for producing phenol from benzene. However, benzene is less reactive with O_2 and must be done in a high-temperature or high-pressure manner, which requires high costs and low safety levels. One way to reduce production costs is to look for alternative methods of phenol synthesis that can be run under mild conditions. Photocatalyst is an environmentally friendly way that can be used to convert benzene into phenol with the help of O_2 in mild conditions. This process involves materials that can absorb photon energy well to produce holes and radicals such as *OH, $*O_2^-$, or $*H_2O$ under aerobic conditions [7]. Due to its reactive nature, this radical compound can later oxidize benzene into our desired phenol product. However, the weakness of this radical compound is its low selectivity, and its reactivity knows no direction. It can oxidize any compound before it, even the phenol product itself. The O–H binding energy in phenolic compounds is smaller than the C–H bonding in benzene, so the phenolic product will be more easily oxidized than its precursor (benzene). The phenolic compounds formed will be oxidized by radical compounds to produce by-products such as by-phenyls or chain hydrocarbons. Therefore, it is essential to develop materials to control the overoxidation process and minimize unwanted pathways. In order to find potential materials, detailed knowledge of the reaction mechanism in the oxidation of benzene to phenol is critical.

5.3.1 Traditional Process in Phenol Production

In industry, phenol is produced by the Cumene or Hock processes. The precursors used are benzene and propylene, adding oxygen from water and a little initiator radical and the help of an acid catalyst, as shown in Eq. (5.1) [76]. The results of this reaction are phenol and acetone. The cumene process is very complex, involving several reaction steps, and hydroperoxide can be formed. This very reactive substance can further oxidize the produced phenol, therefore reducing the yield presentation of the phenol formed.

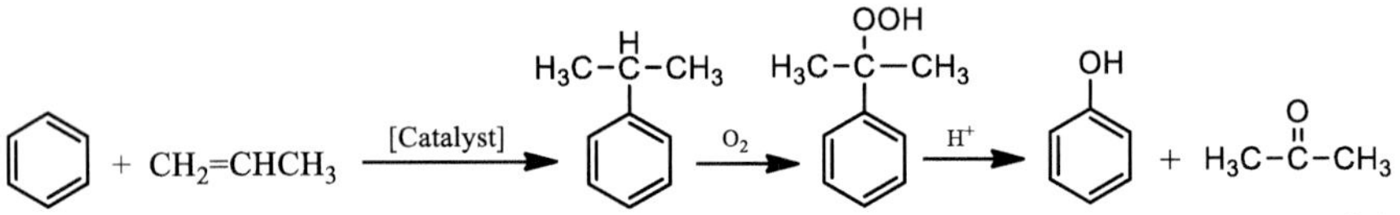

(5.1)

In the first stage of this process, benzene is reacted with propene in a ratio of 3:1 with the help of an acid catalyst. This excess concentration of benzene helps to limit the polyalkylation and oligomerization reactions of propane. The cumene formed is then converted to cumene hydroperoxide by flowing oxygen and adding peroxide. The presence of oxygen helps form radical compounds, which are helpful for binding peroxide groups (–OOH) to the cumene structure. This reaction takes place at a temperature of 77–117 °C and at a high pressure of 1–7 atm to maintain the system in the liquid phase. Furthermore, cumene hydroperoxide is decomposed by adding sulfuric acid at 40–100 °C, then neutralized to form phenol and propanone (acetone). Finally, the product is separated by a distillation column [36].

The biggest drawback of this process is that the percentage of phenol yield obtained is tiny despite phenol products being more profitable targets when compared to propanone. As seen from the reaction above, the weakness of this reaction occurs when the addition of sulfuric acid initiates phenol and propanone formation. The water content in sulfuric acid helps mediate the ongoing oxidation reaction. Increasing the yield of phenol can be done by finding a material with a high level of selectivity in limiting the water release in the system immediately before use to facilitate the formation of new radicals to prolong the oxidation process or by looking for a material that has a high selectivity for phenol and helps stabilize it from excessive free radical attack.

5.3.2 Photocatalyst Process in Phenol Production

Another drawback of the traditional process is that benzene is less reactive to O_2 under mild conditions. This process can only run at the high-temperature or high pressure, which requires high costs and low safety levels. One way to reduce production costs is to look for alternative methods of phenol synthesis that can run under mild conditions. Photocatalyst is a potential phenol synthesis method to reduce production costs and increase the level of reactivity of benzene to oxygen. Highly selective phenol production is desired without sacrificing the photocatalysts at ambient conditions. Thus, stable yet selective enough photocatalysts are a must while maintaining vigorous activity in benzene conversion. Especially in hydroxylation, benzene into phenol, the kinetic reaction is determined by the catalyst ability to activate the C double bond in the benzene structure followed by oxidation of activated benzene by the oxidizing agent (Eqs. 5.1–5.7). A suitable oxidant is needed to avoid over-oxidation of phenol while maintaining good conversion. Reactive oxygen species (ROS) are the most efficient and energy-saving oxidants. Water can form in situ ROS facilitated by photo-excited electron–hole pairs. Many studies reported the primary mechanism of hydroxylation of benzene to phenol by utilizing electron–hole pairs at the surface of the semiconductor to overcome the energy barrier on Sp^2 carbon activation on benzene. At the surface of the semiconductor, an excited electron into the conduction band will leave a hole at the valence band that can react with hydroxyl ions from water. At a certain point, this reaction can form hydroxyl radicals, which subsequently react with benzene to form phenol. Photogenerated holes can use enough energy to convert adsorbed water or surface-attached hydroxyls into hydroxyl radicals (•OH). The exact mechanism is not fully understood, but •OH radicals are known to be the most potent oxidative species for organic oxidation. As a result, they are commonly used in the photocatalytic treatment of organic pollutants. However, in selective organic oxidation, •OH can lead to nonselective oxidation of organics into CO_2 and water instead of producing valuable organic chemicals. Therefore, significant efforts have been made to prevent the formation of •OH to increase reaction selectivity. The oxidation of saturated C–H bonds is more challenging than selective alcohol oxidation because of their higher bond energy. The focus of most researchers in this area has been on two reactions: the conversion of toluene to benzaldehyde and cyclohexane to cyclohexanone. By combining information from the literature, it can be concluded that the mechanism of hydrocarbon photo-oxidation involves the hydrocarbon being initially oxidized to alkyl radicals (•RH) through the generation of photogenerated holes. The following reaction can predict the resume of the hole's involvement in photo-oxidation [46].

$$RH_2 + h^+(VB) \rightarrow {}^{\cdot}RH + H^+ \quad (5.2)$$

$$^{\cdot}RH_2 + O_2 \rightarrow {}^{\cdot}OORH \quad (5.3)$$

$$^{\cdot}OORH + e^-_{(CB)} + H^+ \rightarrow RO + H_2O \quad (5.4)$$

$$^{\cdot}OORH + {}^{\cdot}O_2^- + H^+ \rightarrow RO + O_2H_2O \quad (5.5)$$

$$2\,{}^{\cdot}\mathrm{OORH} \rightarrow \mathrm{RO} + \mathrm{RHOH} + \mathrm{O_2} \quad (5.6)$$

$$\mathrm{RHOH} + 2\,{}^{\cdot}\mathrm{OH} \rightarrow \mathrm{RO} + 2\mathrm{H_2O} \quad (5.7)$$

Other ROS such as superoxide radical ($\cdot O^{2-}$)$_6$ and singlet oxygen also have been reported could be formed during photocatalysis, yet they have limited use, especially in selective oxidation like phenol production from benzene [85]. Therefore, we believe the main ROS that could be useful in the hydroxylation of benzene to phenol will still be dominated by the hydroxyl radical (•OH). Besides the importance of oxidants, the type of catalyst and their intrinsic properties are other things that will be discussed in this section.

Photocatalysts usually use semiconductor materials to capture photon energy and help break benzene into phenol. The most commonly used semiconductor material is TiO_2 because it has good chemical stability, high oxidizing ability, low cost, and safety. When the TiO_2 photocatalyst gets a higher energy exposure than its energy band gap, a positive charge is generated in the valence band. In contrast, a negative charge is promoted in the conductive band. Positive holes and electrons play an essential role in oxidation and reduction reactions. Positive charges can attract water and form hydroxyl radicals, while negative charges can reduce oxygen to produce superoxide ions. The highly reactive hydroxyl radical compounds can be used to transfer benzene into phenol [10].

TiO_2 has two phases, namely, the rutile and anatase phases. The anatase phase has a higher photocatalytic activity because it has a higher surface hydroxyl group. The anatase TiO_2 phase has a wide band gap of 3.2 eV, so it can only be active under exposure to UV light [11]. However, the largest composition of sunlight is visible light. Therefore, to maximize the absorption of sunlight and maximize its use, the researchers then developed TiO_2 materials with several additional modifications, such as complex compounds, mesocellular siliceous foam, metals, transition metals, etc.

Yuzawa et al. developed a Pt–TiO_2-based photocatalyst for the hydroxylation of benzene in water [14]. They compared their Pt–TiO_2 irradiation wavelength with several materials with different wavelength irradiation; they found an effect of the photocatalyst wavelength on the selectivity of phenol. The best phenol selectivity (91%) is obtained if the photocatalyst can absorb solar energy up to more than 385 nm. The mechanism is predicted in four possible pathways, such as reactions in Eqs. (5.8–5.11). Pt–TiO_2 with a longer wavelength can significantly increase the phenol yield. The optimum photon energy helps to oxidize H_2O to *OH radicals. The *OH radical attacks benzene to form phenol and releases H_2 gas (Eqs. 5.8). Pt–TiO_2 with less wavelength produces a small amount of phenol. The mechanism was predicted in three possible pathways, such as reactions in Eqs. (5.9–5.11).

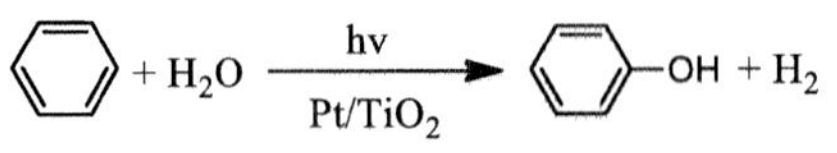

(5.8)

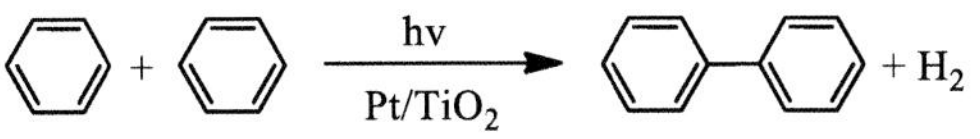
(5.9)

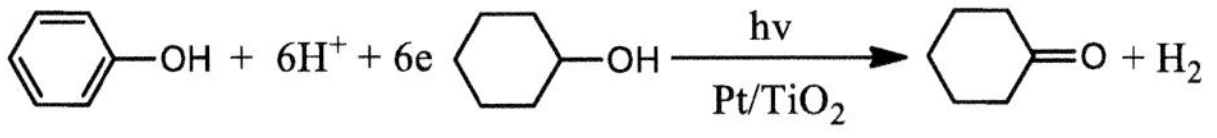
(5.10)

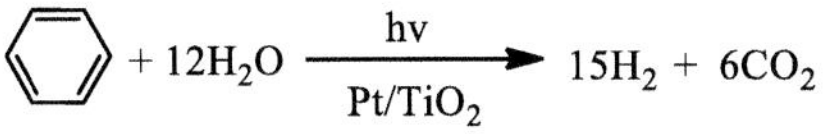
(5.11)

Although using photocatalysts based on noble metals such as Pt and Au showed high efficiency and selectivity for phenol formation, using noble metal-based photocatalysts is costly and not environmentally friendly. Therefore, exploring and developing transition metal-based photocatalysts is an alternative strategy to achieve green energy, low-cost, and high-performance goals to increase productivity and selectivity for phenol. Transition metals, such as Ni, Cu, Co, Fe, etc., can provide vacant d-orbitals (electrophiles) or lone pairs of electrons (nucleophiles) to form intermediate products, thereby reducing the activation energy of the reaction and improving the reaction kinetics.

Gupta et al. used Fe impregnated onto the surface of TiO_2 as photocatalyst hydroxylation of benzene to phenol. The results showed selectivity to 80–86% phenol under UV light irradiation for 1–2 h. It was due to changes in the oxidation stat–e of Fe when irradiated under UV light. Fe^{3+} is formed on the surface and helps to stabilize the intermediate products of benzene oxidation and prevents further oxidation by *OH radicals.

Devaraji and Jo modify Fe–TiO_2 by adding Cr doping into its structure [11]. This modified catalyst was then compared with the basic materials, namely, Fe–TiO_2 and Cr–TiO_2. The results showed a selectivity of 90% for phenol after 12 h of irradiation under UV light with the help of an H_2O_2 oxidizer. The degree of selectivity to phenol in the Fe–Cr–TiO_2 modification increases by about 5–10% compared to the base material. These results prove that the simultaneous presence of Cr and Fe in the TiO_2 lattice can increase the photocatalytic activity of phenol production when compared to TiO_2 doped with Fe or Cr alone. TiO_2 doped with Fe–Cr also shows visible light absorption, but the energy from visible light absorption is not strong enough to convert benzene into phenol. However, visible light energy helps strengthen the absorption of UV energy to increase its conversion power and selectivity to phenol. This invention shows that the conversion of benzene to phenol occurs due to the excitation of electrons in the valence band due to exposure to UV irradiation. The reaction mechanism of the hydroxylation of benzene to phenol with the help of the Fe–Cr–TiO_2 photocatalyst is predicted through two reaction pathways, namely, the A-path, and B-path, as shown in Fig. 5.19.

UV light applied to the surface of the Fe–Cr–TiO_2 photocatalyst helps release electrons from the valence band and is excited to the conductive band on the A-path; electrons trapped at the dopant level can reduce Fe^{3+} to Fe^{2+}, which can then be imagined with H_2O_2 and protons from H_2O to produce hydroxyl radicals and Fe^{3+}.

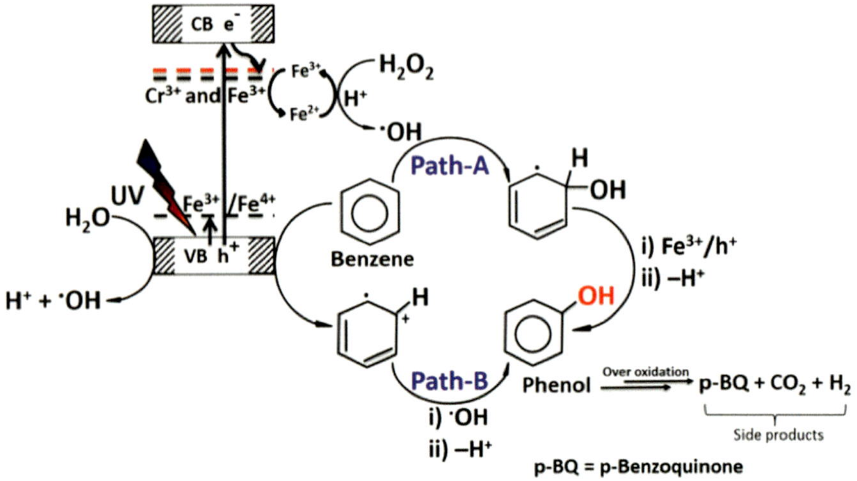

Fig. 5.19 Prediction of Fe–Cr–TiO_2 photocatalyst mechanism pathways. Reprinted with permission from ref. [11]. License number 5531351014879. 2018. Elsevier

This hydroxyl radical attacks the aromatic benzene ring and forms the hydroxycyclohexedienyl radical. The positive photogeneration hole in the valence band (Fe^{3+}) then oxidizes the hydroxycyclohexedienyl radical to phenol through deprotonation. Meanwhile, starting from the positive hole in the valence band, the pathway-B mechanism analyzes it with benzene to produce benzene radical ions. These radical ions are then analyzed with hydroxyl radicals to form phenols via deprotonation of unstable intermediates.

As explained in the previous sub-chapter, iron is a potential material as a catalyst because of its low price and excellent catalyst effectiveness. Iron has been used as a catalyst for a long time in the Fenton process. Nonetheless, the utilization of iron catalysts gives rise to several challenges, including stability and its more hydrophilic nature. Iron is easily oxidized to form Fe^{3+} on the surface; this process runs very fast, so its effectiveness is often limited by time. In addition, Fe^{3+} is also hydrophilic, which is not good for the hydroxylation of benzene to phenol. This is because phenol has a hydroxyl group (OH) which is more electronegative and hydrophilic. Fe^{3+} on the catalyst's surface will easily bind to the hydroxyl group on phenol and oxidize it to form byproducts or hydrocarbon chain compounds. To overcome this problem, materials are developed to bind Fe^{3+} on the surface to reduce the reactivity of Fe^{3+} to the phenolic hydroxyl groups. At the same time, the material must also be hydrophobic to reduce its interaction with phenol. One of the potential materials is the organosilane group. Organosilanes have two side groups: a silane group and a hydrocarbon chain.

The OH group in silanes has a strong electronegativity to form bonds with Fe^{3+}. At the same time, the hydrocarbon chain on the top side tends to be hydrophobic, which will prevent further phenol oxidation. Wei et al. succeeded in synthesizing

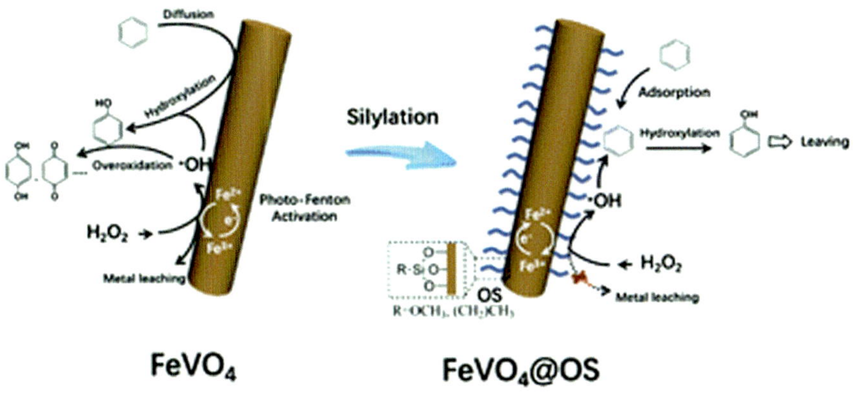

Fig. 5.20 Predict mechanism in $FeVO_4$@OS photocatalyst for reaction of hydroxylation benzene into phenol. Reprinted with permission from ref. 62. License number 135783. 2021 Elsevier

$FeVO_4$, which was deposited on tetramethoxylation, and dodecyltrimethoxysilane has a high selectivity to phenol, which is 98%. The presence of a vanadate group in its structure contributes to lowering the activation energy of Fe, while the organosilane deposited on the $FeVO_4$ surface helps suppress Fe^{3+} activity on the surface. H_2O_2 is used as the oxidizing agent in this system. Initially, the peroxide oxidizes the iron from Fe^{2+} to Fe^{3+}. Benzene is adsorbed by the hydrocarbon groups on the surface and oxidized with the help of Fe^{3+} to form phenol, as shown in Fig. 5.20.

Besides TiO_2, ZnO is also very popular as a semiconductor often used as a photocatalyst. Compared to TiO_2, ZnO has a wider direct band gap of 3.37 eV, is less toxic, and is more stable in a wide-range pH while also providing the opportunity for activation by visible light. Iron-based catalysts are often deposited onto the ZnO surface to help reduce the band gap and increase its catalytic activity. Yang et al. tested the ability of $ZnFe_3O_4$ (ZFO) to hydroxylate benzene into phenol. Although $ZnFe_3O_4$ can absorb photons in the visible light range, its catalytic activity is still very low. It is due to the leaching and overoxidation of phenol by Fe^{3+} on the surface. Yang et al. then developed the $ZnFe_3O_4$ material with modified surface properties by adding a carbon shell on the surface. The principle of the present invention is the same as the previously discussed organosilane addition. The incorporation of a carbon shell offers several advantages, including its rigid structure to cover the Fe surface for leaching issues and increase its catalytic stability. Additionally, carbon is also hydrophobic due to defects on the surface, which can reduce its interaction with phenol so that its selectivity increases and has a π-conjugated electron system that can facilitate the adsorption and activation of benzene in the hydroxylation process. As a result, the catalytic effectiveness of $ZnFe_3O_4$ increased, and its selectivity to phenol increased by about 99.4%. The mechanism is predicted as shown in Fig. 5.21. ZFO absorbs photon energy, electrons in the valence band are excited to the conductance band, and with the help of a peroxide oxidizer, Fe^{2+} is converted to Fe^{3+}, while

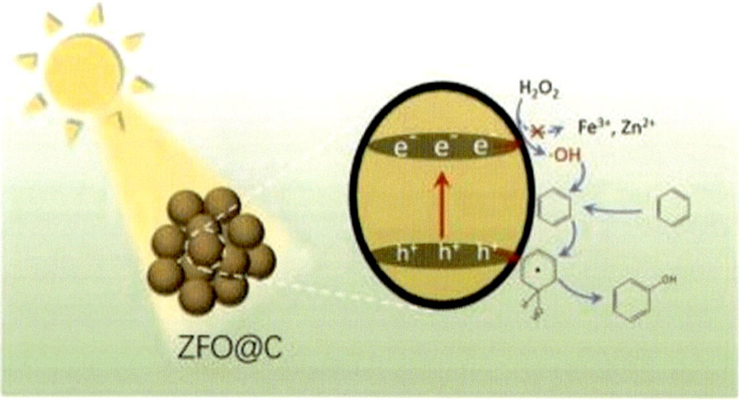

Fig. 5.21 Predict mechanism in ZFO@C photocatalyst for the reaction of hydroxylation benzene into phenol. Reprinted with permission from ref. [24]. License number 5554480814204. 2019 Elsevier

Zn is oxidized to Zn^{2+}. These two metal ions synergistically contribute to the benzene to phenol oxidation process.

Another photocatalyst known as having the ability to convert benzene into phenol is a homogenous catalyst. Although homogenous photocatalysts show promising results, a lack of practice in the separation process of the catalyst at the end of the reaction is unavoidable. As reported, more than 99% phenol can be obtained during hydroxylation of the benzene in the presence of MIL complexes photocatalysis. It has been shown that due to the synergistic effect of functional group in Keggin type phosphor tungsten (PW) structure and its counter ions exhibited high yield (20.9%) and selectivity (> 99%) of phenol. Ions pairing between anion in PW and cation from ionic liquid (IL) promotes the generation of photoinduced charges, giving the crucial effect during hydroxylation reaction. This intermolecular interaction happens inside the photocatalyst structure and has been proved by the mass spectrometry (MS) analysis (Fig. 5.22).

From the MS spectra, it can be seen that demonstrated a dissymmetry signal at −15.9 ppm slight shifting to up-field for C_{16}Qu-PW is attributable to the strong interaction between organic cations and Keggin type POM anions [85]. Furthermore, the FTIR technique confirmed that the functionality of the IL side chain could provide stability of phase transformation of these homogenous photocatalysts into solid form, which solves the common separation problem. Solid C13 NMR and XRD analyses supported the evidence of phase transformation of homogenous photocatalysts into solid structures. This transformation facilitates reusability, a cycle of oxygen radical species formation and transfer. Undeniably, heterogeneous photocatalysts are still a possible option for industrial application. However, the performance is inferior to homogenous; this problem can be solved by adequately modifying their intrinsic properties. For example, upon exposure to visible light, the photocatalysts 5%Cr-CdS/ZnO generate electron–hole pairs. The electrons produced by the

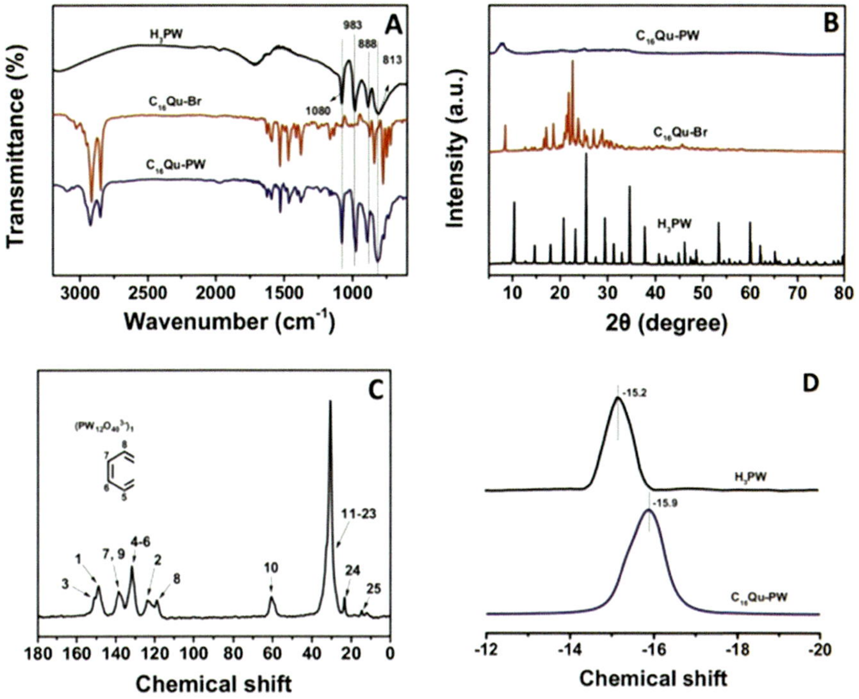

Fig. 5.22 **a** FT-IR spectra; **b** XRD patterns of H_3PW, C_{16}Qu-Br, and C_{16}Qu-PW; **c** 13C; and **d** 31P MAS NMR spectra of H_3PW and C_{16}Qu-PW. Reprinted with permission from ref. 46. License number 5538710685710. 2019. Elsevier

light are elevated from the valence band to a newly created electron trapping level caused by the addition of Cr^{3+} dopants. Two reaction paths of phenol production from benzene by realizing the ·OH radicals were produced on-site and then interacted with the benzene rings to produce hydroxyl cyclohexadiene radicals. Positive holes in the valence band further oxidized these radicals into phenol. This whole process occurred in situ. An alternative route following the VB of 5%Cr-CdS/ZnO contains created voids that have the potential to interact with enhanced available holes in VB3. Phenol could be formed by the liberation of proton radicals from the hydroxycyclohexadienyl radical. These routes can be seen in Fig. 5.23 [86].

Various types of co-catalyst nanoparticles have been tested for their positive impact on preventing the polymerization of phenolic products and the full oxidation of benzene. One particular type of nanoparticle, the Au shell-Pd core variant, which is supported on TiO_2, has shown promising results in this regard, based on the UV–Vis spectra data Au and Pd-rich alloy of TiO_2 nanoparticles, which shows promoted phenol production compared with bare TiO_2 or monometallic counterparts. Surface morphology with alloy composition benefit inhibited hydroquinone formation [87]. This strategy involves the doping of metal nanoparticles onto the surface

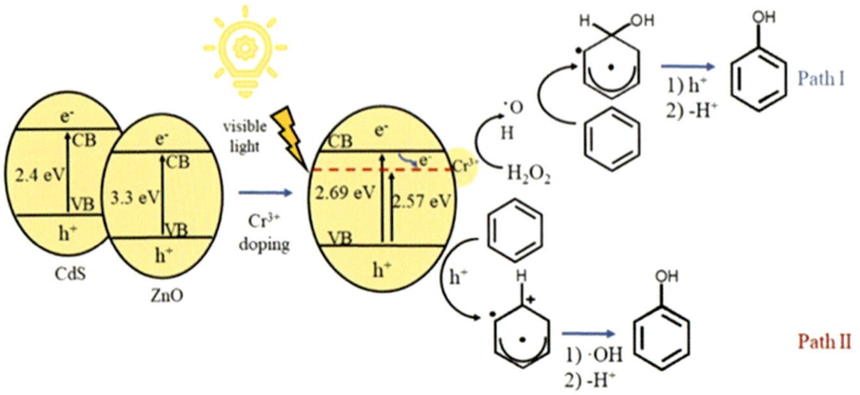

Fig. 5.23 Proposed mechanism for OH radical formation and the routes for the photocatalytic hydroxylation of benzene to phenol. Reprinted with permission from ref. [86]. License number 5538711011882. 2023. Elsevier

of the semiconductor and utilizes the Schottky barrier effect. At the metal–semiconductor interface, the electron given by the nanoparticle due to the SPR effect will not recombine but instead be transferred to the CB of the semiconductor. Subsequentially, the left hole could react with an organic compound that realizes the oxidation reaction. Supported by other results, it was confirmed that loaded metal nanoparticles onto the surface could enhance charge separation. Generally, the highly oxidized condition results from utilizing bare semiconductors for photocatalysis due to the very negative potential of its valence band. Therefore, the appropriate selection of semiconductors from their band gap information is very crucial. For example, it has been reported that WO_3 may photo-catalytically create phenol from benzene with high selectivity (e.g., 74% at 69% benzene conversion), significantly greater than TiO_2, which produced CO_2 as the major product. The electrons generated by light on the Pt/WO_3 surface tended to start a process where two electrons reduce O_2 and create H_2O_2. This H_2O_2 was unable to over-oxidize the produced phenol. On the other hand, the electrons on the conduction band of TiO_2 quickly generated radical oxygen species like $\bullet O^{2-}$, which caused the phenol to undergo subsequent oxidation to CO_2. As a result, this process decreased the selectivity [88]. In addition to the extensive development of controllable nano-architecture syntheses, the development of nanocatalysts with or without the support material has expanded photocatalysts' performance by employing modulation of the band gap through defect engineering [89], nanostructured modification [33], crystal control [47], or by surface manipulation [90]. When benzene is hydroxylated to phenol, it is advantageous for adsorbing the benzene molecule at the nonpolar surface, like graphitic carbon [91]. Carbon materials, including graphene and C60, are conductive and have a band gap of 0 eV around the Fermi level. These materials have been important in both fundamental research and industrial applications. Researchers can leverage the distinct characteristics of these different components by combining photocatalysts with non-metal materials like carbon-based substances or metal oxides. This approach offers a valuable set of techniques to manipulate how light is harvested, how well the substance can adsorb, and ultimately enhance the activity and specificity of the photocatalyst. Thus, when an oxidant, such as hydroxyl radical, is available on the surface of the carbon, the oxidation reaction of benzene can proceed.

However, using carbon materials in photocatalysts is mainly a supporting material in catalytic systems. As demonstrated, $FeCl_3$ was firmly attached to the g-C_3N_4 support through the interaction between Fe^{3+} and the nitrogen atoms of g-C_3N_4. When used in the hydroxylation of benzene to phenol with the presence of H_2O_2, the $FeCl_3$/eg-C_3N_4 materials demonstrated robust catalytic activity even after five consecutive reactions. The catalyst shows remarkable reproducibility by achieving more than 99% selectivity [92]. A method was used to introduce carbon atoms into the BN structure, creating ternary-component materials known as BCN. Two well-known types of BCN are diamond-like structure cubic boron carbon nitride (c-BCN) and hexagonal boron carbon nitride (h-BCN) [93, 94]. These materials were found to have a narrower band gap than BN but wider than carbon materials, resulting in a semiconductor band gap that falls between the range of 0–5.5 eV. Several studies have reported that carbonaceous material could behave like semiconductors and produce oxidant species. For instance, under visible light, RGO was discovered to activate H_2O_2, generating OH via a Fenton-like pathway previously used for phenol degradation. In another report, CuPd bimetallic nanoparticles were uniformly deposited onto the C_3N_4 nanocomposite. By KOH etching, graphitic carbon nitride forms a void called holey-gC_3N_4 (HCN) [57]. It has been confirmed by using ESR (electron spin resonance) and DMPO as a radical trapper, indicating two ROS, i.e. $\bullet O^{2-}$ and •OH, were formed in atomically dispersed CuPd on HCN (HCN-5). It was suggested that two main steps for hydroxyl radicals were formed from peroxide decomposition, followed by selective benzene oxidation by these radicals. In the initial steps, Adsorbed peroxide at CuPd sites will promote stabilization and form CuPd = O. Next, as cumulative peroxide molecules increase at this site has led to the formation of O = CuPd = O. This newly formed active site plays a key role in activating the C–H bond from benzene by forming a low-energy C–O bond. Lastly, at the final step, continuing adsorption of the benzene molecule will generate phenol, which sequentially also regenerates the CuPd = O site. The detailed data for the reaction and simulated energy diagram by DFT are presented in Fig. 5.24.

The ternary h-BCN nanosheets, which have the benefits of both graphene and h-BN, functioned as a heterogeneous catalyst with a unique ability to adsorb benzene and an appropriate energy band structure as a semiconductor [19]. These characteristics of the ternary h-BCN nanosheets resulted in efficient photocatalytic oxidation of benzene into phenol with high yield and selectivity for phenol formation in the one-step hydroxylation of benzene. The process occurred under visible light irradiation and in the presence of $FeCl_3$ and H_2O_2. These findings indicate that h-BCN nanosheets have the potential to become a new type of photocatalytic material. In this process, benzene molecules are attracted to the h-BCN nanosheets through a π-π interaction between the benzene and the conjugated h-BCN. When the nanosheets are exposed to visible light, electron–hole pairs form in the conduction and valence

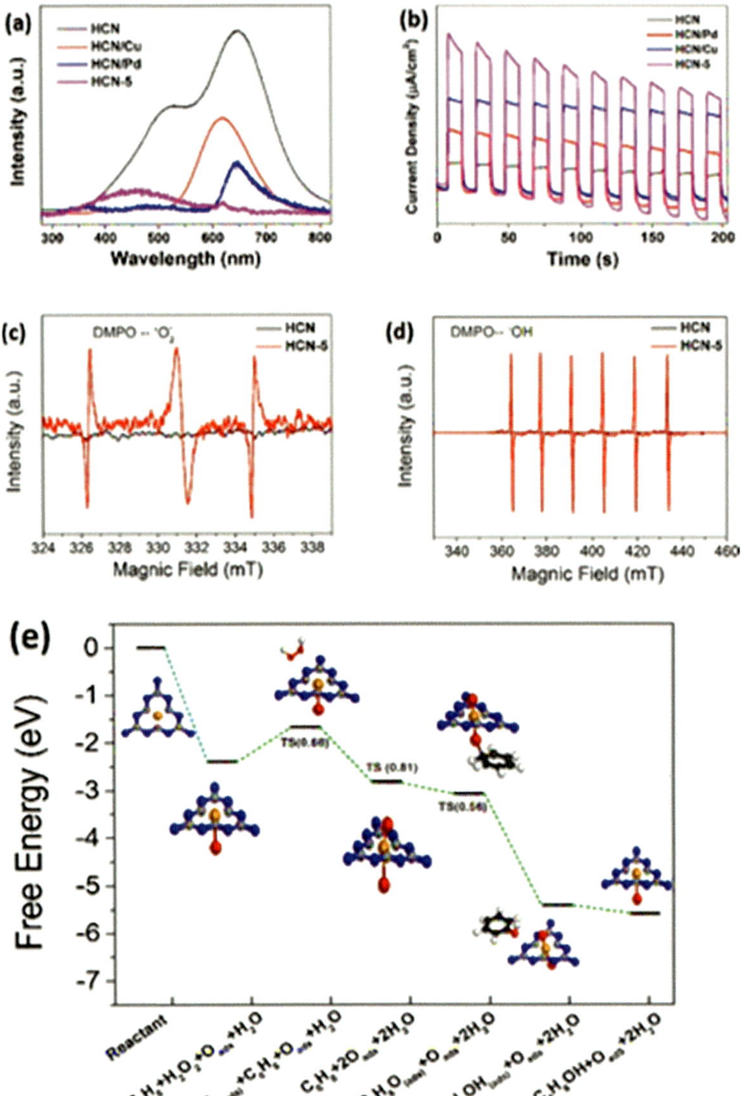

Fig. 5.24 **a** The photoluminescence spectrum of the catalysts synthesized in their original form; **b** the temporary photocurrent reaction of the samples synthesized in their original form; **c** the electron spin resonance spectrum of DMPO spin trapping of $\bullet O^{2-}$ for HCN-5; **d** the electron spin resonance spectrum of DMPO spin trapping of •OH for HCN-5; **e** the mechanism suggested for the oxidation of benzene over HCN-5 through DFT calculation. HCN-5; Reprinted with permission from ref. [57]. License number 5538841433587. 2019. Elsevier

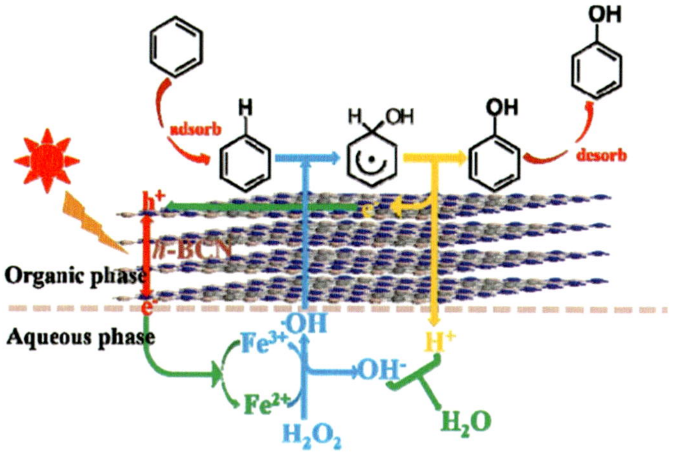

Fig. 5.25 Schematic description of how benzene is hydroxylated using h-BCN in combination with $FeCl_3$, and the steps involved in the reaction. Reprinted with permission from ref. [84]. License number 5538871271401. 2019. Elsevier

bands of h-BCN, respectively. The photo-generated electrons reduce Fe^{3+} to Fe^{2+}. And then, Fe^{2+} reacts with H_2O_2 to produce Fe^{3+}, radical •OH, and OH^-. These •OH radicals then attack the benzene molecules adsorbed on the h-BCN surface, forming hydroxyl-cyclohexadienyl radicals. These radicals release protons (H^+) and electrons (e^-) into the system, which in turn react with OH^- species to form H_2O and recombine with holes (h^+) in the valence band of h-BCN. To summarize, h-BCN nanosheets serve as a photocatalyst by providing both electrons and holes while concurrently acting as a solid surface on which benzene molecules readily adsorb and become activated through the π-π interaction with h-BCN. The schematic mechanism for the reaction is illustrated in Fig. 5.25.

5.4 Application of Photocatalysis in Generating Value-Added Products

5.4.1 Carbon-Based Materials

Photocatalysis using carbon-based materials is an emerging field that has gained significant attention recently due to its potential for generating value-added products. Carbon-based photocatalysts, such as graphene and carbon nanotubes, have unique electronic and structural properties that make them excellent candidates for photocatalytic applications.

5.4.1.1 C_3N_4-based Materials

In recent years, C_3N_4-based materials have garnered considerable attention as potential photocatalysts for various energy and environmental applications. These materials, made up of a carbon nitride compound with a layered structure and narrow bandgap, can absorb visible light and generate photoinduced charge carriers for catalytic reactions. Additionally, they possess desirable properties like chemical and thermal resistance, low density, and biocompatibility [95–97].

One material synthesized using tri-s-triazine units under ambient conditions is graphitic carbon nitride (g-C_3N_4), considered the most stable method of preparation. However, the stability of g-C_3N_4 is yet to be fully understood due to the electronic structure of the tri-s-triazine units, which contain a nitrogen lone pair [98]. Several approaches have been explored, including creating mesoporous graphitic carbon nitride (m-g-C_3N_4). This method results in a material with a significant internal surface area, making it suitable for hosting metal nanoparticles or semiconductors [98].

5.4.1.2 Bare C_3N_4

The study aimed to evaluate the effectiveness of pure C_3N_4 when subjected to visible light irradiation to oxidize aromatic alcohols to aldehydes in the presence of water and oxygen. The results showed that at a neutral pH and a temperature of 60 °C, 4-methoxybenzyl alcohol underwent a chemical transformation to produce the corresponding aldehyde with a selectivity of 89% following a conversion rate of 56%. Introducing HCl into the reaction system resulted in the oxidation of benzyl alcohol to benzaldehyde, with 93% selectivity and 49% conversion [99]. The synthesis of C_3N_4 involved the thermal condensation of dicyandiamide, and the obtained solid was heated to 400, 450, and 500 °C (thermos-exfoliation), treated mechanically for 1.5, 4.0, and 8 h, or stirred in H_2SO_4, HCl, and HNO_3 solution before being used for oxidizing benzyl alcohol to produce benzaldehyde under LED irradiation at a wavelength of approximately 392 nm and pH water condition of 5.6 [100]. The preparation of these materials proved advantageous as it increased the conversion and selectivity of benzyl alcohol to aldehyde, as shown in Fig. 5.26.

The experiment conducted for 4 h revealed that the selectivity reached 90% and conversion was 66%, while benzaldehyde yielded around 59% when using the thermos-exfoliated material at 500 °C, which had a specific surface area of 87 $m^2\ g^{-1}$. This yield was 3–4 times higher than bulk g-C_3N_4 [101]. This discovery suggests that mechanical and chemical treatments could modify the electronic properties of the bulk g-C_3N_4 material, resulting in defects that widen its band gap. This increased band gap, together with a greater surface area, enhances the production of benzaldehyde while still maintaining a selectivity of no less than 80% [100]. Efforts were made to enhance the selectivity towards aldehydes, including investigating the effect of H_2O_2 on C_3N_4 [102] and treating carbon nitride in molten potassium and lithium salts [103]. Photoelectrochemical studies revealed that bulk C_3N_4 (MCN) treated

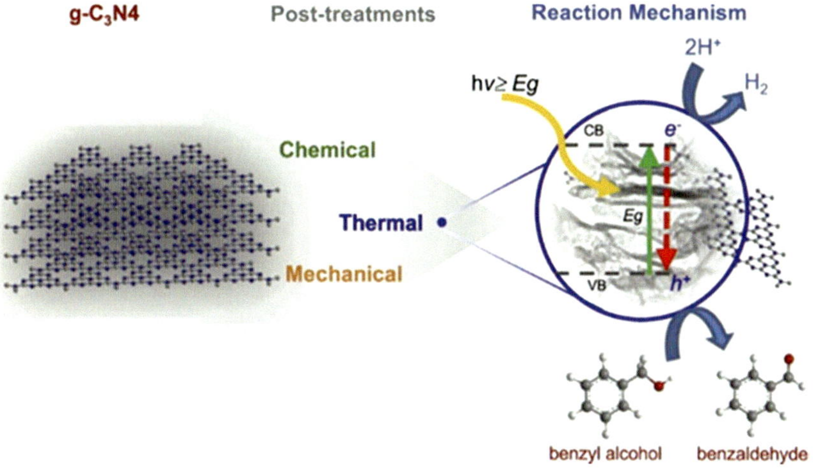

Fig. 5.26 Scheme for the partial oxidation of benzyl alcohol to benzaldehyde. Reprinted with permission from ref. 68. License number 553775153169. 2017 Elsevier

with H_2O_2 had poor mobility of charge carriers, possibly due to defects. The efficiency of generating reactive oxygen species (ROS) was reduced in the solids MCN-TE-H_2O_2 and MCN H_2O_2, as evidenced by their decreased capacity to produce DMPO-O_2/O_2H spin adducts (refer to Fig. 5.27). However, bulk C_3N_4 (MCN-TE) that was thermally etched and treated with H_2O_2 exhibited good results with the highest photoconductivity. It was observed that both treatments exhibited reduced efficacy in generating reactive oxygen species, as demonstrated by their diminished capacity to oxidize organic alcohols partially. Utilizing carbon nitride subjected to treatment in molten potassium and lithium salts was observed to be significantly more effective than employing unprocessed g-C_3N_4. The maximum conversion of benzyl alcohol achieved was 81%, while selectivity to benzaldehyde was 96%. This enhanced efficiency was attributed to reducing the size of photocatalyst particles to the nanometer scale, facilitating better accessibility of the photogenerated charges to the catalyst surface [103]

Numerous techniques have been proposed to augment the photocatalytic potential of carbon nitrides. One such method involves the hydrothermal synthesis of poly(triazine imide), which research has indicated exhibits superior photocatalytic attributes compared to g-C_3N_4 materials [104]. Polytriazine imide carbon nitride photocatalyst successfully achieved the selective production of cyclohexanone from cyclohexane in water at 60 °C with an exceptional selectivity of 100% even at just 6% conversion [105].

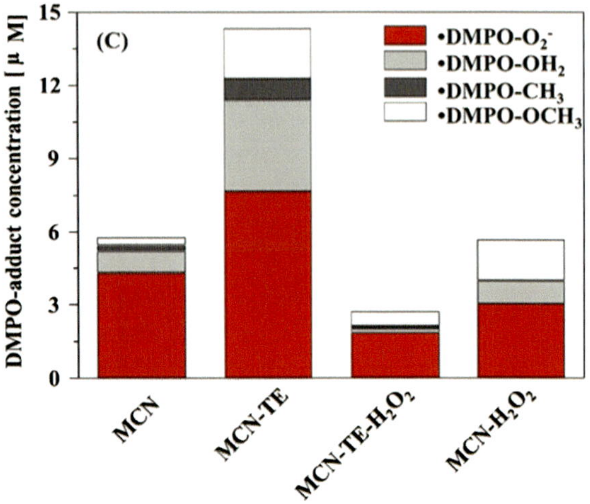

Fig. 5.27 The simulation analysis of experimental EPR spectra conducted in DMSO/water suspensions of the C_3N_4-based sample revealed the concentration of DMPO spin-adducts. Reprinted with permission from ref. 106. License number 55303927277. 2018 Elsevier

5.4.1.3 C_3N_4 Doped with Metal

The placement of gold nanoparticles (Au-NP) on the mesoporous surface of g-C_3N_4 creates a metal–semiconductor rectifying contact, forming a Mott-Schottky heterojunction that elevates the photocatalytic properties of C_3N_4. The selective photooxidation reaction of cyclohexane to cyclohexanone using Au-C_3N_4 materials indicated that H_2O_2-forming C_3N_4 played a crucial role in the reaction. The absence of Au-NP resulted in a lower conversion rate of 2.6%, while their presence increased it to 10.5%. With a 10% Au loading, the catalytic performance and selectivity of the process were enhanced from 72.7% to 100% due to the maximum surface plasma resonance effect [106]. Using Pd-C_3N_4 as a photocatalyst and Pd as a coupling catalyst led to a reaction with a mixed H_2O_2/EtOH solvent, and the Mott-Schottky heterojunction facilitated the reaction to proceed with 100% conversion and 97% selectivity [107]. In addition, using triethanolamine as a sacrificial agent proved effective in achieving the oxidation of amines to imines and hydrogen production from water. The optimum amount of nickel by weight was determined to be 0.41%, resulting in a benzylamine conversion rate of 61.3% and an H_2 production rate of 596 μmol g^{-1} h^{-1} [108].

5.4.1.4 C_3N_4 Doped with Organic and Inorganic Species

Heterogeneous synthesis photocatalysis in water poses a significant challenge due to the low solubility of organic molecules in water. However, this issue could be overcome by incorporating micelles through the addition of a surfactant to the reagent system. Despite this solution, another limitation must be considered [109]. To address this, a proposal has been suggested to utilize an amphiphile-attached photocatalyst that aggregates nanoparticles in water. These nanoparticles could function as a source of photocatalyst, offering lipophilic nuclei suitable for Ir-based photo redox catalysis

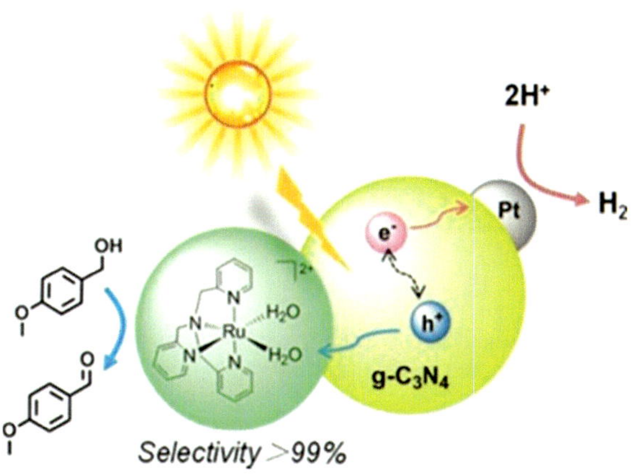

Fig. 5.28 Ru-g-C_3N_4 photocatalyst to produce 4-MBA oxidation and hydrogen production via photocatalytic. Reprinted with permission from ref. 114. License number 5537750379497. 2018 Elsevier

[110, 111]. The primary objective of this study was to investigate the potential role of nitrogen vacancies in graphitic carbon nitride in promoting the aerobic oxidative C–C bond cleavage of vicinal diols under visible light by using cetyltrimethylammonium bromide as a surfactant in water. The micellar semiconductor/catalyst system exhibited consistent activity for up to 10 experiments and demonstrated effectiveness under sunlight irradiation, making it suitable for large-scale applications. Furthermore, the investigation evaluated an alternative composite material comprising Pt-g-C_3N_4 and an organometallic ruthenium complex, demonstrating its potential for partially oxidizing benzyl alcohols to aldehydes and producing H_2. The exceptional selectivity of 99% was attributed to the synergistic influence of both heterogeneous and homogeneous catalysts and the electrostatic interaction between the positively charged ruthenium catalyst and the negatively charged surface of Pt-g-C_3N_4, as illustrated in Fig. 5.28.

Benzyl alcohol, 4-methoxybenzyl alcohol, and piperonyl alcohol were subjected to selective oxidation under UV in the presence of pure and P-doped C_3N_4 photocatalysts, both in an aqueous environment [70]. The experiment's findings indicated that pure and P-doped C_3N_4 photocatalysts achieved 100% selectivity towards benzaldehyde and 4-methoxybenzyl alcohol. However, the selectivity was only 46% towards piperonal. The presence of P in the C_3N_4 structure decreased its oxidizing capabilities, leading to higher selectivity towards the corresponding aldehyde rather than complete alcohol mineralization. In a separate experiment, the condensation of benzyl alcohol to benzoin was conducted using an alkalinized photocatalyst g-C_3N_4 (K-CN), which resulted in a conversion rate of 98% and a remarkable selectivity of 96% [113]. The reaction process illustrated in Fig. 5.29 depends on the initial oxidation of the alcohol to produce benzaldehyde. Subsequently, two benzaldehydes react with each other to create a carbon–carbon bond, leading to the creation of the final product.

The introduction of K + modification has been discovered to enhance the generation of superoxide and hydrogen peroxide, therefore promoting the oxidation of alcohol to aldehyde. Moreover, the K-modification promotes the formation of a

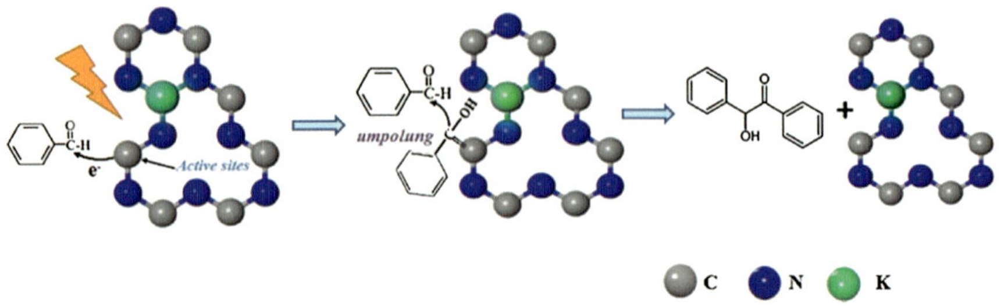

Fig. 5.29 Scheme of the reaction mechanism of photocatalytic benzoin synthesis over K-CN catalyst. Reprinted with permission from ref. 38. License number 555367010697. 2018 Elsevier

carbon–carbon bond between two aldehydes in comparison to the unmodified material. This modification involves substituting amino hydrogen atoms with potassium (K) atoms, leading to deprotonation and an increase in the nucleophilicity of the surface. Consequently, the negatively charged surface is more prone to combine with the carbonium ions present within the carbonyl groups of the intermediate benzaldehydes [24].

5.4.2 *Other Carbon-Based Materials*

The study explored the potential of defective reduced graphene oxide (D-rGO) as a photocatalyst and co-catalyst and its combination with CdS nanoparticles to facilitate the reduction of 4-nitrophenol to 4-aminophenol and H_2 production. The results showed that D-rGO exhibited high charge mobility, and the presence of CdS increased the number of active sites, particularly under sunlight [114]. Recent research utilizing scanning electrochemical microscopy suggests that carbon quantum dots (CDs) ranging from 5 to 10 nm in size, produced via electrochemical ablation of graphite, exhibit remarkable proton properties under visible light in solution. This material is effective as an acid catalyst and catalyzes several organic reactions, resulting in reasonable conversion rates (34.7–46.2%) in water under visible light. The catalytic activity of the CDs exhibits a strong dependence on the intensity of light and temperature [115]. The application of CDs as a photocatalyst was successful for the selective conversion of benzyl alcohols into benzaldehyde in the presence of H_2O_2 when illuminated with near-infrared (NIR) light [116]. The result shows that the conversion and the selectivity are around 92% and 100%, respectively.

The introduction of Cu and N-dopants in CDs led to a significant enhancement of the photocatalytic performance by 3.5 times compared to the unmodified material, primarily due to the improved electron-donating and electron-accepting capacities that play a critical role in the heterogeneous photocatalysis. Additionally, Au/CD heterostructures displayed high efficiency in converting cyclohexane to cyclohexanone with a selectivity of 99.99%, which is 20 times higher than pure CDs [117]. Similar improvement has been observed with metals such as Ag and Cu [118].

5.4.2.1 Sulfides and Selenide Materials

Sulfides and selenides have become of considerable interest due to their possible application as photocatalysts because of their distinctive electronic and structural properties. Their narrow bandgap makes them capable of absorbing visible light and creating charge carriers, initiating chemical reactions upon exposure to light. Zinc sulfide (ZnS) is one of the most well-researched photocatalysts used to break down organic pollutants in water and air [119]. Zinc sulfide (ZnS) can be utilized as a photocatalyst to produce valuable products by reducing CO_2 under visible light irradiation. In this context, ZnS photocatalysts were developed through the deposition of ruthenium nanoparticles. These photocatalysts reveal variations in their electronic band structure, morphology, and photocatalytic activities. Among the prepared ZnS-based photocatalysts, the Ru@ZnS material catalyzed CO_2 reduction to C1 products of two-electron reduction with a small amount of methane being generated, indicating the ability of the materials to catalyze other multielectron reduction processes. However, the presented data showed a high selectivity of carbon dioxide reduction with low hydrogen production efficiency, demonstrating that ZnS is active in photocatalytic CO_2 reduction. moreover, the materials demonstrated a possibility of highly selective formate production in aqueous media [120]. Copper sulfide (Cu_2S) has been identified as a promising photocatalyst with potential applications in organic pollutant degradation, hydrogen gas production, and CO_2 reduction to produce valuable products.

The research investigating the influence of morphology on Cu_2S nanocrystals revealed that those with chalcocite structure had a higher rate of photocatalytic hydrogen production, reaching 234 μmol g^{-1} h^{-1}. The highest hydrogen production rate, reaching 324 μmol g^{-1} h^{-1}, was achieved by a Cu_2S-based composite photocatalyst containing MoS_2, indicating a synergistic effect between the two materials. This combination effectively isolates photoexcited electrons and reduces the recombination of electron–hole pairs, leading to improved photocatalytic activity under visible light irradiation, which can be attributed to the materials' narrow band gap [121]. The Cu_2S photocatalyst that was synthesized can be effectively utilized to reduce CO_2 to carbon monoxide and methane, with a formation rate of 3.02 and 0.13 μmol g^{-1} h^{-1}, respectively. This process is compatible with solar fuel generation photocatalytic reactions and can be achieved by utilizing a band-pass filter to select wavelengths longer than 570 nm [44].

The unique light absorption properties of colloidal semiconductor quantum dots (QDs), particularly those composed of sulfides and selenides, have made them attractive for organic synthesis using visible light due to their high surface-to-volume ratio. While CdSe QDs have been used in various reactions, the challenge of photo

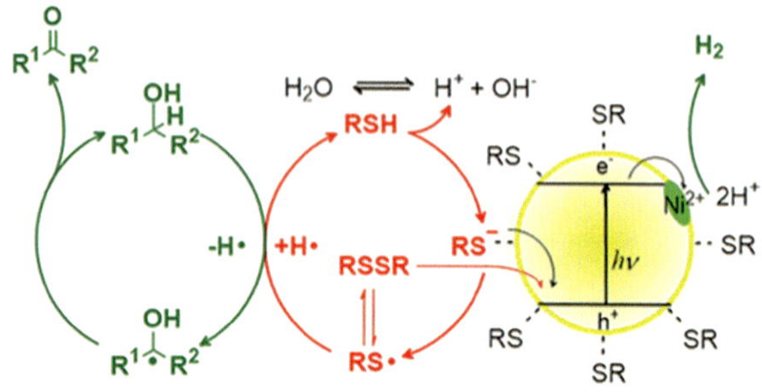

Fig. 5.30 Presence of MPA-CdSe QD in alcohol oxidation. Reprinted with permission from ref. 124. License number 5538201376813. 2017 Wiley

corrosion becomes prominent, primarily when used in water-based reactions. Consequently, the use of CdSe QDs in water-based reactions typically requires the presence of hole scavengers to protect the photocatalyst. Nevertheless, visible light-mediated reactions on CdSe QDs have demonstrated the ability to lead to the specific binding of thiols to disulfides and the production of H_2 without the need for oxidizing or sacrificial agents [123]. Under visible light and in water, CdSe quantum dots can selectively oxidize alcohols into aldehydes and ketones by producing the thiyl radical. This radical is responsible for extracting a hydrogen atom from the carbon atom adjacent to the alcoholic group (Fig. 5.30).

The reaction can be achieved with high efficiency and selectivity [122]. Various experiments have demonstrated that CdS and CdSe nanoparticles can serve as efficient and selective photocatalysts for reducing aromatic azides to aromatic amines. The reaction is a two-electron process, and the quantum yield has reached 0.5 [124]. In a separate study, the combination of CdS nanospheres and graphene in a nanocomposite exhibited higher effectiveness in reducing nitroaromatic compounds under visible light exposure. Adding ammonium formate as a hole scavenger further improved the reusability of the photocatalyst [125]. A recent study has indicated that using CdS materials in oxidizing benzyl alcohol in water has been constrained by photo corrosion. Nevertheless, CdS-graphene nanocomposites have demonstrated the ability to convert benzyl alcohol into benzaldehyde under visible light irradiation and aerobic conditions while sustaining good stability against photo corrosion. This outcome is attributed to the excellent efficiency of electronic transport in graphene [126]. Another recent report describes an innovative approach for converting a-hydroxyl acids, derived from biomass and glucose, into amino acids. The process involves using NH_3 and CdS nanosheets, which are more efficient than other CdS morphologies. The nanosheets can selectively produce reactive oxygen species during hydrogen evolution, leading to higher activity [127]. By employing the NH_3 and CdS nanosheets method, researchers have achieved the conversion of glucose into alanine in one step, as well as the synthesis of multiple other amino acids.

5.4.2.2 TiO_2

TiO_2 is a commonly used material as a photocatalyst because of its abundance, low cost, excellent electronic properties, and high stability. Despite these advantages, the performance of TiO_2 is usually improved by doping, adding a co-catalyst, or pairing it with other materials to form heterojunctions. TiO_2-based photocatalysts are often used for various applications, such as water splitting, CO_2 reduction, and many organic reactions.

TiO_2 is among the few materials that possess more negative CB than the reduction potential of H^+ and a more positive VB than the oxidation potential of H_2O [128]. TiO_2 is often doped with noble metals (Au, Ag, Pt, and Pd), non-noble metals (Cu, Ni, and Co), and non-metals (C, S, F, and N) to improve their performance [129]. For example, Kuntahnkudee et al. studied the effects of dispersing Au nanoparticles onto TiO_2 via photo-deposition at various pH levels [130]. The highest performing material was Au/TiO_2, synthesized at pH 10. The H_2 production rate was 134 μmol g^{-1} h^{-1}, which was 28 times the rate of bare TiO_2.

Another example is a study conducted by Wu et al. Pd nanoparticles are deposited onto TiO_2 via chemical reduction [131]. The highest-performing material in this study had a production rate of 3096 μmol g^{-1} h^{-1}, 40 times the rate of bare TiO_2. In both cases, the improved performance was attributed to a lower band gap, increased charge separation, and a broader wavelength range of light absorption. The metallic state of the metal dopants is also essential, as higher performance was observed for catalysts with the neutral metallic state instead of the ionic one.

An alternative strategy to increase hydrogen gas production is to change the oxygen evolution reaction with the oxidation reaction of organic molecules, which brings the advantage of synthesizing value-added products. For instance, Kozlova et al. compared photocatalytic performance between TiO_2 doped with Pd and CuO_x [132]. In this study, Pd-doped TiO_2 performed better as the glycerol conversion for Pd/TiO_2 was 18%, while the conversion for CuO_x/TiO_2 was only 9%. Both catalysts produced glyceraldehyde, lactic acid, and glycolic aldehyde. However, CuO_x/TiO_2 also produced ethylene glycol.

Another example is a study by Payormhorm and Idem, in which C-doped TiO_2 was synthesized via the sol-microwave method for glycerol oxidation [133]. Using carbon as the dopant brings several advantages, including low cost and reduced toxicity by preventing metal leaching. The highest-performing material in this study had a glycerol conversion of 67.5%. The material highly favors the production of formic acid, where the yield of formic acid reached 49%. Other oxidation products were also detected: glyceraldehyde, dihydroxyacetone, and lactic acid.

Other than glycerol, glucose can also be oxidized using a TiO_2-based photocatalyst. Roongraung et al. synthesized TiO_2 supported on type-Y zeolite (ZeY) and modified with silver [7, 134]. The highest-performing material was the catalyst with a TiO_2:ZeY ratio of 20% mol, which attained a glucose conversion of 75% after 2 h of illumination. The yields of the products were 9% for gluconic acid, 35% for formic acid, 26% for arabinose, and 4% for xylitol. The effect of Ag loading on TiO_2/ZeY was also observed. The best performance was done by the catalyst loaded with

1%w Ag. After modification, glucose conversion increased to 97% compared to the conversion of 68% using bare TiO_2/ZeY. The yields of the products were 3.43% for gluconic acid, 57.79% for formic acid, 9.97% for arabinose, and 11.87% for xylitol. The addition of Ag increased the selectivity of the oxidation towards formic acid. Meanwhile, Zhao et al. studied TiO_2 modified with carbon quantum dots (CQDs) for glucose oxidation [135]. The two CQD types were synthesized: CQD-Na, which used sodium citrate, and CQD-H, which used citric acid. The highest performing material for H_2 production was P25-CQDs-H-2 with a production rate of 2430 μmol g^{-1} h^{-1}. Meanwhile, the highest glucose conversion was obtained using P25-CQDs-Na-2 with a conversion of 27%, which is 2.2 times higher than that of bare P25.

5.4.2.3 $FeVO_4$, $ZnFe_2O_4$ and $CdWO_4$

Iron vanadate ($FeVO_4$) is commonly used for hydrogen production via water splitting [136–138], CO_2 reduction [139], and degradation of dyes and various pollutants [140, 141]. For value-adding reactions, while $FeVO_4$ is mainly limited to benzene hydroxylation to phenol, it also finds applications in other hydrocarbon-converting reactions. Braganza and Salker synthesized Al-doped $FeVO_4$ via the coprecipitation method and studied its performance for phenol methylation [142]. This catalyst highly favors methylation in the ortho position, resulting in the production of 2,6-xylenol. The best-performing material was the one dubbed AF1, which is the catalyst with 2% wt of Al loaded during synthesis. AF1 achieved maximum phenol conversion of 98% and 2,6-xylenol selectivity of 94% after 3 h. However, the conversion and selectivity started to decrease slowly after 6 h. After 10 h, conversion and selectivity dropped to 80% and 82%, respectively. In addition, Liu et al. studied $V_2O_5/FeVO_4$ for epoxidation of cyclooctene [143]. After optimization, the catalyst could achieve a conversion of 96.5% and cyclooctene oxide selectivity of 90.2%. Meanwhile, both spinel zinc ferrite ($ZnFe_2O_4$) and cadmium tungstate ($CdWO_4$) have been used in many applications such as pollutant degradation [144–147], CO_2 reduction [148–150], and hydrogen production via water splitting [151–154]. However, other than benzene hydroxylation, both catalysts were not used for any other value-adding reactions.

5.5 Conclusion

Phenol production using the photocatalyst method shows promise as an environmentally friendly process with a high degree of selectivity. The commonly used photocatalysts are semiconductor materials such as TiO_2, ZnO, WO_3, and metal oxides. These materials possess a suitable bandgap that allows them to absorb photon energy from sunlight in the UV or visible region. This absorption leads to the creation of electron–hole pairs and reactive oxygen species (ROS), such as hydroxyl radicals (•OH), which result from oxidizing agents such as peroxide or water. Benzene molecules adsorb onto the catalyst surface, and under irradiation, the excited electrons or holes participate in a series of electron transfer and redox reactions, leading to the formation of phenol and other by-products. The main problem in this process is the over-oxidation of phenol due to the retirement of the OH group in the phenol structure. Researchers have developed strategies to mitigate this over-oxidation of phenol and enhance the material photocatalytic activity and selectivity through doping, nanostructuring, and surface modification. Additional doping, such as noble metal or transition metal, is known to have the ability to act as a catalyst due to unoccupied electrons on its d-orbital. The size modification of the photocatalyst enhances the surface properties and conductivity, thereby gaining the absorption of photon energy. The hydrophilic properties of photocatalysts negatively influence the selectivity since phenol is more likely to have a hydrophilic group (OH). The surface modification by adding potential materials such as carbon, nitrogen, boron, metal, and organosilanes is known to have the ability to suppress the hydrophilic properties on the surface and increase the selectivity of phenol. Besides being used in the hydroxylation of benzene into phenol, photocatalysts also hold potential in several processes to produce and transform organic compounds such as benzaldehyde, benzene, alcohol, glycerol, and others. In summary, photocatalytic hydroxylation of benzene to phenol presents a promising route for sustainable phenol production. Advancements in material development play a vital role in improving the performance of photocatalysts, enabling efficient light absorption and promoting the desired chemical transformations. Understanding the mechanism of this process aids in optimizing reaction conditions and designing more efficient catalysts. The applications of photocatalytic hydroxylation extend to various industries, offering eco-friendly alternatives for phenol synthesis and contributing to developing a more sustainable chemical landscape.

References

1. Frollini, E.; Silva, C. G.; Ramires, E. C. Phenolic Resins as a Matrix Material in Advanced Fiber-Reinforced Polymer (FRP) Composites. *Advanced Fibre-Reinforced Polymer (FRP) Composites for Structural Applications*. Elsevier Ltd 2013, pp 7–43. https://doi.org/10.1533/9780857098641.1.7.
2. Li, J.; Xu, Y.; Ding, Z.; Mahadi, A. H.; Zhao, Y.; Song, Y. F. Photocatalytic Selective Oxidation of Benzene to Phenol in Water over Layered Double Hydroxide: A Thermodynamic and Kinetic Perspective. *Chem. Eng. J.* **2020**, *388*. https://doi.org/10.1016/j.cej.2020.124248.
3. Navarro S, Sherman E, Colmer-Hamood JA, Nelius T, Myntti M, Hamood AN (2022) Urinary Catheters Coated with a Novel Biofilm Preventative Agent Inhibit Biofilm Development by Diverse Bacterial Uropathogens. Antibiotics 11(11):1514
4. Fortuin, J. P.; Waterman, H. I. *Production of Phenol from Cumene*; Pasalmll Pren Ltd, 1058.
5. Hu L, Yue B, Wang C, Chen X, He H (2014) Enhanced Catalytic Activity over Vanadium-Containing Silylated SBA-15 Catalysts for Styrene Epoxidation and Benzene Hydroxylation. Appl Catal A Gen 477:141–146
6. Wang F, Chen Y, Zhu R, Sun J (2017) Novel Synthesis of Magnetic, Porous C/$ZnFe_2O_4$ Photocatalyst with Enhanced Activity under Visible Light Based on the Fenton-like Reaction. Dalt. Trans. 46(34):11306–11317. https://doi.org/10.1039/c7dt01528c

7. Zhang, S.; Jin, M.; Xu, H.; Li, W.; Ye, Y.; Shi, T.; Zhou, H.; Chen, C.; Wang, G.; Zhang, Y.; Lin, Y.; Zheng, L.; Zhang, H.; Zhao, H. Hydrogen Peroxide Assisted Electrooxidation of Benzene to Phenol over Bifunctional Ni–(O–C_2)$_4$ Sites. *Adv. Sci.* **2022**, *9* (35). https://doi.org/10.1002/advs.202204043.
8. Devaraji, P.; Sathu, N. K.; Gopinath, C. S. Ambient Oxidation of Benzene to Phenol by Photocatalysis on Au/$Ti_{0.98}V_{0.02}O_2$: Role of Holes. *ACS Catal.* **2014**, *4* (9), 2844–2853. https://doi.org/10.1021/cs500724z.
9. Morimoto Y, Bunno S, Fujieda N, Sugimoto H, Itoh S (2015) Direct Hydroxylation of Benzene to Phenol Using Hydrogen Peroxide Catalyzed by Nickel Complexes Supported by Pyridylalkylamine Ligands. J Am Chem Soc 137(18):5867–5870. https://doi.org/10.1021/jacs.5b01814
10. Gupta N, Bansal P, Pal B (2015) Metal Ion-TiO_2 Nanocomposites for the Selective Photooxidation of Benzene to Phenol and Cycloalkanol to Cycloalkanone. J Exp Nanosci 10(2):148–160
11. Devaraji P, Jo W-K (2018) Noble Metal Free Fe and Cr Dual-Doped Nanocrystalline Titania (Ti_1- X− YMx+ YO_2) for High Selective Photocatalytic Conversion of Benzene to Phenol at Ambient Temperature. Appl Catal A Gen 565:1–12
12. Goto T, Ogawa M (2016) Efficient Photocatalytic Oxidation of Benzene to Phenol by Metal Complex-Clay/TiO_2 Hybrid Photocatalyst. RSC Adv 6(28):23794–23797
13. Marino T, Molinari R, García H (2013) Selectivity of Gold Nanoparticles on the Photocatalytic Activity of TiO_2 for the Hydroxylation of Benzene by Water. Catal Today 206:40–45. https://doi.org/10.1016/j.cattod.2012.01.030
14. Yuzawa H, Aoki M, Otake K, Hattori T, Itoh H, Yoshida H (2012) Reaction Mechanism of Aromatic Ring Hydroxylation by Water over Platinum-Loaded Titanium Oxide Photocatalyst. J Phys Chem C 116(48):25376–25387
15. Huy TH, Bui DP, Kang F, Wang YF, Liu SH, Thi CM, You SJ, Chang GM, Pham VV (2019) SnO_2/TiO_2 Nanotube Heterojunction: The First Investigation of NO Degradation by Visible Light-Driven Photocatalysis. Chemosphere 215:323–332. https://doi.org/10.1016/j.chemosphere.2018.10.033
16. Shi Q, Ping G, Wang X, Xu H, Li J, Cui J, Abroshan H, Ding H, Li G (2019) CuO/TiO_2 Heterojunction Composites: An Efficient Photocatalyst for Selective Oxidation of Methanol to Methyl Formate. J. Mater. Chem. A 7(5):2253–2260. https://doi.org/10.1039/c8ta09439j
17. Wu, C.; Dai, J.; Ma, J.; Zhang, T.; Qiang, L.; Xue, J. Mechanistic Study of B-TiO2/BiVO4 S-Scheme Heterojunction Photocatalyst for Tetracycline Hydrochloride Removal and H_2 Production. *Sep. Purif. Technol.* **2023**, *312*. https://doi.org/10.1016/j.seppur.2023.123398.
18. Nam, Y.; Lim, J. H.; Ko, K. C.; Lee, J. Y. Photocatalytic Activity of TiO_2 Nanoparticles: A Theoretical Aspect. *Journal of Materials Chemistry A*. Royal Society of Chemistry 2019, pp 13833–13859. https://doi.org/10.1039/c9ta03385h.
19. Ide Y, Nakamura N, Hattori H, Ogino R, Ogawa M, Sadakane M, Sano T (2011) Sunlight-Induced Efficient and Selective Photocatalytic Benzene Oxidation on TiO_2-Supported Gold Nanoparticles under CO_2 Atmosphere. Chem Commun 47(41):11531–11533
20. Zhang, K.; Huang, D.; Zhang, Y.; El Houda Bouroubi, N.; Chen, P.; Ganbold, N.; He, P.; Liu, J.; Fang, Y.; Gan, M.; Zhu, J.; Yang, B. Natural Mineral-Derived Fe/Mn-BC as Efficient Peroxydisulfate Activator for 2,4-Dichlorophenol Removal from Wastewater: Performance and Sustainable Catalytic Mechanism. *J. Environ. Manage.* **2023**, *335*. https://doi.org/10.1016/j.jenvman.2023.117540.
21. Zaarour M, Cazemier J, Ruiz-Martínez J (2020) Recent Developments in the Control of Selectivity in Hydrogenation Reactions by Confined Metal Functionalities. Catal Sci Technol 10(24):8140–8172
22. Otor HO, Steiner JB, García-Sancho C, Alba-Rubio AC (2020) Encapsulation Methods for Control of Catalyst Deactivation: A Review. ACS Catal 10(14):7630–7656
23. Deng J, Jiang J, Zhang Y, Lin X, Du C, Xiong Y (2008) $FeVO_4$ as a Highly Active Heterogeneous Fenton-like Catalyst towards the Degradation of Orange II. Appl Catal B Environ 84(3–4):468–473

24. Wei D, Huang L, Liang H, Zou J, Chen W, Yang C, Hou Y, Zheng D, Zhang J (2021) Photocatalytic Hydroxylation of Benzene to Phenol over Organosilane-Functionalized $FeVO_4$ Nanorods. Catal Sci Technol 11(17):5931–5937
25. Cao Y, Zhao B, Bao X, Wang Y (2018) Fabricating Metal@N-Doped Carbon Catalysts via a Thermal Method. ACS Catal 8(8):7077–7085. https://doi.org/10.1021/acscatal.8b01644
26. Sun L, Zhan W, Shang J, Chen G, Wang S, Chen Y, Long Z (2019) Carbon-Encapsulated Fe_3O_4 for Catalyzing the Aerobic Oxidation of Benzyl Alcohol and Benzene. React Kinet Mech Catal 126(2):1055–1065. https://doi.org/10.1007/s11144-018-01529-x
27. Zeng, L.; Liang, H.; An, P.; Yu, D.; Yang, C.; Hou, Y.; Zhang, J. Carbon Encapsulated Bimetallic FeCo Nanoalloys for One-Step Hydroxylation of Benzene to Phenol. *Appl. Catal. A Gen.* **2022**, *633*. https://doi.org/10.1016/j.apcata.2022.118499.
28. Yang, B.; Zhang, S.; Gao, Y.; Huang, L.; Yang, C.; Hou, Y.; Zhang, J. Unique Functionalities of Carbon Shells Coating on $ZnFe_2O_4$ for Enhanced Photocatalytic Hydroxylation of Benzene to Phenol. *Appl. Catal. B Environ.* **2022**, *304*. https://doi.org/10.1016/j.apcatb.2021.120999.
29. Zeng Q, Bai J, Li J, Li L, Xia L, Zhou B, Sun Y (2018) Highly-Stable and Efficient Photocatalytic Fuel Cell Based on an Epitaxial $TiO_2/WO_3/W$ Nanothorn Photoanode and Enhanced Radical Reactions for Simultaneous Electricity Production and Wastewater Treatment. Appl Energy 220:127–137. https://doi.org/10.1016/j.apenergy.2018.03.042
30. Jiang, T.; Cheng, L.; Han, Y.; Feng, J.; Zhang, J. One-Pot Hydrothermal Synthesis of Bi_2O_3-WO_3 p-n Heterojunction Film for Photoelectrocatalytic Degradation of Norfloxacin. *Sep. Purif. Technol.* **2020**, *238*. https://doi.org/10.1016/j.seppur.2019.116428.
31. Liao, M.; Su, L.; Deng, Y.; Xiong, S.; Tang, R.; Wu, Z.; Ding, C.; Yang, L.; Gong, D. Strategies to Improve WO_3-Based Photocatalysts for Wastewater Treatment: A Review. *Journal of Materials Science*. Springer 2021, pp 14416–14447. https://doi.org/10.1007/s10853-021-06202-8.
32. Higashimoto, S.; Kurikawa, Y.; Tanabe, Y.; Fukushima, T.; Harada, A.; Murata, M.; Sakata, Y.; Kobayashi, H. Photocatalytic Property of WO_3 Modified with Noble Metal Co-Catalysts towards Selective Hydroxylation of Benzene to Phenol under Visible Light Irradiation. *Appl. Catal. B Environ.* **2023**, *325*. https://doi.org/10.1016/j.apcatb.2022.122289.
33. Szilágyi IM, Fórizs B, Rosseler O, Szegedi Á, Németh P, Király P, Tárkányi G, Vajna B, Varga-Josepovits K, László K, Tóth AL, Baranyai P, Leskelä M (2012) WO_3 Photocatalysts: Influence of Structure and Composition. J Catal 294:119–127. https://doi.org/10.1016/j.jcat.2012.07.013
34. Chen P, Chen L, Zeng Y, Ding F, Jiang X, Liu N, Au CT, Yin SF (2018) Three-Dimension Hierarchical Heterostructure of $CdWO_4$ Microrods Decorated with Bi2WO6 Nanoplates for High-Selectivity Photocatalytic Benzene Hydroxylation to Phenol. Appl Catal B Environ 234:311–317. https://doi.org/10.1016/j.apcatb.2018.04.028
35. Lin, M.; Chen, H.; Zhang, Z.; Wang, X. Engineering Interface Structures for Heterojunction Photocatalysts. *Physical Chemistry Chemical Physics*. Royal Society of Chemistry 2023, pp 4388–4407. https://doi.org/10.1039/d2cp05281d.
36. Chen X, Zhang J, Fu X, Antonietti M, Wang X (2009) Fe-g-C_3N_4-Catalyzed Oxidation of Benzene to Phenol Using Hydrogen Peroxide and Visible Light. J Am Chem Soc 131(33):11658–11659
37. Guo, H.; Chen, M.; Zhong, Q.; Wang, Y.; Ma, W.; Ding, J. Synthesis of Z-Scheme α-Fe_2O_3/g-C_3N_4 Composite with Enhanced Visible-Light Photocatalytic Reduction of CO_2 to CH_3OH. *J. CO_2 Util.* **2019**, *33*, 233–241.
38. Guo, R. tang; Wang, J.; Bi, Z. xu; Chen, X.; Hu, X.; Pan, W. guo. Recent Advances and Perspectives of g–C_3N_4–Based Materials for Photocatalytic Dyes Degradation. *Chemosphere*. Elsevier Ltd 2022. https://doi.org/10.1016/j.chemosphere.2022.133834.
39. Fang Y, Wang X (2017) Metal-free Boron-containing Heterogeneous Catalysts. Angew. Chemie Int. Ed. 56(49):15506–15518
40. Huang, C.; Chen, C.; Zhang, M.; Lin, L.; Ye, X.; Lin, S.; Antonietti, M.; Wang, X. Carbon-Doped BN Nanosheets for Metal-Free Photoredox Catalysis. *Nat. Commun.* **2015**, *6*. https://doi.org/10.1038/ncomms8698.

41. Zheng M, Yuan T, Shi J, Cai W, Wang X (2019) Photocatalytic Oxygenation and Deoxygenation Transformations over BCN Nanosheets. ACS Catal 9(9):8068–8072. https://doi.org/10.1021/acscatal.9b02550
42. Zheng, M.; Cai, W.; Fang, Y.; Wang, X. Nanoscale Boron Carbonitride Semiconductors for Photoredox Catalysis. *Nanoscale*. Royal Society of Chemistry 2020, pp 3593–3604. https://doi.org/10.1039/c9nr09333h.
43. Liu Y, Zheng Y, Dong P, Zhang W, Wu W, Mao J (2021) Atomically Dispersed Cu Anchored on Nitrogen and Boron Co-doped Carbon Nanosheets for Enhancing Catalytic Performance. ACS Appl Mater Interfaces 13(51):61047–61054. https://doi.org/10.1021/acsami.1c17205
44. Gu Y, Li Q, Zang D, Huang Y, Yu H, Wei Y (2021) Light-Induced Efficient Hydroxylation of Benzene to Phenol by Quinolinium and Polyoxovanadate-Based Supramolecular Catalysts. Angew. Chemie - Int. Ed. 60(24):13310–13316. https://doi.org/10.1002/anie.202011164
45. Long Z, Sun L, Zhang M, Zhang Y, Zong C, Xue Z, Wang T, Chen G (2019) Metal-Free Photocatalytic Aerobic Hydroxylation of Benzene Catalyzed by the Commercially Available Quinoline Sulfate. Catal Commun 121:1–4. https://doi.org/10.1016/j.catcom.2018.11.018
46. Zhang, L.; Hou, Q.; Zhou, Y.; Wang, J. Phosphotungstic Anion-Paired Quinoline Salt for Heterogeneous Photocatalytic Hydroxylation of Benzene to Phenol with Air. *Mol. Catal.* **2019**, *473*. https://doi.org/10.1016/j.mcat.2019.110397.
47. Zhang, Y.; Fu, F.; Li, Y.; Zhang, D.; Chen, Y. One-Step Synthesis of Ag@TiO_2 Nanoparticles for Enhanced Photocatalytic Performance. *Nanomaterials* **2018**, *8* (12). https://doi.org/10.3390/NANO8121032.
48. Bi, X.; Du, G.; Kalam, A.; Sun, D.; Yu, Y.; Su, Q.; Xu, B.; Al-Sehemi, A. G. Tuning Oxygen Vacancy Content in TiO_2 Nanoparticles to Enhance the Photocatalytic Performance. *Chem. Eng. Sci.* **2021**, *234*. https://doi.org/10.1016/j.ces.2021.116440.
49. Kamarulzaman, N.; Kasim, M. F.; Rusdi, R. Band Gap Narrowing and Widening of ZnO Nanostructures and Doped Materials. *Nanoscale Res. Lett.* **2015**, *10* (1). https://doi.org/10.1186/s11671-015-1034-9.
50. Rajamanickam D, Shanthi M (2016) Photocatalytic Degradation of an Organic Pollutant by Zinc Oxide – Solar Process. Arab J Chem 9:S1858–S1868. https://doi.org/10.1016/j.arabjc.2012.05.006
51. Ong, C. B.; Ng, L. Y.; Mohammad, A. W. A Review of ZnO Nanoparticles as Solar Photocatalysts: Synthesis, Mechanisms, and Applications. *Renewable and Sustainable Energy Reviews*. Elsevier Ltd 2018, pp 536–551. https://doi.org/10.1016/j.rser.2017.08.020.
52. Yu L, Li M, Huang C, Zhang Y, He J, Zhou X, Zhu H (2018) Photoelectrochemical Properties of N Doped Black TiO_2 Nanotube Arrays. Mater Lett 216:239–242. https://doi.org/10.1016/j.matlet.2018.01.126
53. Kim K, Choe H, Jeong Y, Lee JH, Hong S (2015) Ru (II)-Catalyzed Site-Selective Hydroxylation of Flavone and Chromone Derivatives: The Importance of the 5-Hydroxyl Motif for the Inhibition of Aurora Kinases. Org Lett 17(10):2550–2553
54. D'Amato EM, Neumann CN, Ritter T (2015) Selective Aromatic C-H Hydroxylation Enabled by H6-Coordination to Iridium (III). Organometallics 34(18):4626–4631
55. Verma S, Nasir Baig RB, Nadagouda MN, Varma RS (2017) Hydroxylation of Benzene via C-H Activation Using Bimetallic CuAg@ g-C_3N_4. ACS Sustain. Chem. Eng. 5(5):3637–3640
56. Bhuyan B, Devi M, Bora D, Dhar SS, Newar R (2018) Design of a Photoactive Bimetallic Cu-Au@ G-C_3N_4 Catalyst for Visible Light Driven Hydroxylation of the Benzene Reaction through C-H Activation. Eur J Inorg Chem 2018(34):3849–3858
57. Zhang Y, Park SJ (2019) Stabilizing CuPd Bimetallic Alloy Nanoparticles Deposited on Holey Carbon Nitride for Selective Hydroxylation of Benzene to Phenol. J Catal 379:154–163. https://doi.org/10.1016/j.jcat.2019.09.032
58. Hosseini SM, Ghiaci M, Kulinich SA, Wunderlich W, Monjezi BH, Ghorbani Y, Ghaziaskar HS, Koupaei AJ (2020) Au-Pd Nanoparticles Enfolded in Coil-like TiO_2 Immobilized on Carbon Fibers Felt as Recyclable Nanocatalyst for Benzene Oxidation under Mild Conditions. Appl Surf Sci 506:144644

59. Nishikawa M, Shiroishi W, Honghao H, Suizu H, Nagai H, Saito N (2017) Probability of Two-Step Photoexcitation of Electron from Valence Band to Conduction Band through Doping Level in TiO_2. J Phys Chem A 121(32):5991–5997
60. Borah P, Ma X, Nguyen KT, Zhao Y (2012) A Vanadyl Complex Grafted to Periodic Mesoporous Organosilica: A Green Catalyst for Selective Hydroxylation of Benzene to Phenol. Angew. Chemie 124(31):7876–7881
61. Zhang J, Qiao Z, Mahurin SM, Jiang X, Chai S, Lu H, Nelson K, Dai S (2015) Hypercrosslinked Phenolic Polymers with Well-developed Mesoporous Frameworks. Angew. Chemie Int. Ed. 54(15):4582–4586
62. Wang J, Xu Y, Li J, Ma X, Xu S-M, Gao R, Zhao Y, Song Y-F (2020) Highly Selective Photo-Hydroxylation of Phenol Using Ultrathin NiFe-Layered Double Hydroxide Nanosheets under Visible-Light up to 550 nm. Green Chem 22(24):8604–8613
63. Lu E, Wu J, Yang B, Yu D, Yu Z, Hou Y, Zhang J (2020) Selective Hydroxylation of Benzene to Phenol over Fe Nanoparticles Encapsulated within N-Doped Carbon Shells. ACS Appl. Nano Mater. 3(9):9192–9199
64. Fujishima A, Rao TN, Tryk DA (2000) Titanium Dioxide Photocatalysis. J Photochem Photobiol C Photochem Rev 1(1):1–21
65. Kudo, A. Y. Received.“. *Renew. energy issue energy Res. Chem. Soc. Rev* **2008**, *38*, 253–278.
66. Shiraishi Y, Saito N, Hirai T (2005) Adsorption-Driven Photocatalytic Activity of Mesoporous Titanium Dioxide. J Am Chem Soc 127(37):12820–12822
67. Ide Y, Torii M, Sano T (2013) Layered Silicate as an Excellent Partner of a TiO_2 Photocatalyst for Efficient and Selective Green Fine-Chemical Synthesis. J Am Chem Soc 135(32):11784–11786
68. Ide Y, Matsuoka M, Ogawa M (2010) Efficient Visible-Light-Induced Photocatalytic Activity on Gold-Nanoparticle-Supported Layered Titanate. J Am Chem Soc 132(47):16762–16764
69. Guo Y, Zhang N, Wang X, Qian Q, Zhang S, Li Z, Zou Z (2017) A Facile Spray Pyrolysis Method to Prepare Ti-Doped $ZnFe_2O_4$ for Boosting Photoelectrochemical Water Splitting. J. Mater. Chem. A 5(16):7571–7577. https://doi.org/10.1039/c6ta11134c
70. Su R, Kesavan L, Jensen MM, Tiruvalam R, He Q, Dimitratos N, Wendt S, Glasius M, Kiely CJ, Hutchings GJ (2014) Selective Photocatalytic Oxidation of Benzene for the Synthesis of Phenol Using Engineered Au–Pd Alloy Nanoparticles Supported on Titanium Dioxide. Chem Commun 50(84):12612–12614
71. Park H, Choi W (2005) Photocatalytic Conversion of Benzene to Phenol Using Modified TiO_2 and Polyoxometalates. Catal Today 101(3–4):291–297
72. Yoshida H, Yuzawa H, Aoki M, Otake K, Itoh H, Hattori T (2008) Photocatalytic Hydroxylation of Aromatic Ring by Using Water as an Oxidant. Chem Commun 38:4634–4636. https://doi.org/10.1039/b811555a
73. Singha A, Bhaduri K, Kothari AC, Chowdhury B (2022) Selective Hydroxylation of Benzene to Phenol via CH Activation over Mesoporous Fe_2O_3-TiO_2 Using H_2O_2. Mol Catal 533:112800
74. Ma X, Dang R, Liu Z, Yang F, Li H, Guo T, Luo J (2020) Facile Synthesis of Heterogeneous Recyclable Pd/CeO2/TiO2 Nanostructured Catalyst for the One Pot Hydroxylation of Benzene to Phenol. Chem Eng Sci 211:115274
75. Sun M-H, Huang S-Z, Chen L-H, Li Y, Yang X-Y, Yuan Z-Y, Su B-L (2016) Applications of Hierarchically Structured Porous Materials from Energy Storage and Conversion, Catalysis, Photocatalysis, Adsorption, Separation, and Sensing to Biomedicine. Chem Soc Rev 45(12):3479–3563
76. Mancuso A, Sacco O, Sannino D, Venditto V, Vaiano V (2020) One-Step Catalytic or Photocatalytic Oxidation of Benzene to Phenol: Possible Alternative Routes for Phenol Synthesis? Catalysts 10(12):1424
77. Schmidt RJ (2005) Industrial Catalytic Processes—Phenol Production. Appl Catal A Gen 280(1):89–103. https://doi.org/10.1016/j.apcata.2004.08.030
78. Espinosa JC, Navalón S, Álvaro M, García H (2016) Reduced Graphene Oxide as a Metal-Free Catalyst for the Light-Assisted Fenton-Like Reaction. ChemCatChem 8(16):2642–2648. https://doi.org/10.1002/cctc.201600364

79. Cai J, Zhang M, Wang D, Li Z (2018) Engineering Surface Wettability of Reduced Graphene Oxide to Realize Efficient Interfacial Photocatalytic Benzene Hydroxylation in Water. ACS Sustain. Chem. Eng. 6(11):15682–15687. https://doi.org/10.1021/acssuschemeng.8b04175
80. Ye X, Cui Y, Wang X (2014) Ferrocene-Modified Carbon Nitride for Direct Oxidation of Benzene to Phenol with Visible Light. Chemsuschem 7(3):738–742. https://doi.org/10.1002/cssc.201301128
81. Ci L, Song L, Jin C, Jariwala D, Wu D, Li Y, Srivastava A, Wang ZF, Storr K, Balicas L, Liu F, Ajayan PM (2010) Atomic Layers of Hybridized Boron Nitride and Graphene Domains. Nat Mater 9(5):430–435. https://doi.org/10.1038/nmat2711
82. Chen L, Zhou M, Luo Z, Wakeel M, Asiri AM, Wang X (2019) Template-Free Synthesis of Carbon-Doped Boron Nitride Nanosheets for Enhanced Photocatalytic Hydrogen Evolution. Appl Catal B Environ 241:246–255
83. Wang X, Liang F, Gu H, Wu S, Cao Y, Lv G, Zhang H, Jia Q, Zhang S (2022) In Situ Synthesized α-Fe_2O_3/BCN Heterojunction for Promoting Photocatalytic CO2 Reduction Performance. J Colloid Interface Sci 621:311–320
84. Wang B, Anpo M, Lin J, Yang C, Zhang Y, Wang X (2019) Direct Hydroxylation of Benzene to Phenol on H-BCN Nanosheets in the Presence of FeCl3 and H2O2 under Visible Light. Catal Today 324:73–82
85. Chen Y-Z, Wang ZU, Wang H, Lu J, Yu S-H, Jiang H-L (2017) Singlet Oxygen-Engaged Selective Photo-Oxidation over Pt Nanocrystals/Porphyrinic MOF: The Roles of Photothermal Effect and Pt Electronic State. J Am Chem Soc 139(5):2035–2044
86. Han, W.; Xiang, W.; Meng, Z.; Dong, S.; Lv, Y. A Novel Cr-Doped CdS/ZnO Nanocomposite for Efficient Photocatalytic Hydroxylation of Benzene to Phenol. *Colloids Surfaces A Physicochem. Eng. Asp.* **2023**, 131529.
87. Sun L, Wang D, Li Y, Wu B, Li Q, Wang C, Wang S, Jiang B (2023) Fully Conversing and Highly Selective Oxidation of Benzene to Phenol Based on MOFs-Derived CuO@CN Photocatalyst. Chinese Chem. Lett. 34(4):107490
88. Tomita O, Ohtani B, Abe R (2014) Highly Selective Phenol Production from Benzene on a Platinum-Loaded Tungsten Oxide Photocatalyst with Water and Molecular Oxygen: Selective Oxidation of Water by Holes for Generating Hydroxyl Radical as the Predominant Source of the Hydroxyl Group. Catal Sci Technol 4(11):3850–3860
89. Wang H, Sun X, Li D, Zhang X, Chen S, Shao W, Tian Y, Xie Y (2017) Boosting Hot-Electron Generation: Exciton Dissociation at the Order-Disorder Interfaces in Polymeric Photocatalysts. J Am Chem Soc 139(6):2468–2473
90. McCreery, R. L.; Bergren, A. J. Surface Functionalization in the Nanoscale Domain. In *Nanofabrication: Techniques and Principles*; Springer, 2011; pp 163–190.
91. Groszek AJ (1987) Graphitic and Polar Surface Sites in Carbonaceous Solids. Carbon N. Y. 25(6):717–722
92. Yu Z-H, Gan Y-L, Xu J, Xue B (2020) Direct Catalytic Hydroxylation of Benzene to Phenol Catalyzed by $FeCl_3$ Supported on Exfoliated Graphitic Carbon Nitride. Catal. Letters 150:301–311
93. Nakano S, Akaishi M, Sasaki T, Yamaoka S (1996) Characterizations of Several Cubic Phases Directly Transformed from the Graphitic BC_2N. Mater Sci Eng A 209(1–2):26–29
94. Yadav VK, Chakraborty H, Klein ML, Waghmare UV, Rao CNR (2019) Defect-Enriched Tunability of Electronic and Charge-Carrier Transport Characteristics of 2D Boron Carbonitride (BCN) Monolayers from Ab Initio Calculations. Nanoscale 11(41):19398–19407
95. Zhang, Y.; Pan, Q.; Chai, G.; Liang, M.; Dong, G.; Zhang, Q.; Qiu, J. Synthesis and Luminescence Mechanism of Multicolor-Emitting g-C_3N_4 Nanopowders by Low Temperature Thermal Condensation of Melamine. *Sci. Rep.* **2013**, *3*. https://doi.org/10.1038/srep01943.
96. Wang, Y.; Wang, X.; Antonietti, M. Polymeric Graphitic Carbon Nitride as a Heterogeneous Organocatalyst: From Photochemistry to Multipurpose Catalysis to Sustainable Chemistry. *Angewandte Chemie - International Edition*. 2012, pp 68–89. https://doi.org/10.1002/anie.201101182.

97. Wang X, Maeda K, Thomas A, Takanabe K, Xin G, Carlsson JM, Domen K, Antonietti M (2009) A Metal-Free Polymeric Photocatalyst for Hydrogen Production from Water under Visible Light. Nat Mater 8(1):76–80. https://doi.org/10.1038/nmat2317
98. Fanchini, G.; Tagliaferro, A.; Conway, N. M. J.; Godet, C. Role of Lone-Pair Interactions and Local Disorder in Determining the Interdependency of Optical Constants of (Formula Presented) Thin Films. *Phys. Rev. B - Condens. Matter Mater. Phys.* **2002**, *66* (19), 1–9. https://doi.org/10.1103/PhysRevB.66.195415.
99. Long B, Ding Z, Wang X (2013) Carbon Nitride for the Selective Oxidation of Aromatic Alcohols in Water under Visible Light. Chemsuschem 6(11):2074–2078. https://doi.org/10.1002/cssc.201300360
100. Lima MJ, Silva AMT, Silva CG, Faria JL (2017) Graphitic Carbon Nitride Modified by Thermal, Chemical and Mechanical Processes as Metal-Free Photocatalyst for the Selective Synthesis of Benzaldehyde from Benzyl Alcohol. J Catal 353:44–53. https://doi.org/10.1016/j.jcat.2017.06.030
101. Akhundi A, García-López EI, Marcì G, Habibi-Yangjeh A, Palmisano L (2017) Comparison between Preparative Methodologies of Nanostructured Carbon Nitride and Their Use as Selective Photocatalysts in Water Suspension. Res Chem Intermed 43(9):5153–5168. https://doi.org/10.1007/s11164-017-3046-9
102. Marcì G, García-López EI, Pomilla FR, Palmisano L, Zaffora A, Santamaria M, Krivtsov I, Ilkaeva M, Barbieriková Z, Brezová V (2019) Photoelectrochemical and EPR Features of Polymeric C_3N_4 and O-Modified C_3N_4 Employed for Selective Photocatalytic Oxidation of Alcohols to Aldehydes. Catal Today 328:21–28. https://doi.org/10.1016/j.cattod.2019.01.075
103. Kindt S, Wicht K, Heinrich MR (2016) Thermally Induced Carbohydroxylation of Styrenes with Aryldiazonium Salts. Angew. Chemie 128(30):8886–8889. https://doi.org/10.1002/ange.201601656
104. Wang, L.; Wang, K.; He, T.; Zhao, Y.; Song, H.; Wang, H. Graphitic Carbon Nitride-Based Photocatalytic Materials: Preparation Strategy and Application. *ACS Sustainable Chemistry and Engineering*. American Chemical Society 2020, pp 16048–16085. https://doi.org/10.1021/acssuschemeng.0c05246.
105. Zhang Y, Hu L, Zhu C, Liu J, Huang H, Liu Y, Kang Z (2016) Air Activation by a Metal-Free Photocatalyst for "Totally-Green" Hydrocarbon Selective Oxidation. Catal Sci Technol 6(19):7252–7258. https://doi.org/10.1039/c6cy01066k
106. Liu J, Yang Y, Liu N, Liu Y, Huang H, Kang Z (2014) Total Photocatalysis Conversion from Cyclohexane to Cyclohexanone by C_3N_4/Au Nanocomposites. Green Chem 16(10):4559–4565. https://doi.org/10.1039/c4gc01126k
107. Li, X. H.; Baar, M.; Blechert, S.; Antonietti, M. Facilitating Room-Temperature Suzuki Coupling Reaction with Light: Mott-Schottky Photocatalyst for C-C-Coupling. *Sci. Rep.* **2013**, *3*. https://doi.org/10.1038/srep01743.
108. Zhang S, Qian X, Yan J, Chen K, Huang J (2020) Nickel-Decorated g-C_3N_4 hollow Spheres as an Efficient Photocatalyst for Hydrogen Evolution and Oxidation of Amines to Imines. New J Chem 44(27):11710–11719. https://doi.org/10.1039/d0nj01218a
109. Lipshutz BH, Isley NA, Fennewald JC, Slack ED (2013) On the Way towards Greener Transition-Metal-Catalyzed Processes as Quantified by e Factors. Angew. Chemie - Int. Ed. 52(42):10952–10958. https://doi.org/10.1002/anie.201302020
110. Bu MJ, Cai C, Gallou F, Lipshutz BH (2018) PQS-Enabled Visible-Light Iridium Photoredox Catalysis in Water at Room Temperature. Green Chem 20(6):1233–1237. https://doi.org/10.1039/c7gc03866f
111. Bu MJ, Lu GP, Jiang J, Cai C (2018) Merging Visible-Light Photoredox and Micellar Catalysis: Arylation Reactions with Anilines Nitrosated in Situ. Catal Sci Technol 8(15):3728–3732. https://doi.org/10.1039/c8cy01221k
112. Li F, Wang Y, Du J, Zhu Y, Xu C, Sun L (2018) Simultaneous Oxidation of Alcohols and Hydrogen Evolution in a Hybrid System under Visible Light Irradiation. Appl Catal B Environ 225:258–263. https://doi.org/10.1016/j.apcatb.2017.11.072

113. Sun X, Jiang D, Zhang L, Wang W (2018) Alkaline Modified g-C_3N_4 Photocatalyst for High Selective Oxide Coupling of Benzyl Alcohol to Benzoin. Appl Catal B Environ 220:553–560. https://doi.org/10.1016/j.apcatb.2017.08.057
114. Chen Y, Wang Y, Zhou X, Zhao Y, Peng W (2020) Defected Graphene as Effective Co-Catalyst of CdS for Enhanced Photocatalytic Activities. Environ Sci Pollut Res 27(21):26810–26816. https://doi.org/10.1007/s11356-020-09066-8
115. Li H, Liu R, Kong W, Liu J, Liu Y, Zhou L, Zhang X, Lee ST, Kang Z (2014) Carbon Quantum Dots with Photo-Generated Proton Property as Efficient Visible Light Controlled Acid Catalyst. Nanoscale 6(2):867–873. https://doi.org/10.1039/c3nr03996j
116. Li H, Liu R, Lian S, Liu Y, Huang H, Kang Z (2013) Near-Infrared Light Controlled Photocatalytic Activity of Carbon Quantum Dots for Highly Selective Oxidation Reaction. Nanoscale 5(8):3289–3297. https://doi.org/10.1039/c3nr00092c
117. Liu R, Huang H, Li H, Liu Y, Zhong J, Li Y, Zhang S, Kang Z (2014) Metal Nanoparticle/Carbon Quantum Dot Composite as a Photocatalyst for High-Efficiency Cyclohexane Oxidation. ACS Catal 4(1):328–336. https://doi.org/10.1021/cs400913h
118. Yang Y, Liu N, Qiao S, Liu R, Huang H, Liu Y (2015) Silver Modified Carbon Quantum Dots for Solvent-Free Selective Oxidation of Cyclohexane. New J Chem 39(4):2815–2821. https://doi.org/10.1039/c4nj02256d
119. Isac, L.; Enesca, A. Recent Developments in ZnS-Based Nanostructures Photocatalysts for Wastewater Treatment. International Journal of Molecular Sciences. MDPI 2022.https://doi.org/10.3390/ijms232415668
120. Baran T, Wojtyła S, Dibenedetto A, Aresta M, Macyk W (2015) Zinc Sulfide Functionalized with Ruthenium Nanoparticles for Photocatalytic Reduction of CO_2. Appl Catal B Environ 178:170–176. https://doi.org/10.1016/j.apcatb.2014.09.052
121. Ha, E.; Xin, Z.; Li, D.; Zhang, J.; Ji, T.; Hu, X.; Wang, L.; Hu, J. Dual-Modified Cu_2S with MoS_2 and Reduced Graphene Oxides as Efficient Photocatalysts for H_2 Evolution Reaction. *Catalysts* **2021**, *11* (11). https://doi.org/10.3390/catal11111278.
122. Zhao L-M, Meng Q-Y, Fan X-B, Ye C, Li X-B, Chen B, Ramamurthy V, Tung C-H, Wu L-Z (2017) Photocatalysis with Quantum Dots and Visible Light: Selective and Efficient Oxidation of Alcohols to Carbonyl Compounds through a Radical Relay Process in Water. Angew. Chemie 129(11):3066–3070. https://doi.org/10.1002/ange.201700243
123. Li XB, Li ZJ, Gao YJ, Meng QY, Yu S, Weiss RG, Tung CH, Wu LZ (2014) Mechanistic Insights into the Interface-Directed Transformation of Thiols into Disulfides and Molecular Hydrogen by Visible-Light Irradiation of Quantum Dots. Angew. Chemie - Int. Ed. 53(8):2085–2089. https://doi.org/10.1002/anie.201310249
124. Warrier M, Lo MKF, Monbouquette H, Garcia-Garibay MA (2004) Photocatalytic Reduction of Aromatic Azides to Amines Using CdS and CdSe Nanoparticles. Photochem Photobiol Sci 3(9):859–863. https://doi.org/10.1039/b404268a
125. Chen Z, Liu S, Yang MQ, Xu YJ (2013) Synthesis of Uniform CdS Nanospheres/Graphene Hybrid Nanocomposites and Their Application as Visible Light Photocatalyst for Selective Reduction of Nitro Organics in Water. ACS Appl Mater Interfaces 5(10):4309–4319. https://doi.org/10.1021/am4010286
126. Zhang N, Yang MQ, Tang ZR, Xu YJ (2013) CdS-Graphene Nanocomposites as Visible Light Photocatalyst for Redox Reactions in Water: A Green Route for Selective Transformation and Environmental Remediation. J Catal 303:60–69. https://doi.org/10.1016/j.jcat.2013.02.026
127. Song, S.; Qu, J.; Han, P.; Hülsey, M. J.; Zhang, G.; Wang, Y.; Wang, S.; Chen, D.; Lu, J.; Yan, N. Visible-Light-Driven Amino Acids Production from Biomass-Based Feedstocks over Ultrathin CdS Nanosheets. *Nat. Commun.* **2020**, *11* (1). https://doi.org/10.1038/s41467-020-18532-3.
128. Wang Q, Lian J, Bai Y, Hui J, Zhong J, Li J, An N, Yu J, Wang F (2015) Photocatalytic Activity of Hydrogen Production from Water over TiO_2 with Different Crystal Structures. Mater Sci Semicond Process 40:418–423
129. Kudo A, Miseki Y (2009) Heterogeneous Photocatalyst Materials for Water Splitting. Chem Soc Rev 38(1):253–278. https://doi.org/10.1039/b800489g

130. Zheng Z, Huang B, Qin X, Zhang X, Dai Y, Whangbo MH (2011) Facile in Situ Synthesis of Visible-Light Plasmonic Photocatalysts M@TiO_2 (M = Au, Pt, Ag) and Evaluation of Their Photocatalytic Oxidation of Benzene to Phenol. J Mater Chem 21(25):9079–9087. https://doi.org/10.1039/c1jm10983a
131. Wu J, Lu S, Ge D, Zhang L, Chen W, Gu H (2016) Photocatalytic Properties of Pd/TiO_2 Nanosheets for Hydrogen Evolution from Water Splitting. RSC Adv 6(72):67502–67508
132. Kozlova EA, Kurenkova AY, Gerasimov EY, Gromov NV, Medvedeva TB, Saraev AA, Kaichev VV (2021) Comparative Study of Photoreforming of Glycerol on Pt/TiO_2 and CuO_x/TiO_2 Photocatalysts under UV Light. Mater Lett 283:128901
133. Payormhorm J, Chuangchote S, Laosiripojana N (2017) CTAB-Assisted Sol-Microwave Method for Fast Synthesis of Mesoporous TiO_2 Photocatalysts for Photocatalytic Conversion of Glucose to Value-Added Sugars. Mater Res Bull 95:546–555
134. Roongraung K, Chuangchote S, Laosiripojana N (2020) Enhancement of Photocatalytic Oxidation of Glucose to Value-Added Chemicals on TiO_2 Photocatalysts by a Zeolite (Type y) Support and Metal Loading. Catalysts 10(4):423
135. Zhao H, Yu X, Li C-F, Yu W, Wang A, Hu Z-Y, Larter S, Li Y, Kibria MG, Hu J (2022) Carbon Quantum Dots Modified TiO_2 Composites for Hydrogen Production and Selective Glucose Photoreforming. J Energy Chem 64:201–208
136. Wang W, Zhang Y, Wang L, Bi Y (2017) Facile Synthesis of Fe_{3+}/Fe_{2+} Self-Doped Nanoporous $FeVO_4$ Photoanodes for Efficient Solar Water Splitting. J. Mater. Chem. A 5(6):2478–2482. https://doi.org/10.1039/c6ta10308a
137. Li N, Wu X, Wang M, Huang K, He J, Ma W, Chen H, Li Y, Feng S (2019) Facile Preparation of $BiVO_4$/$FeVO_4$ Heterostructure for Efficient Water-Splitting Applications. Int J Hydrogen Energy 44(41):23046–23053. https://doi.org/10.1016/j.ijhydene.2019.07.063
138. Zhang L, Jin Z, Ma X, Zhang Y, Wang H (2019) Properties of Iron Vanadate over CdS Nanorods for Efficient Photocatalytic Hydrogen Production. New J Chem 43(8):3609–3618. https://doi.org/10.1039/C8NJ06110F
139. Li Y, Zeng Z, Chen Y, Zhang Y, Wang W, Xu X, Du M, Li Z, Zou Z (2022) $FeVO_4$ Nanowires for Efficient Photocatalytic CO2 Reduction. Catal Sci Technol 4:3289–3294. https://doi.org/10.1039/d2cy00324d
140. Sharma S, Dutta V, Raizada P, Hosseini-Bandegharaei A, Thakur VK, Singh A, Le Q, Van; Nguyen, V. H., Singh, P. (2021) Constructing a Novel All-Solid-State Z-Scheme $BiVO_4$/CQDs/$FeVO_4$ Photocatalyst and Its Enhancement to the Photocatalytic Activity. Mater Lett 297:129940. https://doi.org/10.1016/j.matlet.2021.129940
141. Li J, Zhao W, Guo Y, Wei Z, Han M, He H, Yang S, Sun C (2015) Facile Synthesis and High Activity of Novel $BiVO_4$/$FeVO_4$ Heterojunction Photocatalyst for Degradation of Metronidazole. Appl Surf Sci 351:270–279. https://doi.org/10.1016/j.apsusc.2015.05.134
142. Braganza CF, Salker AV (2018) Al-Doped $FeVO_4$ Nanoparticles for Vapour Phase Methylation of Phenol. ChemistrySelect 3(26):7602–7607. https://doi.org/10.1002/slct.201800231
143. Liu J, Wang W, Wang L, Jian P (2023) Heterostructured V_2O_5/$FeVO_4$ for Enhanced Liquid-Phase Epoxidation of Cyclooctene. J Colloid Interface Sci 630:804–812. https://doi.org/10.1016/j.jcis.2022.10.076
144. Aslam I, Cao C, Tanveer M, Farooq MH, Khan WS, Tahir M, Idrees F, Khalid S (2015) A Novel Z-Scheme WO_3/$CdWO_4$ Photocatalyst with Enhanced Visible-Light Photocatalytic Activity for the Degradation of Organic Pollutants. RSC Adv 5(8):6019–6026. https://doi.org/10.1039/c4ra15847d
145. Huang Y, Liang Y, Rao Y, Zhu D, Cao JJ, Shen Z, Ho W, Lee SC (2017) Environment-Friendly Carbon Quantum Dots/$ZnFe2O_4$ Photocatalysts: Characterization, Biocompatibility, and Mechanisms for NO Removal. Environ Sci Technol 51(5):2924–2933. https://doi.org/10.1021/acs.est.6b04460
146. Huang K, Hong Y, Yan X, Huang C, Chen J, Chen M, Shi W, Liu C (2016) Hydrothermal Synthesis of g-C_3N_4/$CdWO_4$ Nanocomposite and Enhanced Photocatalytic Activity for Tetracycline Degradation under Visible Light. CrystEngComm 18(34):6453–6463. https://doi.org/10.1039/c6ce01019a

147. Nguyen, L. T. T.; Vo, D. V. N.; Nguyen, L. T. H.; Duong, A. T. T.; Nguyen, H. Q.; Chu, N. M.; Nguyen, D. T. C.; Tran, T. Van. Synthesis, Characterization, and Application of $ZnFe_2O_4$@ZnO Nanoparticles for Photocatalytic Degradation of Rhodamine B under Visible-Light Illumination. *Environ. Technol. Innov.* **2022**, *25*. https://doi.org/10.1016/j.eti.2021.102130.
148. Tahir, M. Well-Designed $ZnFe_2O_4$/Ag/TiO_2 Nanorods Heterojunction with Ag as Electron Mediator for Photocatalytic CO_2 Reduction to Fuels under UV/Visible Light. *J. CO2 Util.* **2020**, *37* (December 2019), 134–146. https://doi.org/10.1016/j.jcou.2019.12.004.
149. Li, Y. Y.; Wei, Z. H.; Fan, J. Bin; Li, Z. J.; Yao, H. C. Photocatalytic CO_2 Reduction Activity of Z-Scheme CdS/$CdWO_4$ Catalysts Constructed by Surface Charge Directed Selective Deposition of CdS. *Appl. Surf. Sci.* **2019**, *483* (February), 442–452. https://doi.org/10.1016/j.apsusc.2019.03.333.
150. Iqbal F, Mumtaz A, Shahabuddin S, Abd Mutalib MI, Shaharun MS, Nguyen TD, Khan MR, Abdullah B (2020) Photocatalytic Reduction of CO_2 to Methanol over $ZnFe_2O_4$/TiO_2 (p–n) Heterojunctions under Visible Light Irradiation. J Chem Technol Biotechnol 95(8):2208–2221. https://doi.org/10.1002/jctb.6408
151. Nagakawa H, Ochiai T, Konuma S, Nagata M (2018) Visible-Light Overall Water Splitting by CdS/WO_3/$CdWO_4$ Tricomposite Photocatalyst Suppressing Photocorrosion. ACS Appl. Energy Mater. 1(12):6730–6735. https://doi.org/10.1021/acsaem.8b01600
152. Althabaiti, S. A.; Malik, M. A.; Kumar Khanna, M.; Bawaked, S. M.; Narasimharao, K.; Al-Sheheri, S. Z.; Fatima, B.; Siddiqui, S. I. One-Pot Facile Synthesis of CuO–$CdWO_4$ Nanocomposite for Photocatalytic Hydrogen Production. *Nanomaterials* **2022**, *12* (24). https://doi.org/10.3390/nano12244472.
153. Dai F, Zhao R, Huai X, Han J, Wang L, Feng S (2018) Magnetic $ZnFe_2O_4$@ZnSe Hollow Nanospheres for Photocatalytic Hydrogen Production Application. Compos Part B Eng 2019(173):106891. https://doi.org/10.1016/j.compositesb.2019.05.102
154. Nagajyothi PC, Devarayapalli KC, Shim J, Prabhakar Vattikuti SV (2020) Highly Efficient White-LED-Light-Driven Photocatalytic Hydrogen Production Using Highly Crystalline $ZnFe_2O_4$/MoS_2 Nanocomposites. Int J Hydrogen Energy 45(57):32756–32769. https://doi.org/10.1016/j.ijhydene.2020.03.047

GPSR Compliance

The European Union's (EU) General Product Safety Regulation (GPSR) is a set of rules that requires consumer products to be safe and our obligations to ensure this.

If you have any concerns about our products, you can contact us on ProductSafety@springernature.com

In case Publisher is established outside the EU, the EU authorized representative is:

Springer Nature Customer Service Center GmbH
Europaplatz 3
69115 Heidelberg, Germany

Batch number: 10756376

Printed by Printforce, the Netherlands